AF361987

# Recent Advances in Machine Learning and Computational Intelligence

# Recent Advances in Machine Learning and Computational Intelligence

Editors

**Yue Wu**
**Xinglong Zhang**
**Pengfei Jia**

MDPI • Basel • Beijing • Wuhan • Barcelona • Belgrade • Manchester • Tokyo • Cluj • Tianjin

*Editors*
Yue Wu
Xidian University
China

Xinglong Zhang
National University of
Defense Technology
China

Pengfei Jia
Guangxi University
China

*Editorial Office*
MDPI
St. Alban-Anlage 66
4052 Basel, Switzerland

This is a reprint of articles from the Special Issue published online in the open access journal *Applied Sciences* (ISSN 2076-3417) (available at: https://www.mdpi.com/journal/applsci/special_issues/2IEDK599AN).

For citation purposes, cite each article independently as indicated on the article page online and as indicated below:

LastName, A.A.; LastName, B.B.; LastName, C.C. Article Title. *Journal Name* **Year**, *Volume Number*, Page Range.

**ISBN 978-3-0365-7482-0 (Hbk)**
**ISBN 978-3-0365-7483-7 (PDF)**

# Contents


**About the Editors** . . . . . . . . . . . . . . . . . . . . . . . . . . . . . . . . . . . . . . . . . . . . . . . . . . **vii**

**Preface to "Recent Advances in Machine Learning and Computational Intelligence"** . . . . . . **ix**

**Yue Wu, Xinglong Zhang and Pengfei Jia**
Special Issue on Recent Advances in Machine Learning and Computational Intelligence
Reprinted from: *Appl. Sci.* **2023**, *13*, 5078, doi:10.3390/app13085078 . . . . . . . . . . . . . . . . **1**

**Lixin Zhao and Hui Jin**
IDEINFO: An Improved Vector-Weighted Optimization Algorithm
Reprinted from: *Appl. Sci.* **2023**, *13*, 2336, doi:10.3390/app13042336 . . . . . . . . . . . . . . . . **5**

**Ayman Mohamed Mostafa, Meeaad Aljasir, Meshrif Alruily, Ahmed Alsayat
and Mohamed Ezz**
Innovative Forward Fusion Feature Selection Algorithm for Sentiment Analysis Using
Supervised Classification
Reprinted from: *Appl. Sci.* **2023**, *13*, 2074, doi:10.3390/app13042074 . . . . . . . . . . . **31**

**Jian Li, Weijian Zhang, Yating Hu, Shengliang Fu, Changyi Liao and Weilin Yu**
RJA-Star Algorithm for UAV Path Planning Based on Improved R5DOS Model
Reprinted from: *Appl. Sci.* **2023**, *13*, 1105, doi:10.3390/app13021105 . . . . . . . . . . . . . . . . **53**

**Mira Kim and Myeong Ho Song**
High Performing Facial Skin Problem Diagnosis with Enhanced Mask R-CNN and Super
Resolution GAN
Reprinted from: *Appl. Sci.* **2023**, *13*, 989, doi:10.3390/app13020989 . . . . . . . . . . . . . . . . **69**

**Zhitong Zhang, Xu Chang, Hongxu Ma, Honglei An and Lin Lang**
Model Predictive Control of Quadruped Robot Based on Reinforcement Learning
Reprinted from: *Appl. Sci.* **2023**, *13*, 154, doi:10.3390/app13010154 . . . . . . . . . . . . . . . . **93**

**Chongren Wang and Zhuoyi Xiao**
A Deep Learning Approach for Credit Scoring Using Feature Embedded Transformer
Reprinted from: *Appl. Sci.* **2022**, *12*, 10995, doi:10.3390/app122110995 . . . . . . . . . . . . . . . **107**

**Yong Huang, Xin Xu, Yong Li, Xing-Long Zhang, Yao Liu and Xiao-Chuan Zhang**
Vehicle-Following Control Based on Deep Reinforcement Learning
Reprinted from: *Appl. Sci.* **2022**, *12*, 10648, doi:10.3390/app122010648 . . . . . . . . . . . . . . . **121**

**Guiliang Ou, Yulin He, Philippe Fournier-Viger and Joshua Zhexue Huang**
A Novel Mixed-Attribute Fusion-Based Naive Bayesian Classifier
Reprinted from: *Appl. Sci.* **2022**, *12*, 10443, doi:10.3390/app122010443 . . . . . . . . . . . . . . . **133**

**Ahlem Aboud, Nizar Rokbani, Bilel Neji, Zaher Al Barakeh, Seyedali Mirjalili
and Adel M. Alimi**
A Distributed Bi-Behaviors Crow Search Algorithm for Dynamic Multi-Objective Optimization
and Many-Objective Optimization Problems
Reprinted from: *Appl. Sci.* **2022**, *12*, 9627, doi:10.3390/app12199627 . . . . . . . . . . . . . . . **149**

**Zhi Liu, Xuelin He and Yunhua Lu**
Combining UNet 3+ and Transformer for Left Ventricle Segmentation via Signed Distance and
Focal Loss
Reprinted from: *Appl. Sci.* **2022**, *12*, 9208, doi:10.3390/app12189208 . . . . . . . . . . . . . . . **193**

# About the Editors

**Yue Wu**

Yue Wu received B.Eng. and Ph.D. degrees from Xidian University, Xi'an, China, in 2011 and 2016, respectively. Since 2016, he has been a teacher with Xidian University. He is currently an associate professor, doctoral supervisor, and deputy director of the Institute of Computational Intelligence at Xidian University. He has published more than 100 papers in high-level international journals and conferences and applied for and authorized more than 30 patents. His research interests include computational intelligence and its applications. He is the registered chairman of BIC-TA 2016 and ECOLE 2017, the publishing vice-chairman of the 6th CCF Big Data Academic Conference, the chairman of the IEEE CCIS2021 Organizing Committee, a Senior Member of the Chinese Computer Federation, etc. He is an Editorial Board Member for over four journals, including Remote Sensing, Applied Sciences, Electronics, and Mathematics.

**Xinglong Zhang**

Xinglong Zhang was born in Anhui, China, in 1990. He received a B.E. degree and M.S. degree in mechanical engineering from Zhejiang University, Hangzhou, China, in 2011 and PLA University of Science and Technology, Nanjing, China, in 2014, respectively, and Ph.D. in system and control from the Politecnico di Milano, Italy, 2018. He is presently an associate professor at the College of Intelligence Science and Technology, National University of Defense Technology, Changsha, China. He is a Member of IEEE CIS Technical Committee on Adaptive Dynamic Programming and Reinforcement Learning. His research interests include learning-based model predictive control, adaptive dynamic programming, and their applications in automotive systems.

**Pengfei Jia**

Pengfei Jia received a BS degree from Tianjin Medical University, Tianjin, China in 2010, and a PhD degree from Chongqing University, Chongqing, China in 2014. He currently works as a Researcher in School of Electrical Engineering, Guangxi University, Nanning City, Guangxi Province, China. His research interests include chip-level gas sensor array, intelligent olfactory models and gas sensing materials, etc. He is an IEEE Member, a Committee Member of Youth Working Committee of the Chinese Association for Artificial Intelligence and a Senior Member of the Chinese Association for Electrical Technology.

# Preface to "Recent Advances in Machine Learning and Computational Intelligence"

Machine learning and computational intelligence are currently among the hottest research directions and have been applied to various areas with many successes, such as fields of image processing, point cloud processing, and natural language processing. Researchers explore many intelligent algorithms which are characterized by computational adaptability, robustness, and high-level performance. These algorithms facilitate intelligent behavior in complex and dynamic environments and the development of technology that enables machines to think, behave, or act more humanely. This book, consisting of 10 articles written by research experts on topics of interest, reports on the latest research in machine learning and computational intelligence. Many novel and interesting methods are introduced, which we hope will provide guiding significance for the further development of machine learning and computational intelligence.

**Yue Wu , Xinglong Zhang, and Pengfei Jia**
*Editors*

*Editorial*

# Special Issue on Recent Advances in Machine Learning and Computational Intelligence

Yue Wu [1,*], Xinglong Zhang [2] and Pengfei Jia [3]

[1]   Department of Computer Science and Technology, Xidian University, Xi'an 710071, China
[2]   College of Intelligence Science and Technology, National University of Defense Technology, Changsha 410073, China; zhangxinglong18@nudt.edu.cn
[3]   School of Electrical Engineering, Guangxi University, Nanning 530004, China; jiapengfei200609@126.com
*   Correspondence: ywu@xidian.edu.cn

**Citation:** Wu, Y.; Zhang, X.; Jia, P. Special Issue on Recent Advances in Machine Learning and Computational Intelligence. *Appl. Sci.* **2023**, *13*, 5078. https://doi.org/10.3390/app13085078

Received: 11 April 2023
Accepted: 12 April 2023
Published: 19 April 2023

## 1. Introduction

Machine learning and computational intelligence are currently high-profile research areas attracting the attention of many researchers. They have achieved remarkable results in various fields such as computer vision and natural language processing, showing their strong advantages. Researchers have explored many intelligent algorithms, characterized by computational adaptability, robustness, and high performance. These algorithms facilitate intelligent behavior in complex and dynamic environments and the development of technology that enables machines to think, behave, or act more humanely. This not only promotes the further development of machine learning and computational intelligence, but also provides richer ideas for applications.

## 2. Recent Advances

In view of the above, this Special Issue was introduced to collect the latest research on this topic. These latest research works have addressed various practical application scenarios by utilizing machine learning and computational intelligence techniques. This Special Issue contains 10 papers written by research experts on related topics of interest. In reviewing this Special Issue, various topics have been addressed, predominantly machine learning techniques and heuristic search algorithms. The following seven papers utilize machine learning techniques to solve problems in the fields of computer vision, natural language processing, classification, and so on. The first paper, authored by M. Kim and M.H. Song, studies the machine learning-based diagnosis of facial skin problems. They used enhanced mask R-CNN and super-resolution GAN to successfully solve this problem [1]. The second paper was written by Z. Liu, X. He and Y. Lu. This paper addresses the problem of left ventricle (LV) segmentation of cardiac magnetic resonance (MR) images, which could help doctors in the clinical diagnosis of cardiovascular diseases (CVDs). They provided an effective solution by combining the strengths of UNet 3+ and Transformer [2]. The authors A.M. Mostafa, M. Aljasir, M. Alruily, A. Alsayat, and M. Ezz provided a comprehensive review of recent sentiment analysis methods based on lexicon or machine learning. They proposed a forward fusion feature selection algorithm for the sentiment analysis problem of Arabic reviews [3]. The fourth paper, by G. Ou, Y. He, P. Fournier-Viger and J.Z. Huang, proposed a new naive Bayesian classifier (NBC) construction method for mixed attribute data classification problems [4]. This method is mainly intended to solve two limitations of the NBC: one is the assumption of strong independence; while the other is that it cannot effectively solve continuous attributes. The fifth paper, by Z. Zhang, X. Chang, H. Ma, H. An, and L. Lang, proposed a new locomotion control algorithm for quadruped robots by combining the advantages of model predictive control (MPC) and reinforcement learning (RL) [5]. It is an adaptive approach that achieves a better locomotion performance and balance stability. The sixth paper, by C. Wang and Z. Xiao, used deep learning methods to study credit scoring problems in the financial industry, and designed an end-to-end feature-embedded transformer (FE-Transformer) credit scoring method [6]. The seventh paper, authored by Y. Huang, X. Xu, Y. Li, X. Zhang, Y. Liu, and X. Zhang, used deep reinforcement learning techniques to solve the vehicle-following control problem [7]. They proposed a subsection proximal policy optimization method (subsection-PPO) to improve the efficiency and safety of the vehicle-following control method.

In addition, three papers studied heuristic search algorithms in the field of computational intelligence. In the first paper, L. Zhao and H. Jin improved the traditional vector-weighted optimization algorithm (INFO) and designed a promising optimization algorithm (IDEINFO) [8]. The algorithm further improves the global search ability and achieves an excellent optimization performance. The second paper showed a new UAV path planning algorithm (RJA-Star), proposed by J. Li, W. Zhang, Y. Hu, S. Fu, C. Liao, and W. Yu [9]. This method significantly reduced the moving distance, computation time, number of nodes, number of corners, and maximum angles, and effectively improved the obstacle avoidance ability of agricultural drones. The final paper, by A. Aboud, N. Rokbani, B. Neji, Z. Al Barakeh, S. Mirjalili, and A.M. Alimi, studied the use of the crow search algorithm (CSA) for dynamic multi-objective optimization and multi-objective optimization problems [10]. The authors designed a distributed bi-behaviors crow search algorithm (DB-CSA) with two new mechanisms.

## 3. Future Outlook

This Special Issue introduces many novel and interesting methods, providing guidance for the further development of machine learning and computational intelligence. Looking forward, there are still many thought-provoking issues worthy of further in-depth exploration by researchers. In the future, it will be necessary to apply machine learning and computational intelligence technologies to solve more challenging problems in various fields and propose more robust, accurate, and efficient solutions.

**Acknowledgments:** Thanks are due to all the authors and peer reviewers for their valuable contributions to this Special Issue. Thanks the reviewers and editors for their valuable comments and feedback to help the authors improve the papers included in the Special Issue. Furthermore, congratulations to all the authors for their outstanding achievements on their topics. Finally, we would like to express our sincere appreciation to the editorial team of *Applied Sciences*.

**Conflicts of Interest:** The authors declare no conflict of interest.

## References

1. Kim, M.; Song, M.H. High Performing Facial Skin Problem Diagnosis with Enhanced Mask R-CNN and Super Resolution GAN. *Appl. Sci.* **2023**, *13*, 989. [CrossRef]
2. Liu, Z.; He, X.; Lu, Y. Combining UNet 3+ and Transformer for Left Ventricle Segmentation via Signed Distance and Focal Loss. *Appl. Sci.* **2022**, *12*, 9208. [CrossRef]
3. Mostafa, A.M.; Aljasir, M.; Alruily, M.; Alsayat, A.; Ezz, M. Innovative Forward Fusion Feature Selection Algorithm for Sentiment Analysis Using Supervised Classification. *Appl. Sci.* **2023**, *13*, 2074. [CrossRef]
4. Ou, G.; He, Y.; Fournier-Viger, P.; Huang, J.Z. A Novel Mixed-Attribute Fusion-Based Naive Bayesian Classifier. *Appl. Sci.* **2022**, *12*, 10443. [CrossRef]
5. Zhang, Z.; Chang, X.; Ma, H.; An, H.; Lang, L. Model Predictive Control of Quadruped Robot Based on Reinforcement Learning. *Appl. Sci.* **2023**, *13*, 154. [CrossRef]
6. Wang, C.; Xiao, Z. A Deep Learning Approach for Credit Scoring Using Feature Embedded Transformer. *Appl. Sci.* **2022**, *12*, 10995. [CrossRef]
7. Huang, Y.; Xu, X.; Li, Y.; Zhang, X.; Liu, Y.; Zhang, X. Vehicle-Following Control Based on Deep Reinforcement Learning. *Appl. Sci.* **2022**, *12*, 10648. [CrossRef]
8. Zhao, L.; Jin, H. IDEINFO: An Improved Vector-Weighted Optimization Algorithm. *Appl. Sci.* **2023**, *13*, 2336. [CrossRef]

9.    Li, J.; Zhang, W.; Hu, Y.; Fu, S.; Liao, C.; Yu, W. RJA-Star Algorithm for UAV Path Planning Based on Improved R5DOS Model. *Appl. Sci.* **2023**, *13*, 1105. [CrossRef]
10.   Aboud, A.; Rokbani, N.; Neji, B.; Al Barakeh, Z.; Mirjalili, S.; Alimi, A.M. A Distributed Bi-Behaviors Crow Search Algorithm for Dynamic Multi-Objective Optimization and Many-Objective Optimization Problems. *Appl. Sci.* **2022**, *12*, 9627. [CrossRef]

*applied sciences*

MDPI

*Article*

# IDEINFO: An Improved Vector-Weighted Optimization Algorithm

**Lixin Zhao and Hui Jin ***

School of Automotive and Traffic Engineering, Liaoning University of Technology, Jinzhou 121000, China
* Correspondence: jinhui5868@edu.email.cn

**Abstract:** This study proposes an improved vector-weighted averaging algorithm (IDEINFO) for the optimization of different problems. The original vector-weighted optimization algorithm (INFO) uses weighted averaging for entity structures and uses three core procedures to update the positions of the vectors. First, the update rule phase is based on the law of averaging and convergence acceleration to generate new vectors. Second, the vector combination phase combines the obtained vectors with the update rules to achieve a promising solution. Third, the local search phase helps the algorithm eliminate low-precision solutions and improve exploitability and convergence. However, this approach pseudo-randomly initializes candidate solutions, and therefore risks falling into local optima. We, therefore, optimize the initial distribution uniformity of potential solutions by using a two-stage backward learning strategy to initialize the candidate solutions, and a difference evolution strategy to perturb these vectors in the combination stage to produce improved candidate solutions. In the search phase, the search range of the algorithm is expanded according to the probability values combined with the $t$-distribution strategy, to improve the global search results. The IDEINFO algorithm is, therefore, a promising tool for optimal design based on the considerable efficiency of the algorithm in the case of optimization constraints.

**Keywords:** differential evolution strategy; global search optimization; optimization algorithm; search accuracy; weighted mean of vectors

**Citation:** Zhao, L.; Jin, H. IDEINFO: An Improved Vector-Weighted Optimization Algorithm. *Appl. Sci.* **2023**, *13*, 2336. https://doi.org/10.3390/app13042336

Academic Editors: Yue Wu, Xinglong Zhang and Pengfei Jia

Received: 6 January 2023
Revised: 8 February 2023
Accepted: 9 February 2023
Published: 11 February 2023

## 1. Introduction

As society develops, so will the complexity of its problems. Solving these increasingly complex problems is a key part of promoting ongoing development. Traditional algorithms no longer meet the necessary performance requirements for such problems. However, extensive research on intelligent algorithms has led to successful industrial applications within the engineering domain, where global optimization of nonlinear and complex objective functions is particularly difficult. Metaheuristic algorithms provide the simplicity needed to solve complex path planning [1,2], engineering optimization [3,4], medical diagnosis [5], intelligent control [6], image engineering [7], and network structure optimization problems [8].

Although many traditional numerical analysis methods have been studied in this regard, some deterministic methods are still not fit to solve challenging problems in the field of highly nonlinear search, due to their complexity. The optimization of problems through the application of deterministic methods, such as Lagrangian or simplex methods, requires both initial information about the problem and complex computations. Therefore, it is not always possible or feasible to use such methods, for problems of this level, to explore the global optimal solution problem, and hence, there remains an urgent need to develop an effective method to solve increasingly complex optimization problems. In fact, optimization methods can take various forms and formulations, perhaps without the formal restrictions they necessitate for core development in stochastic class exploration. Problems dealing with these forms, such as multi-objective optimization, fuzzy optimization, robust optimization, modulo optimization, large-scale optimization, and single-objective optimization, can therefore be utilized.

In recent years, several population-based optimization algorithms have been applied as simple and reliable methods to solve problems in computer science and industry. Many researchers have demonstrated that these population-based approaches are promising for solving many challenging problems. Some algorithms use methods that mimic natural evolutionary mechanisms and basic genetic rules such as selection, reproduction, mutation, and migration [9]. One of the most popular evolutionary methods is the introduced genetic algorithm (GA) [10]. With their unique three core operations of crossover, mutation, and selection, genetic algorithms have achieved excellent performance on many optimization problems. Other popular evolutionary algorithms include differential evolution (DE) [11] and genetic programming (GP) [12]. Such evolutionary algorithms simulate the way organisms evolve in nature and are highly adaptable to optimization problems. Moreover, some methods are developed from the laws of physics, such as the simulated annealing algorithm (SA) [13], which simulates the annealing mechanism in physical materials science. Moreover, with its excellent local search capability, SA is used to optimize nonlinear and linearized problems such as multilayer perceptron (MLP) training for motor speed regulation, and proportional-integral differential controller design [14]. The Grey Wolf Optimization (GWO) algorithm [15,16] is a new population intelligence optimization algorithm widely used in many important fields. It primarily mimics the stratification pattern and hunting behavior of gray wolf packs, and optimizes through wolf stalking, encircling, and pouncing behaviors. Compared with traditional optimization algorithms such as PSO and GA, GWO has the advantages of fewer parameters, simple principles, and easy implementation. However, GWO also has disadvantages, such as a slow convergence speed, low solution accuracy, and easily falling into local optimality. One of the latest mature methods is the gradient-based optimizer (GBO) [17], which considers Newtonian logic to explore suitable regions and achieve a global solution. This method has been applied in many fields, including sentiment recognition [18] and parameter evaluation [19]. Most population approaches model the particle swarm optimization (PSO) equation by changing the heuristic basis of collective social behavior around a population of animals [20]. Particle Swarm Optimization (PSO) is one of the most successful algorithms of this type, inspired by the individual and collective intelligence of birds when they flock together. Specifically, PSO has some parameters that need to be adjusted and, unlike other methods, PSO has a memory machine that retains the knowledge of the better performing particles, which helps the algorithm to find the optimal solution faster. Currently, PSO has been engaged with the large-scale optimization problem. In addition, there are many newly developed improved intelligent optimization algorithms, for example, ref. [21] proposed an improved GWO to solve the instability and convergence accuracy problems when GWO is applied to mobile robot path planning as a metaheuristic algorithm with a powerful optimization search capability. In [22], an improved crawler search algorithm (IRSA), based on the sine cosine algorithm and Levy flight, was proposed. The improved sine cosine algorithm has an enhanced global search capability, avoids local minima traps by a comprehensive search of the solution space, and the Levy flight operator, with a jump size control factor, improves the exploitation capability of the search agent. A new metaheuristic optimization algorithm based on ancient warfare strategies was proposed in [23], introducing a new weight update mechanism and a weak troop migration strategy. The proposed warfare strategy algorithm achieves a good balance between the exploration and development phases.

Although the abovementioned optimization methods can solve a variety of challenging practical engineering optimization problems, according to the No Free Lunch (NFL) theorem [24], no single optimization method can be the best tool to solve all problems. The case where some form of structure exists over the set of objective values, rather than the typical total ordering. It is shown that in such cases, when attention is restricted to natural measures of performance and optimization algorithms that measure performance and optimize based on such structure, the "no free lunch" result holds for any class of problems with closed structure under permutations [25]. In contrast, the INFO algorithm, proposed in [26], is a forward-looking algorithm that provides a promising platform for the future of the optimization literature in computer science through innovative attempts to target this approach. Our goal is to apply this improved vector-weighted averaging method to various optimization problems and make it a scalable optimizer.

In [26], a new optimizer (INFO) is designed, that can form a more stable structure by modifying the weighted averaging and updating the position of the vectors. The update phase, the combination phase, and the local search are the three core steps of INFO. Unlike other methods, the mean-based update rule is used in INFO to generate new vectors, which will speed up the convergence. In the vector combination phase, the two vectors obtained in the vector update phase are combined to generate a new vector to improve the local search capability. This operation ensures the diversity of the population to some extent. Considering the global optimal position and the mean-based rule, the local operation can effectively improve the vulnerability of the information material to local optima. The focus introduces the three core procedures mentioned above to optimize various optimization cases and engineering problems, such as structural and mechanical engineering problems and water resource systems. The INFO algorithm uses the concept of weighted averages to move agents toward better positions. The main motivation for INFO is to emphasize its performance aspects, potentially solving some optimization problems that cannot be solved by other methods.

In general, evolutionary algorithms can be divided into two types: single-solution-based and population-based algorithms [27]. In the first case, the search process of the algorithm starts with a single solution and updates its position during the optimization process. The best-known single-solution-based algorithms include single-solution-based simulation algorithms, including simulated annealing (SA) [13]. However, their drawbacks are the plausibility of high positions captured in the local optimum and the failure of information exchange, as these methods have only a single trend in the opposite direction. GA, DE, PSO, Ant Colony Optimization (ACO) [28], Artificial Bee Colony (ABC) [29], Harris Hawkeye Optimization (HHO) [30], Hunger Games Search (HGS) [31], Rungakuta Optimizer (RUN) [32], Sticky Fungus Algorithm (SMA) [33], and Whale Optimization (WOA) [34] are some examples of population-based algorithms. These methods can eliminate local optimization because they use a set of solutions in the optimization process. In addition, information exchange can be shared between solutions, which helps them to search better in difficult search spaces. However, these algorithms require significant computational costs for function evaluation and high-dimensional computation during optimization. Based on the above discussion, population-based algorithms are considered more reliable and robust optimization methods than single-solution-based algorithms.

In general, the best formulation of the algorithm is investigated by evaluating different types of benchmark and engineering problems. Typically, the optimizer uses one or more operators to perform two phases: exploration and exploitation. An optimization algorithm requires a search mechanism to find promising regions in the search space, which is done in the exploration phase. The exploitation phase improves the local search capability and the speed of convergence to promising regions. Balancing these two phases is a challenging problem for any optimization algorithm. According to previous studies, no precise rules have been established so far to distinguish the most appropriate transition time from the exploration to the development phase, or the stochastic nature of this type of optimizer, due to its unexplored form [35]. Therefore, addressing this problem is crucial for developing and designing a stable and reliable optimization algorithm. Concerning the main challenges, in order to create a high-performance optimization algorithm, we focused on vector-weighted optimization algorithms with efficient optimization performance, which is based on the principle of vector-weighted averaging. By avoiding nature-inspired thinking, INFO can provide a promising way to avoid and reduce the challenges of other optimization algorithms, thus taking a step forward in having strong optimization capabilities for practical problems in the field of complex unknown search.

In this paper, we report on our improvements to the INFO algorithm, which provides the following contributions:

- A two-stage backward learning strategy that initializes candidate solutions, resulting in improved distribution uniformity and enhanced search capability.
- A DE strategy that perturbs vector individuals to iteratively generate candidate solutions via greedy selection, which eliminates poorly adapted vectors and improves the local search ability.
- A combined $t$-distribution and probabilistic strategy that expands the search range and avoids local optima traps.

To evaluate our algorithm's performance, 14 sets of test functions are applied, improvement metrics are tested separately, and we compare our improved INFO model with the SSA, GWO, and baseline INFO algorithms.

## 2. Materials and Methods

### 2.1. INFO

INFO is a population-based optimizer that applies weighted averaging rules to vectors in a search space to find the optimal solution after several consecutive generations. The baseline model has the advantages of strong search optimization and fast convergence [26]. The following subsections describe the phases of the algorithm's operation.

### 2.2. Initialization Phase

The INFO algorithm comprises a population of $D$ vectors in an $Np$-dimensional search domain. In this step, the algorithm applies two main control parameters: weighted average $\delta$ and proportionality $\sigma$. Generally, the scale factor is increased by updating the regular operator to obtain a vector, which depends on the size of the search domain. $\sigma$ is used to calculate the exponential weighted averages of vectors. These parameters do not require fine-tuning and are dynamically updated during generation.

### 2.3. Update-Rule Phase

The update-rule operation increases the diversity of the population during the search using the weighted average of vectors to create new vectors. This phase consists of two main activities. The first starts with a random initial solution and extracts the weighted mean of a set of random vectors to move to the next candidate solution. The second activity accelerates convergence. This process is defined by Equations (1)–(4).

If $rand < 0.5$,

$$z1_l^g = x_l^g + \sigma \times MeanRule + randn \times \frac{\left(x_{bs} - x_{a1}^g\right)}{\left(f(x_{bs}) - f\left(x_{a1}^g\right) + 1\right)}, \tag{1}$$

$$z2_l^g = x_{bs} + \sigma \times MeanRule + randn \times \frac{\left(x_{a1}^g - x_b^g\right)}{\left(f\left(x_{a1}^g\right) - f\left(x_{a2}^g\right) + 1\right)}; \tag{2}$$

otherwise,

$$z1_l^g = x_a^g + \sigma \times MeanRule + randn \times \frac{\left(x_{a2}^g - x_{a3}^g\right)}{\left(f\left(x_{a2}^g\right) - f\left(x_{a3}^g\right) + 1\right)}, \tag{3}$$

$$z2_l^g = x_{bt} + \sigma \times MeanRule + randn \times \frac{\left(x_{a1}^g - x_{a2}^g\right)}{\left(f\left(x_{a1}^g\right) - f\left(x_{a2}^g\right) + 1\right)}, \tag{4}$$

where $z1_l^g$ and $z2_l^g$ are the new position vectors for the $g$th iteration, $\sigma$ represents the scaling factor of vectors, and $\alpha1 \neq \alpha2 \neq \alpha3 \neq 1$. $\sigma$ and $\alpha$ are calculated using Equations (5) and (6), respectively, where $\alpha$ is a random integer in $[1, Np]$; *randn* is a random value in a standard positive terrestrial distribution.

$$\sigma = 2\alpha \times rand - \alpha, \tag{5}$$

$$\alpha = 2 \times exp\left(-4 \times \frac{g}{Maxg}\right), \tag{6}$$

$$MeanRule = r \times WM1_l^g + (1 - r) \times WM2_l^g, \tag{7}$$

where $r$ is a random number in $[0, 0.5]$, and $l = 1, 2, \ldots, Np$. $WM2_l^g$ is defined as follows:

$$WM2_l^g = \delta \times \frac{w_1(x_{a1} - x_{a2}) + w_2(x_{a1} - x_{a3}) + w_3(x_{a2} - x_{a3})}{w_1 + w_2 + w_3 + \varepsilon} + \varepsilon \times rand. \tag{8}$$

For $WM1_l^g, l = 1, 2, \ldots, Np$, and its $w_1, w_2, w_3$, and $\omega$ are expressed as follows:

$$w_1 = cos((f(x_{a1}) - f(x_{a2})) + \pi) \times exp\left(-\left|\frac{f(x_{a1}) - f(x_{a2})}{\omega}\right|\right), \tag{9}$$

$$w_2 = cos((f(x_{a1}) - f(x_{a3})) + \pi) \times exp\left(-\left|\frac{f(x_{a1}) - f(x_{a3})}{\omega}\right|\right), \tag{10}$$

$$w_3 = cos((f(x_{a2}) - f(x_{a3})) + \pi) \times exp\left(-\left|\frac{f(x_{a2}) - f(x_{a3})}{\omega}\right|\right), \tag{11}$$

$$\omega = max(f(x_{a1}), f(x_{a2}), f(x_{a3})), \tag{12}$$

$$M2_l^g = \delta \times \frac{w_1(x_{bs} - x_{bt}) + w_2(x_{bs} - x_{ws}) + w_3(x_{bt} - x_{ws})}{w_1 + w_2 + w_3 + \varepsilon} + \varepsilon \times rand. \tag{13}$$

For $WM2_l^g, l = 1, 2, \ldots, Np$, and its $w_1, w_2, w_3$, and $\omega$ are expressed as follows:

$$w_1 = cos((f(x_{bs}) - f(x_{bt})) + \pi) \times exp\left(-\left|\frac{f(x_{bs}) - f(x_{bt})}{\omega}\right|\right), \tag{14}$$

$$w_2 = cos((f(x_{bs}) - f(x_{ws})) + \pi) \times exp\left(-\left|\frac{f(x_{bs}) - f(x_{ws})}{\omega}\right|\right), \tag{15}$$

$$w_3 = cos((f(x_{bt}) - f(x_{ws})) + \pi) \times exp\left(-\left|\frac{f(x_{bt}) - f(x_{ws})}{\omega}\right|\right), \tag{16}$$

$$\omega = f(x_{ws}), \tag{17}$$

$$\delta = 2\beta \times rand - \beta, \tag{18}$$

$$\alpha = \beta = 2exp\left(-4 \times \frac{g}{Maxg}\right). \tag{19}$$

Weighting functions $w_1, w_2$, and $w_3$ are used to calculate the weighted averages of the vectors, and $x_{bs}, x_{bt}$, and $x_{ws}$ are the optimal, suboptimal, and worst solution vectors in the first $g$ generations of the population, respectively.

Convergence acceleration (*CA*) is added to the update-rule operator to improve the global search capability and find the best vector in the search space. With the INFO algorithm, the best solution is assumed to be the one closest to the global optimum. Hence, *CA* moves the vector in that direction. The *CA* presented in the equation is multiplied by a random number in the range $[0, 1]$ at each step, and new vectors are computed as follows:

$$CA = randn \times \frac{(x_{bs} - x_{a1})}{(f(x_{bs}) - f(x_{a1}) + \varepsilon)}, \tag{20}$$

$$z_l^g = x_l^g + \delta \times MeanRule + CA. \tag{21}$$

### 2.4. Vector-Merging Stage

In this stage, INFO combines vectors $z1_l^g$ and $z2_l^g$ with $rand < 0.5$ to generate the new vector, $u_1^g$.

If $rand1 < 0.5$ and $rand2 < 0.5$,

$$u_l^g = z1_l^g + \mu \left| z1_l^g - z2_l^g \right|; \tag{22}$$

otherwise, if $rand1 < 0.5$ and $rand2 \geq 0.5$,

$$u_l^g = z2_l^g + \mu \left| z1_l^g - z2_l^g \right|. \tag{23}$$

In either case, when $rand1 < 0.5$,

$$u_l^g = x_l^g, \tag{24}$$

where $u_l^g$ is the new vector formed by combining the first $g$ generational vectors, and $\mu = 0.05 \times randn$.

### 2.5. Local Search Phase

The local search phase aims to avoid local optima and generate a new vector when $rand < 0.5$.

$$xrnd = \phi \times x_{avg} + (1 - \phi) \times (\phi \times x_{bt} + (1 - \phi) \times x_{bs}), \tag{25}$$

$$x_{avg} = \frac{(xa + xb + x3)}{3}, \tag{26}$$

If $rand1 < 0.5$ and $rand2 < 0.5$,

$$u_1^g = xbs + randn \times \left( MeanRule + randn \times \left( x_{bs}^g - x_{a1}^g \right) \right); \tag{27}$$

otherwise, if $rand1 < 0.5$ and $rand2 \geq 0.5$,

$$u_1^g = xrnd + randn \times (MeanRule + randn \times (v1 \times x_{bs} - v2 \times x_{rnd})), \tag{28}$$

where $\phi$ is a random value in $[0, 1]$, $x_{rnd}$ is a new solution made by combining $x_{avg}$, $x_{bt}$, and $x_{bs}$, and $v_1$ and $v_2$ are two random numbers defined as follows:

$$v1 = \begin{cases} 2 \times rand & p > 0.5 \\ 1 & p \leq 0.5 \end{cases}, \tag{29}$$

$$v2 = \begin{cases} rand & p < 0.5 \\ 1 & p \geq 0.5 \end{cases}. \tag{30}$$

### 2.6. Improved INFO Algorithm Design

#### 2.6.1. Two-Stage Backward Learning Strategy

Because the baseline INFO algorithm randomly initializes the positions of candidate solutions using pseudo-random numbers, local extrema traps are possible. Thus, it is expected that a reverse learning approach to initializing the population, combined with a greedy search, will improve the model's global performance.

The reverse point learning model is defined as follows. Let the initial solution $X(x_1, x_2, \ldots, x_d)$, be located at a point where $x_i \in [a_i, b_i]$, $i \in [1, d]$. $a_i, b_i$ denotes the lower and upper limits of the $i$th dimensional coordinate, respectively, taking $d$ as the dimension of the search space. The coordinates of the reverse point, $X_1\left(\hat{x}_1, \hat{x}_2, ..., \hat{x}_d\right)$, are then calculated as follows:

$$\hat{x}_i = a_i + b_i - x_i. \tag{31}$$

The two-stage backward learning strategy is divided into random and basic backward learning types. The size of the random number in $[0, 1]$ is compared to the switching probability $(0.5)$ to select the backward learning stage. If $P$ is less than this random number, basic backward learning is selected:

$$\overrightarrow{Cnew} = lbi + (ubi - \vec{C}); \tag{32}$$

otherwise, random reverse learning is applied:

$$\overrightarrow{Cnew} = lbi + R(ubi - \vec{C}). \tag{33}$$

$R$ is a random number in $(0, 1)$, $lbi$ and $ubi$ are the upper and lower bounds, respectively, $\vec{C}$ denotes the position updated by concentration only, and $\overrightarrow{Cnew}$ denotes the new position obtained after the two learning stages.

Although this process accelerates convergence to a certain extent, there is no guarantee that the new solution will necessarily be better than the original. Hence, the greedy algorithm is applied for merit:

$$\vec{C} = \begin{cases} \overrightarrow{Cnew}, f(\overrightarrow{Cnew}) \leq f(\vec{C}) \\ \vec{C}, f(\overrightarrow{Cnew}) > f(\vec{C}) \end{cases}. \tag{34}$$

### 2.6.2. DE Strategy

To diversify the population, and expand the search range during each iteration, new child sparks are generated via the differential algorithm to potentially improve the vectors of the next generation [36,37]. DE performs vector synthesis with individuals to be mutated by randomly selecting three from the population and scaling their vector differences:

$$\begin{cases} p_i^{g+1} = x_{d1}^g + F \cdot \left(x_{d2}^g - x_{d3}^g\right), \\ i \neq d1 \neq d2 \neq d3 \end{cases} \tag{35}$$

where $x_{di}^g$ is the $g$th individual in the $di$th generation population, $F$ is the scaling factor that increases the operator with adaptive variation, and

$$F = F0 \cdot 2^\tau, \tag{36}$$

where $F0$ is the variation operator, which takes a value in $[0, 2]$, which is usually 0.5. $\tau = e^{1 - \frac{Gm}{Gm-G}}$, $Gm$ is the maximum evolutionary generation, and $G$ is the current evolutionary generation. To increase the diversity of new populations, crossover operations are introduced as follows:

$$w_{i,j}^{g+1} = \begin{cases} p_{i,j}^{g+1}, rand(0,1) \leq R, or, j = jrand \\ x_{i,j}^g, other \end{cases}, \tag{37}$$

where $R \in [0, 1]$ is the crossover probability, $jrand$ is the random number, $p_{i,j}^{g+1}$ is the intermediate generated by the first $g$ generation population variation, $\left\{ p_i^{g+1} \middle| p_{i,j}^{min_{i,j}^{g+1} max_{i,j}} \right\}$, and $x_{i,j}^g$ is the individual prior to variation. To determine whether $w_i^{g+1}$ can become an individual of the first $g + 1$ generation populations, $p_i^{g+1}$ and $p_i^g$ are compared in terms of fitness, and the optimal value is selected as follows:

$$p_i^{g+1} = \begin{cases} w_i^{g+1}, f\left(w_i^{g+1}\right) \leq f\left(p_i^g\right) \\ p_i^g, other \end{cases}.$$

(38)

### 2.6.3. *t*-Distribution Strategy with the Number of Iterations as a Parameter

The $T(n)$ distribution (i.e., student distribution) contains parametric degrees of freedom, $n$, which determine its curve shape. The smaller the value, the flatter the curve and lower the middle [38]. The $t-$ distribution algorithm perturbs the positions of the vectors to achieve population variation, as follows:

$$x_i^t = x_i + x_i \cdot t(\ iter\ ),$$

(39)

where $x_i^t$ is the new position of the *i*th vector in the population after mutation, $x_i$ is the position of the individual before mutation, and $t(iter)$ is the value of the $t$-distribution using the number of iterations as the degrees of freedom.

In the early iterations, the value of *iter* is small, and the results generated by the $t$-distribution are similar to the Coasean variant in economics, which has a strong global search capability. In later periods, the value of *iter* grows, becoming more similar to the Gaussian variant, which has a strong local search capability. Thus, INFO's algorithmic accuracy is improved.

### 2.6.4. Improving the INFO Algorithm with Refined Two-Stage Backward Learning DE and *t*-Distribution Strategies

The steps for improving the INFO algorithm based on two-stage backward learning DE and *t*-distribution strategies are presented as a flowchart in Figure 1. The stepwise descriptions are as follows:

Step 1: Initialize the algorithm parameters, including the population size, maximum number of iterations $T$, and variable dimensionality.

Step 2: The two-stage backward learning strategy initializes the vector positions and calculates the fitness of the individuals.

Step 3: Start iterations ($t < T$).

Step 4: Update the vector positions using the mean-value rule, calculate the average position of the population, and update the positions.

Step 5: Vector merging and position updating.

Step 6: Local searching and position updating.

Step 7: Calculate the current vector fitness and compare it to obtain the current optimal individual.

Step 8: Generate probability $p$ and perturb the position of the vector using the DE strategy if $p > 0.5$, and vice versa, perturbing the position of the vector according to the $t$-distribution strategy.

Step 9: Calculate the vector fitness values after perturbation and compare and update the positions of the optimal vectors.

Step 10: The individual positions and fitness values of the best vectors are recorded.

Step 11: Repeat Steps 4–10, and when the maximum number of iterations is reached, the optimal vector position and fitness are output.

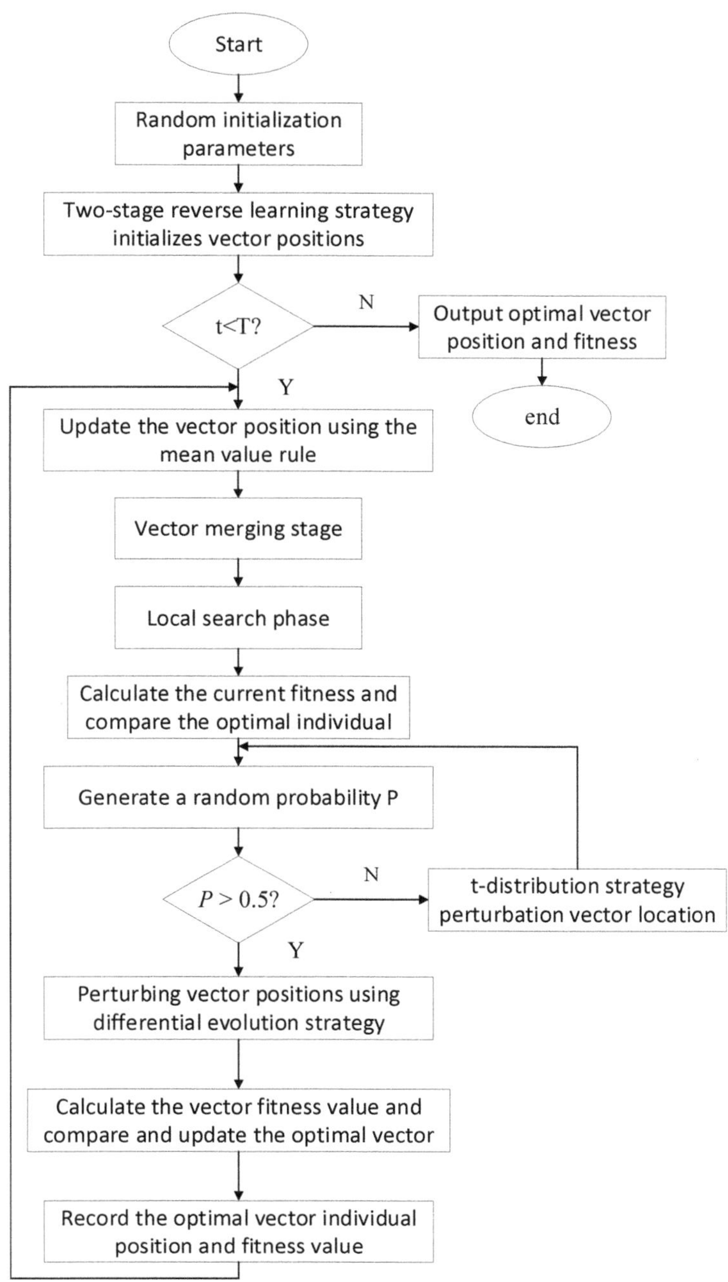


**Figure 1.** Steps of improving the INFO algorithm based on two-stage backward learning DE and *t*-distribution strategies.

### 3. Results

*3.1. Experimental Design and Test Functions*

The performance evaluation in this article is based on an Intel(R) Core(TM) i5-4590 CPU, a 3.30 GHz main frequency, 8 GB of memory, and Windows 10 (64-bit)operating system. The programming software is MATLAB2022(a). Table 1 lists the 14 benchmark functions, and Table 2 lists their dimensions, search ranges, and optimal objective function solutions. The parameter settings for each algorithm are given in Table 3. For a fair comparison, the population size of all algorithms was set to 30, the maximum number of iterations was 1000, and each group of experiments was repeated 50 times to determine the final results [39].

**Table 1.** Test function equations.

| Function | Equation |
| --- | --- |
| F1 | $F1(x) = \sum_{i=1}^{n} \lvert xi \rvert + \prod_{i=1}^{n} \lvert xi \rvert$ |
| F2 | $F2(x) = \sum_{i=1}^{n} \left( \sum_{j=1}^{i} xj \right)^2$ |
| F3 | $F3(x) = max\{xi \mid 1 \leq i \leq n\}$ |
| F4 | $F4(x) = \sum_{i=1}^{n-1} \left[ 100(xi+1 - xi^2)^2 + (xi-1)^2 \right]$ |
| F5 | $F5(x) = \sum_{i=1}^{n} ([xi+0.5])^2$ |
| F6 | $F6(x) = \sum_{i=1}^{n} ix_i^4 + random[0,1]$ |
| F7 | $F7(x) = \sum_{i=1}^{n} -xisin(\sqrt{\lvert xi \rvert})$ |
| F8 | $F8(x) = -20exp(-0.2\sqrt{\frac{1}{n}\sum_{i=1}^{n} xi^2}) - exp(\frac{1}{n}\sum_{i=1}^{n} cos(2\pi xi)) + 20 + e$ |
| F9 | $F9(x) = \frac{\pi}{n}\left\{ 10sin(\pi y1) + \sum_{i=1} \left(yi-1\right)^2 \left[1 + 10sin^2(\pi yi+1)\right] + (yn-1)^2 \right\}$ $+\sum_{i=1}^{n} u(xi,10,100,4) yi = 1 + \frac{xi+1}{4}$ $u(xi,a,k,m) = \begin{cases} k(xi-a)^m & xi > a \\ 0 & -a < xi < a \\ k(-xi-a)^m & xi < -a \end{cases}$ |
| F10 | $F10(x) = 0.1\left\{ sin^2(3\pi x1) + \sum_{i=1}^{n} \left(xi-1\right)^2 \left[1 + sin^2(3\pi xi+1)\right] \right.$ $\left. + \left(xn-1\right)^2 \left[1 + sin^2(2\pi xn)\right] \right\} + \sum_{i=1}^{n} u(xi,5,100,4)$ |
| F11 | $F11(x) = \sum_{i=1}^{11} \left[ ai - \frac{x1(bi^2 + bix2)}{bi^2 + bix3 + x4} \right]^2$ |
| F12 | $F12(x) = 4x1^2 - 2.1x1^4 + \frac{1}{3}x1^6 + x1x2 - 4x2^2 + 4x2^4$ |
| F13 | $F13(x) = -\sum_{i=1}^{7} \left[ (x-ai)(x-ai)^T + ci \right]^{-1}$ |
| F14 | $F14(x) = -\sum_{i=1}^{10} \left[ (x-ai)(x-ai)^T + ci \right]^{-1}$ |

**Table 2.** Test function information.

| Function | Dimensionality | Scope | Theoretical Minimum Value |
|---|---|---|---|
| F1 | 30 | $[-10, 10]$ | 0 |
| F2 | 30 | $[-100, 100]$ | 0 |
| F3 | 30 | $[-100, 100]$ | 0 |
| F4 | 30 | $[-30, 30]$ | 0 |
| F5 | 30 | $[-100, 100]$ | 0 |
| F6 | 30 | $[-1.28, 1.28]$ | 0 |
| F7 | 30 | $[-500, 500]$ | $-12{,}569.5$ |
| F8 | 30 | $[-32, 32]$ | 0 |
| F9 | 30 | $[-50, 50]$ | 0 |
| F10 | 30 | $[-50, 50]$ | 0 |
| F11 | 4 | $[-5, 5]$ | 0.1484 |
| F12 | 2 | $[-5, 5]$ | $-1$ |
| F13 | 4 | $[0, 10]$ | $-1$ |
| F14 | 4 | $[0, 10]$ | $-1$ |

**Table 3.** Algorithm parameter setting.

| Name of the Algorithm | Parameters | Max-Fes |
|---|---|---|
| GWO | $A \in [2, 0]$ | 30,000 |
| WOA | $a1 \in [2, 0], a2 \in [-2, -1], b = 1$ | 30,000 |
| SSA | $c \in [2, 0]$ | 30,000 |
| INFO | $c = 2, d = 4$ | 30,000 |
| INFO1 | $c = 2, d = 4$ | 30,000 |
| INFO2 | $c = 2, d = 4$ | 30,000 |
| IDEINFO | $c = 2, d = 4$ | 30,000 |

*3.2. Experimental Results*

3.2.1. High-Dimensional Single-Objective Test Function

A high-latitude test function is one whose search space dimensionality is relatively high. To demonstrate the optimization-seeking effects of the improved INFO algorithm on high-latitude single-objective test functions, six of the 14 single-objective types were selected for testing. Their convergence curves are shown in Figure 2, and their optimization-seeking space diagrams are shown in Figure 3. INFO1 represents an improved strategy that introduces two-stage backward learning and a greedy mechanism in the original INFO algorithm. INFO2 represents an improved strategy that introduces a differential evolution algorithm and an adaptive $t$-distribution in the original INFO algorithm. The results for INFO1 and INFO2 are illustrated in Figures 4 and 5, respectively.

To verify the efficacy of IDEINFO, ablation experiments based on INFO1 and INFO2 were conducted for comparison. In this test, INFO1 represents an improved strategy that introduces two-stage backward learning and a greedy mechanism in the original INFO algorithm. INFO2 represents an improved strategy that introduces a differential evolution algorithm and an adaptive $t$-distribution in the original INFO algorithm.

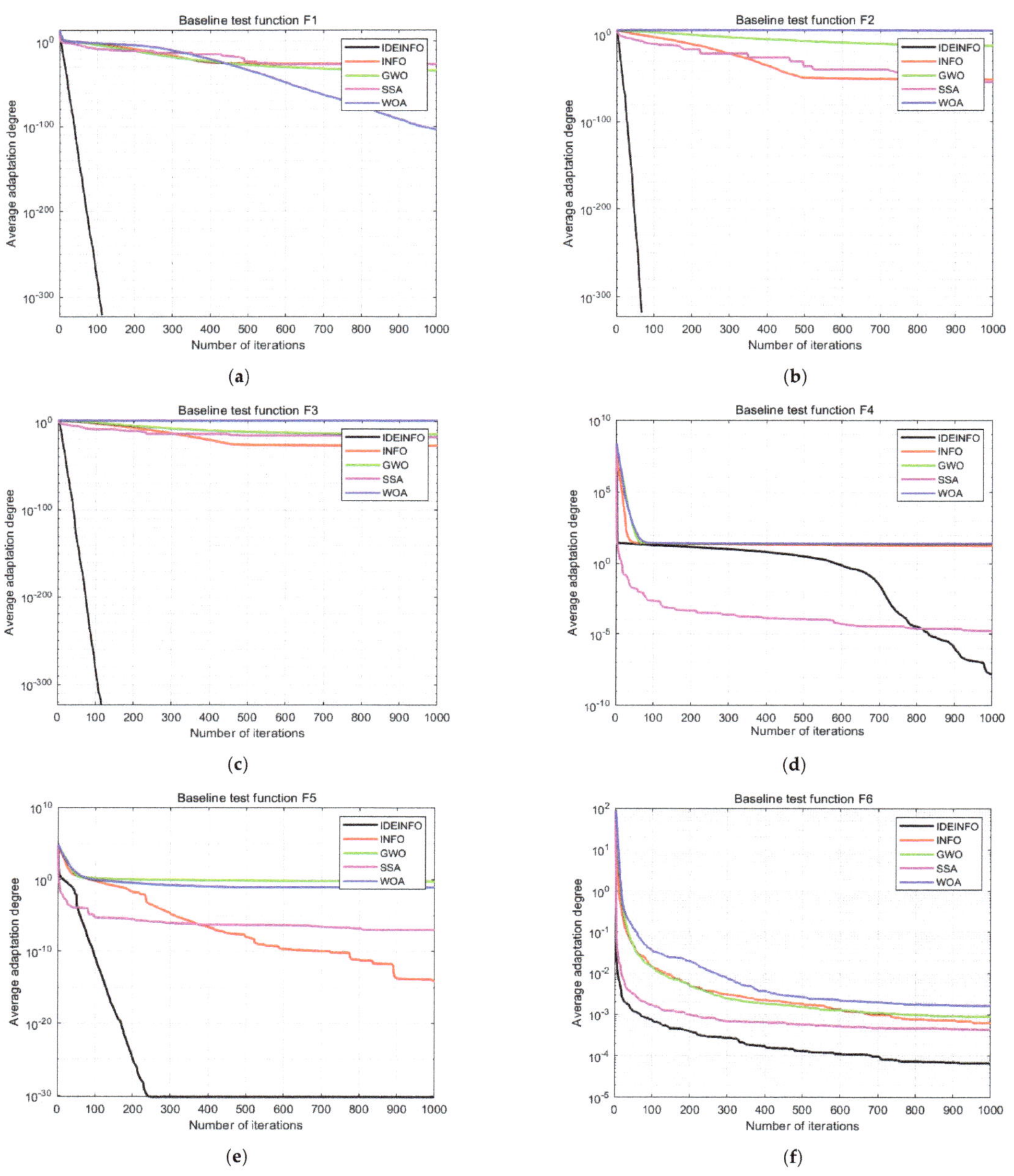

**Figure 2.** Convergence curves of six selected high-dimensional single-objective test functions: (**a**) F1, (**b**) F2, (**c**) F3, (**d**) F4, (**e**) F5, and (**f**) F6.

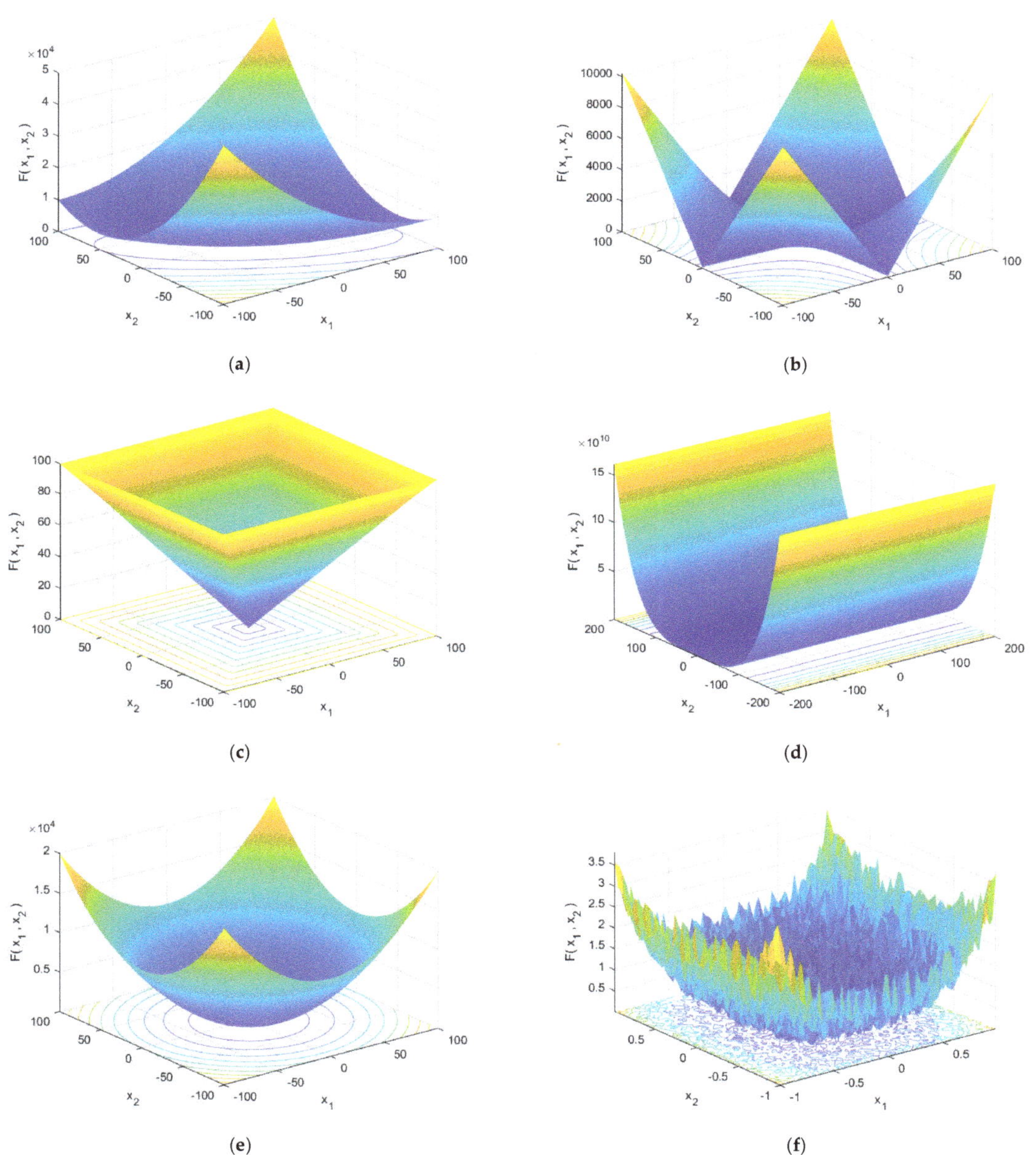

**Figure 3.** Optimization-seeking space diagrams for six selected high-dimensional single-objective test functions: (**a**) F1, (**b**) F2, (**c**) F3, (**d**) F4, (**e**) F5, and (**f**) F6.

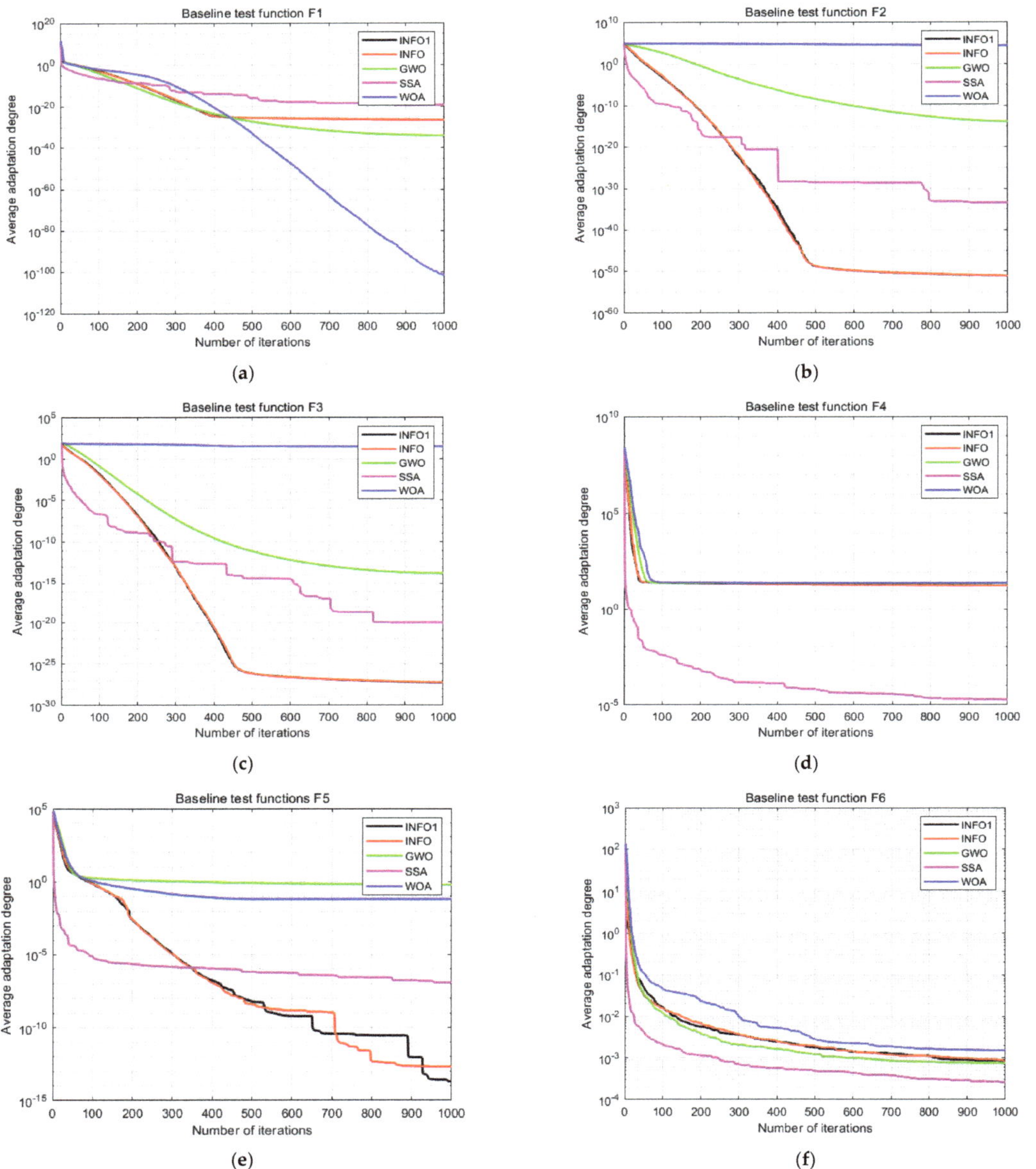

**Figure 4.** Results for six selected high-dimensional single-objective test functions: (**a**) F1, (**b**) F2, (**c**) F3, (**d**) F4, (**e**) F5, and (**f**) F6.

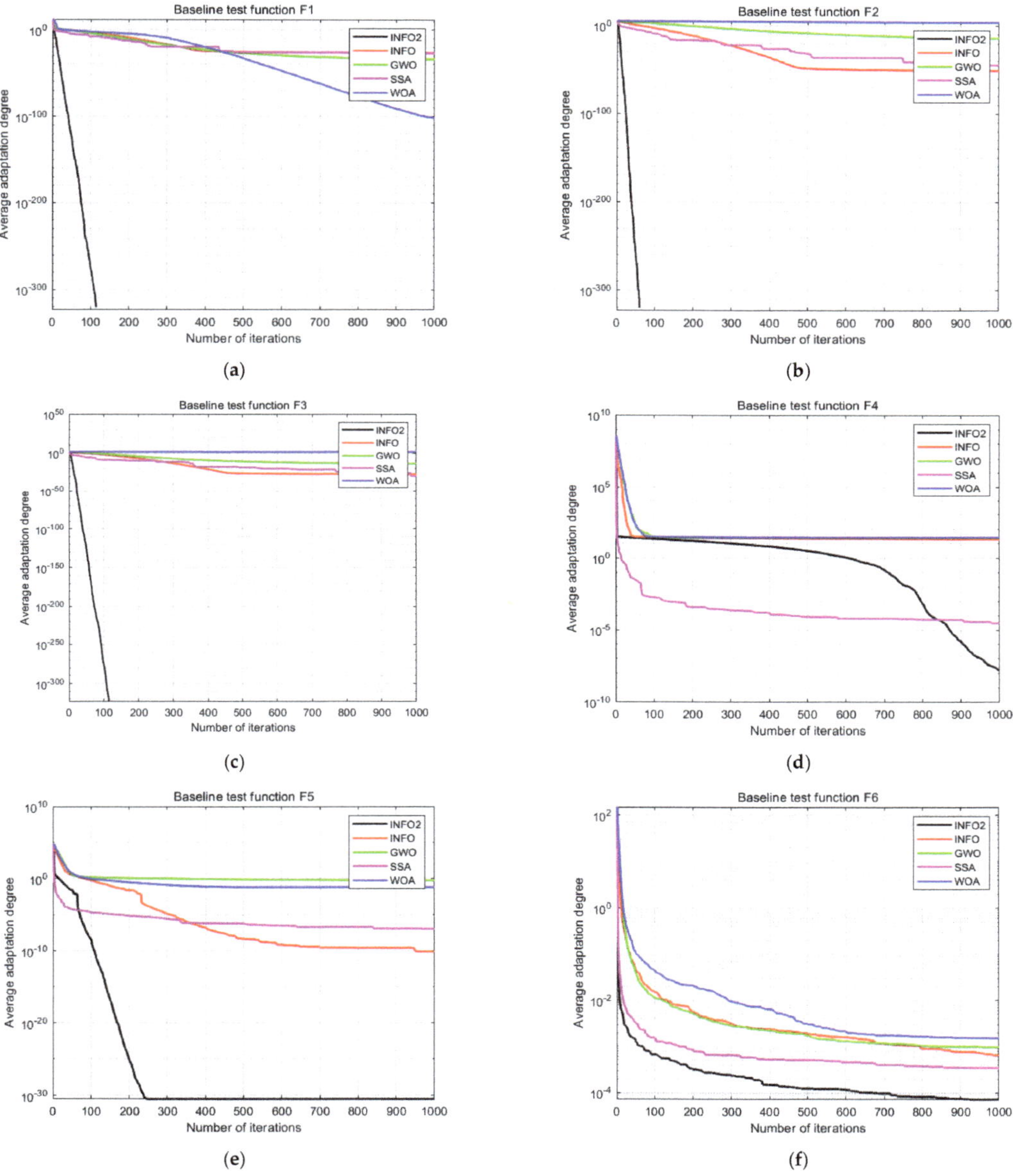

**Figure 5.** Results for six selected high-dimensional single-objective test functions: (**a**) F1, (**b**) F2, (**c**) F3, (**d**) F4, (**e**) F5, and (**f**) F6.

### 3.2.2. High-Dimensional Multi-Objective Test Function

To demonstrate the optimization-seeking effects of the improved INFO algorithm on high-latitude multi-objective test functions, four more of the 14 test functions were selected for testing. Their convergence curves are shown in Figure 6, and their optimization-seeking space diagrams are shown in Figure 7. The results for INFO1 and INFO2 are shown in Figures 8 and 9, respectively. As above, in this test, INFO1 represents an improved strategy that introduces two-stage backward learning and a greedy mechanism in the original INFO algorithm, and INFO2 represents an improved strategy that introduces a differential evolution algorithm and an adaptive $t$-distribution in the original INFO algorithm.

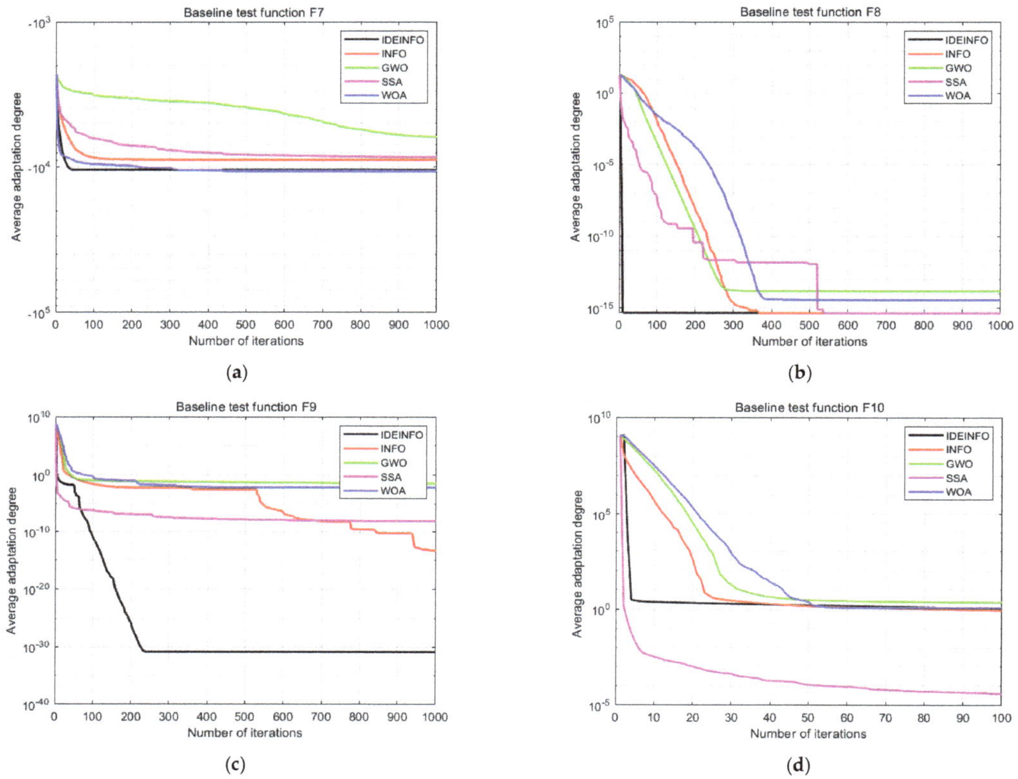

**Figure 6.** Convergence curves of four selected high-dimensional multi-objective test functions: (**a**) F7, (**b**) F8, (**c**) F9, and (**d**) F10.

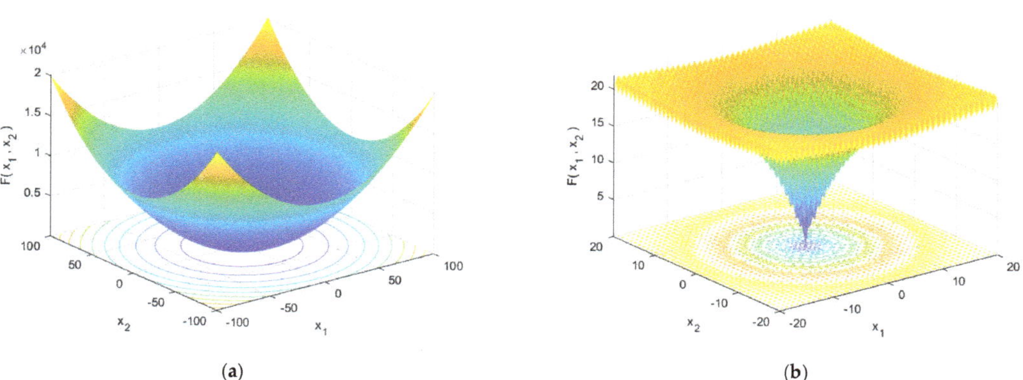

(**a**)  (**b**)

**Figure 7.** *Cont.*

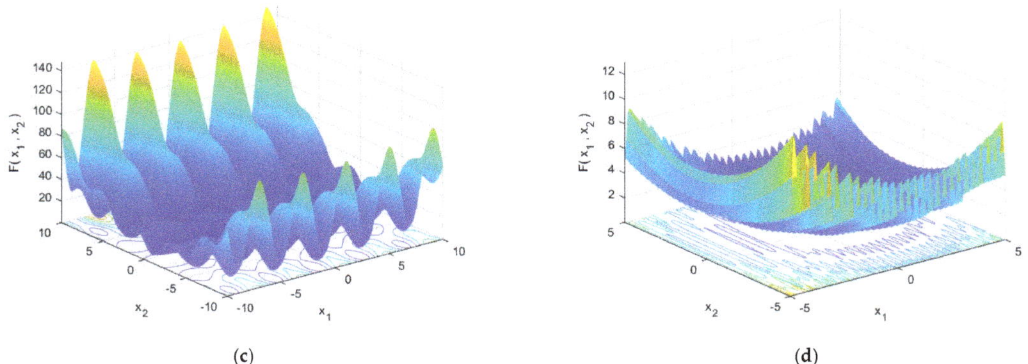

**Figure 7.** Optimization-seeking space diagrams for four selected high-dimensional multi-objective test functions: (**a**) F7, (**b**) F8, (**c**) F9, and (**d**) F10.

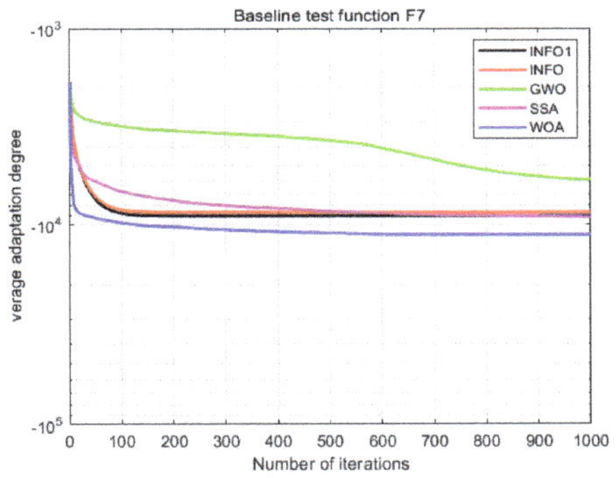

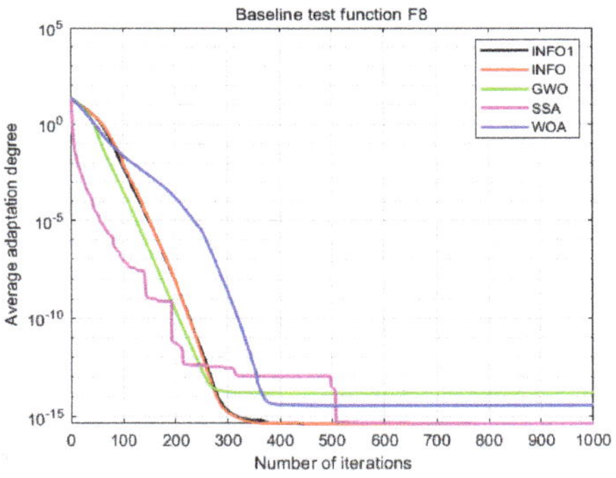

**Figure 8.** Results for four selected high-dimensional multi-objective test functions (**a**) F7, (**b**) F8, (**c**) F9, and (**d**) F10.

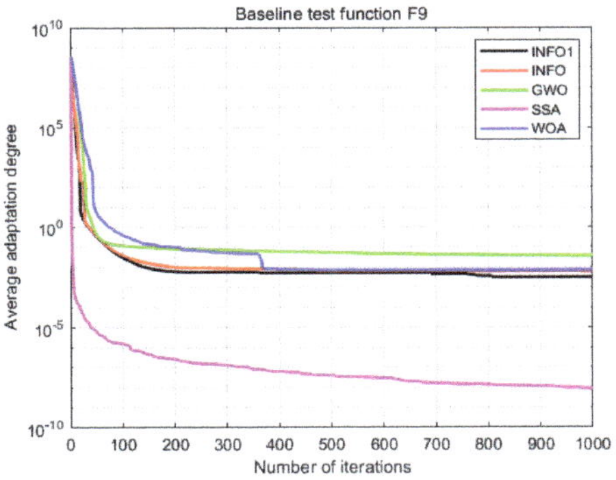

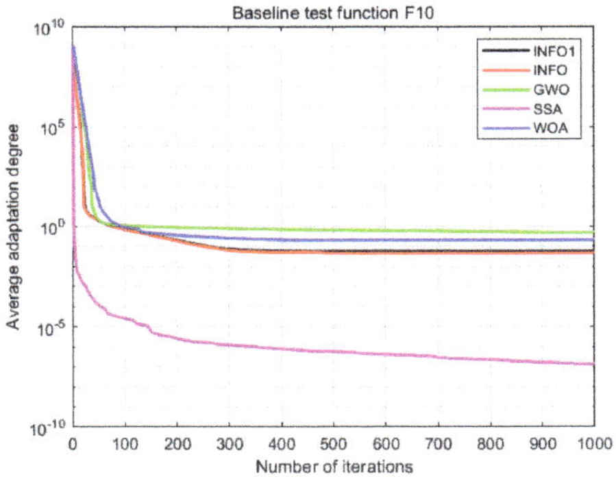

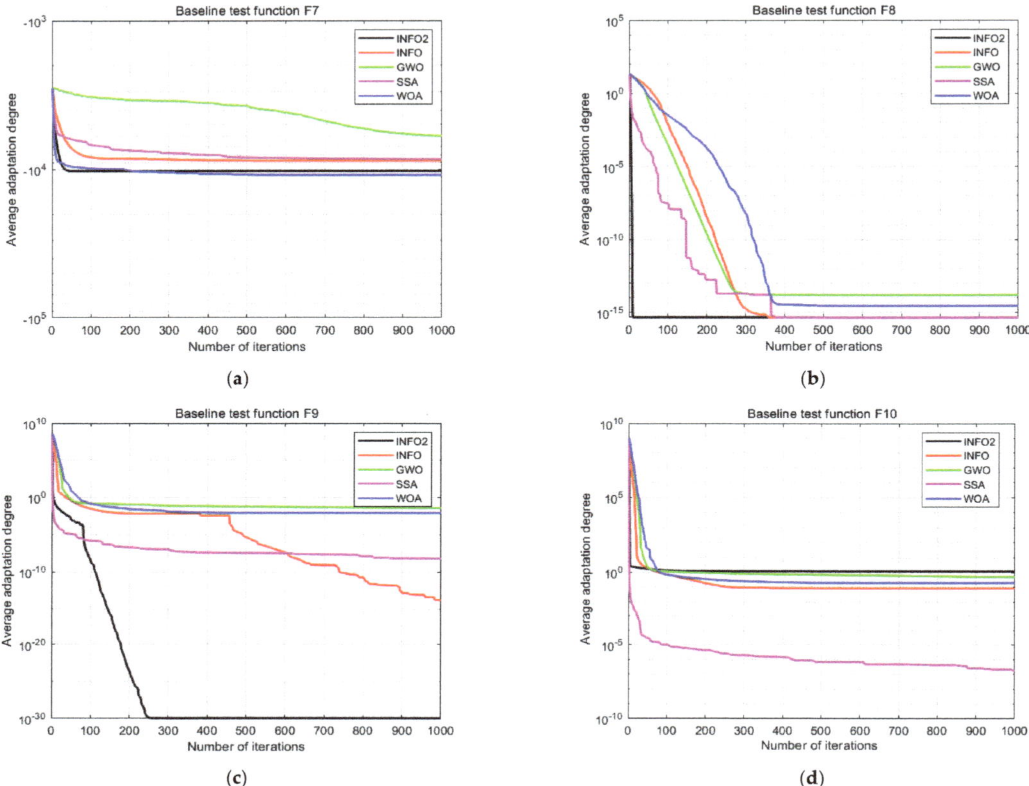

**Figure 9.** Results for four selected high-dimensional multi-objective test functions (**a**) F7, (**b**) F8, (**c**) F9, and (**d**) F10.

To verify the efficacy of IDEINFO, ablation experiments based on INFO1 and INFO2 were conducted for comparison. INFO1 represents an improved strategy that introduces two-stage backward learning and a greedy mechanism in the original INFO algorithm. INFO2 represents an improved strategy that introduces a differential evolution algorithm and an adaptive *t*-distribution in the original INFO algorithm.

### 3.2.3. Low-Dimensional Test Functions

A low-latitude test function is one whose search space dimensionality is relatively low. To demonstrate the optimization-seeking effects of the improved INFO algorithm on low-latitude test functions, the last four of the 14 functions were selected for testing. Their convergence curves are shown in Figure 10, and their optimization-seeking space diagrams are presented in Figure 11. As above, INFO1 represents an improved strategy that introduces two-stage backward learning and a greedy mechanism in the original INFO algorithm, while INFO2 represents an improved strategy that introduces a differential evolution algorithm and an adaptive *t*-distribution in the original INFO algorithm. The results graphs for INFO1 and INFO2 are illustrated in Figures 12 and 13, respectively.

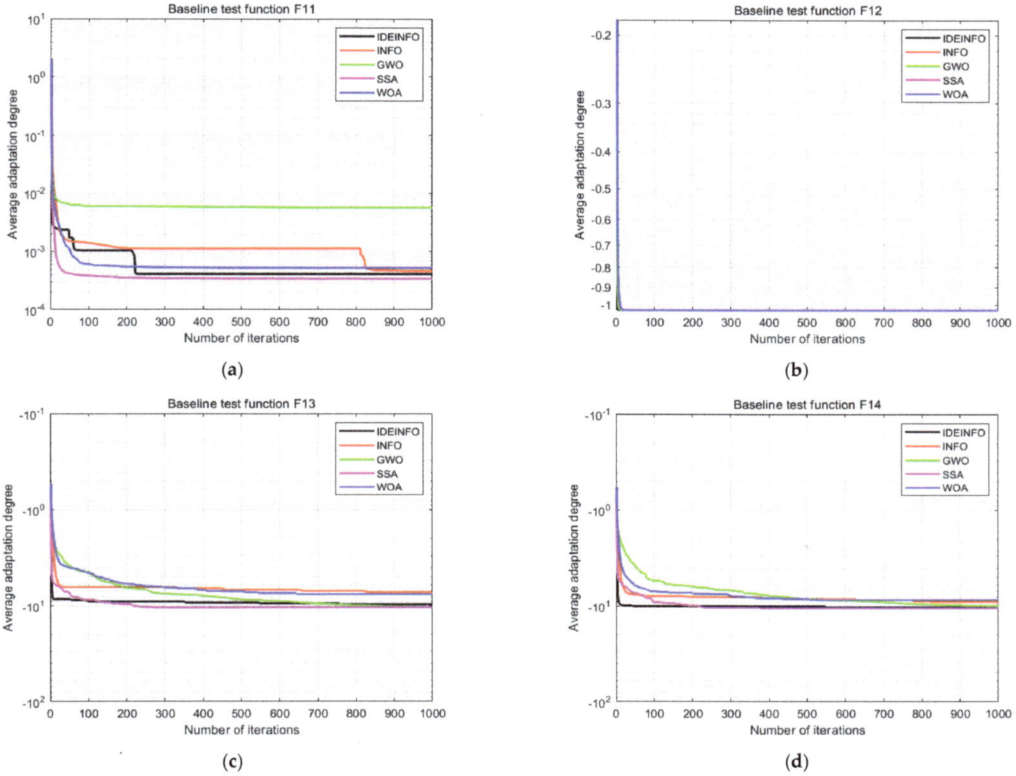

**Figure 10.** Convergence curves of four selected low-dimensional test functions: (**a**) F11, (**b**) F12, (**c**) F13, and (**d**) F14.

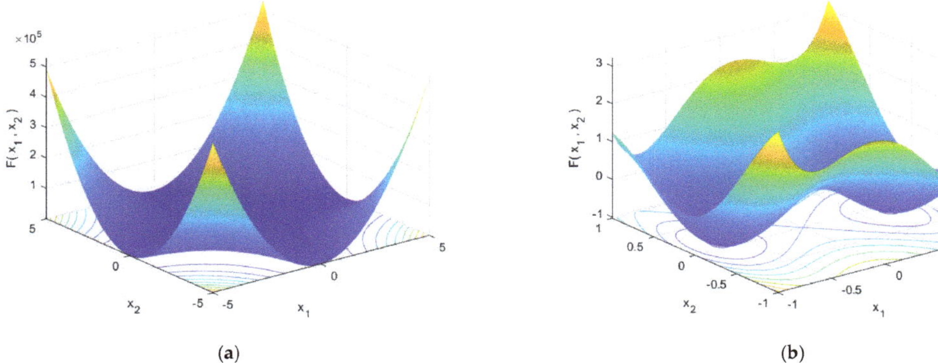

**Figure 11.** *Cont.*

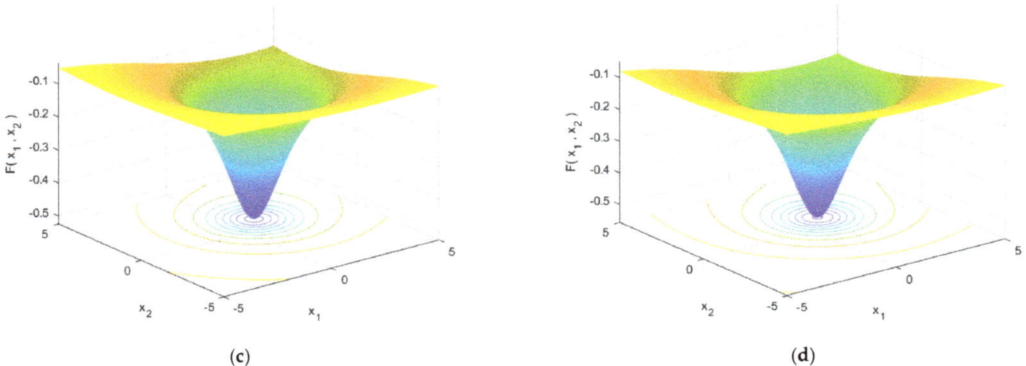

**Figure 11.** Optimization-seeking space diagrams for four selected low-dimensional test functions: (**a**) F11, (**b**) F12, (**c**) F13, and (**d**) F14.

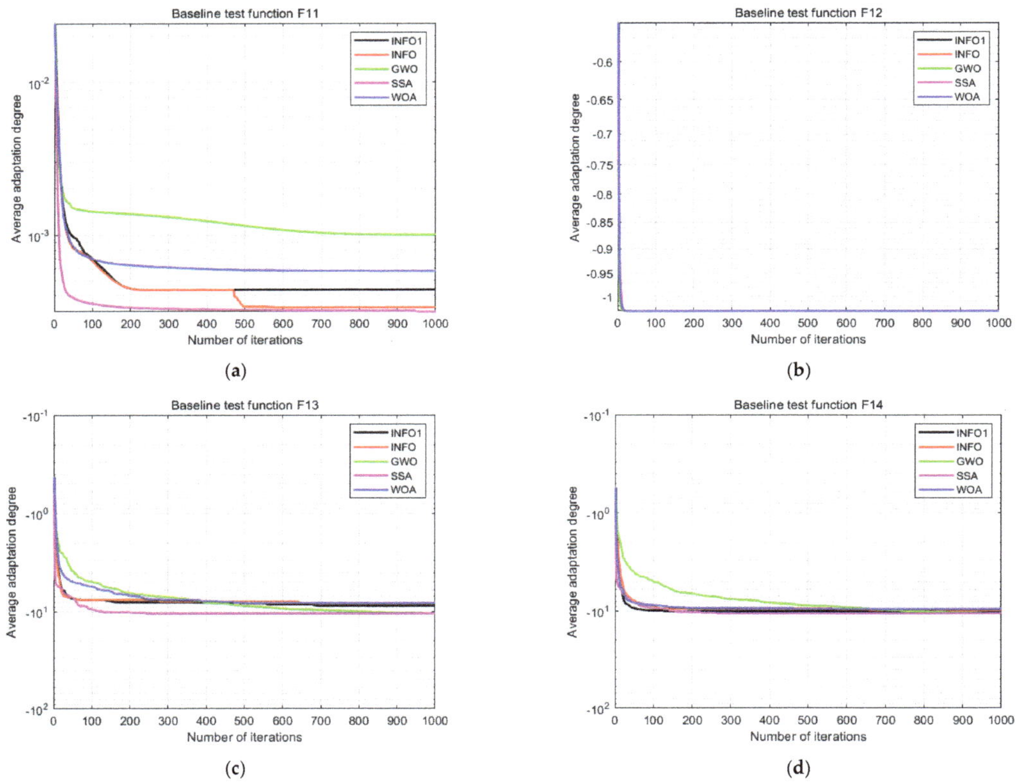

**Figure 12.** Results for four selected low-dimensional test functions (**a**) F11, (**b**) F12, (**c**) F13, and (**d**) F14.

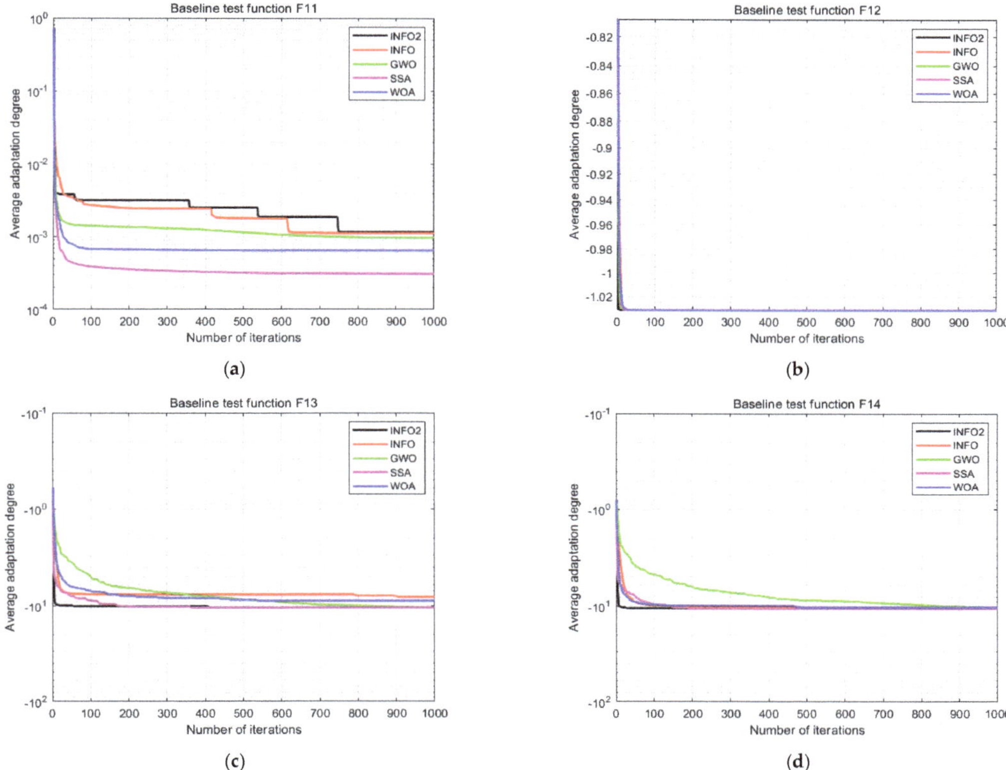

**Figure 13.** Results for four selected low-dimensional test functions (**a**) F11, (**b**) F12, (**c**) F13, and (**d**) F14.

To verify the efficacy of IDEINFO, ablation experiments based on INFO1 and INFO2 were conducted for comparison. INFO1 represents an improved strategy that introduces two-stage backward learning and a greedy mechanism in the original INFO algorithm.INFO2 represents an improved strategy that introduces a differential evolution algorithm and an adaptive $t$-distribution in the original INFO algorithm.

## 4. Discussion

### 4.1. Comparison of Convergence Results

Tables 4–6 present the summary analysis of the results of testing high-latitude single-target, high-latitude multi-target, and low-latitude benchmark functions. From these tables, it can be seen that the improved INFO algorithm has the best results, and comparison tests with INFO1 and INFO2 show that the improved method provides superior accuracy and robustness. For the high-dimensional test function, the experimental results in Table 4 show that, in general, the difference between IDEINFO and INFO2 is not significant, but IDEINFO has the highest optimization accuracy among the six compared algorithms, and the solution accuracy and stability of IDEINFO are significantly better than the five comparison algorithms. The experimental results in Table 5 show that IDEINFO has the highest overall. However, among the F9 functions, INFO2 outperforms the other comparative algorithms.

**Table 4.** Experimental results for six high-latitude single-target benchmark test functions.

| F | Index | IDEINFO | INFO | SSA | GWO | WOA | INFO1 | INFO2 |
|---|---|---|---|---|---|---|---|---|
| F1 | Best | **0** | $6.7057 \times 10^{-28}$ | **0** | $2.7058 \times 10^{-36}$ | $1.3810 \times 10^{-115}$ | $4.2970 \times 10^{-28}$ | **0** |
| | Mean | **0** | $2.8115 \times 10^{-27}$ | $6.4300 \times 10^{-20}$ | $1.0693 \times 10^{-34}$ | $2.5704 \times 10^{-102}$ | $2.8729 \times 10^{-27}$ | **0** |
| | Std | **0** | $6.8192 \times 10^{-28}$ | $7.2766 \times 10^{-19}$ | $1.0672^{-34}$ | $2.1129 \times 10^{-101}$ | $6.3544 \times 10^{-28}$ | **0** |
| F2 | Best | **0** | $1.2106 \times 10^{-53}$ | **0** | $6.3897 \times 10^{-21}$ | 9569.1641 | $9.2088 \times 10^{-54}$ | **0** |
| | Mean | **0** | $7.8038 \times 10^{-52}$ | $4.2017 \times 10^{-34}$ | $1.4629 \times 10^{-14}$ | 20,730.2593 | $7.6727 \times 10^{-52}$ | **0** |
| | Std | **0** | $7.0452 \times 10^{-52}$ | $5.9418 \times 10^{-33}$ | $5.7738 \times 10^{-14}$ | 2569.6795 | $8.3436 \times 10^{-52}$ | **0** |
| F3 | Best | **0** | $3.4184 \times 10^{-29}$ | 0 | $2.9501 \times 10^{-16}$ | 0.00011002 | $5.7132 \times 10^{-29}$ | **0** |
| | Mean | **0** | $5.7937 \times 10^{-28}$ | $1.1195 \times 10^{-20}$ | $1.7655 \times 10^{-14}$ | 32.8655 | $5.2304 \times 10^{-28}$ | **0** |
| | Std | **0** | $3.4028 \times 10^{-28}$ | $1.4256 \times 10^{-19}$ | $3.4399 \times 10^{-14}$ | 28.2881 | $2.51 \times 10^{-28}$ | **0** |
| F4 | Best | $\mathbf{4.6904 \times 10^{-15}}$ | 16.9501 | $2.2361 \times 10^{-11}$ | 25.0954 | 26.1798 | 16.5694 | $3.1549 \times 10^{-14}$ |
| | Mean | $\mathbf{1.5759 \times 10^{-8}}$ | 19.4498 | $2.5536 \times 10^{-5}$ | 26.7773 | 27.1942 | 19.4832 | $1.5887 \times 10^{-8}$ |
| | Std | $\mathbf{3.6914 \times 10^{-8}}$ | 0.7041 | $6.0245 \times 10^{-5}$ | 0.7764 | 0.5283 | 0.7499 | $6.1874 \times 10^{-8}$ |
| F5 | Best | $\mathbf{1.2326 \times 10^{-32}}$ | $2.6038 \times 10^{-20}$ | $4.009 \times 10^{-11}$ | $1.2798 \times 10^{-5}$ | 0.0092902 | $1.1449 \times 10^{-19}$ | $1.5407 \times 10^{-32}$ |
| | Mean | $\mathbf{3.2376 \times 10^{-31}}$ | $2.0788 \times 10^{-13}$ | $1.2121 \times 10^{-7}$ | 0.6228 | 0.068798 | $1.9366 \times 10^{-14}$ | $7.0196 \times 10^{-31}$ |
| | Std | $\mathbf{5.1236 \times 10^{-31}}$ | $1.0668 \times 10^{-12}$ | $3.522 \times 10^{-7}$ | 0.42264 | 0.078427 | $6.2517 \times 10^{-14}$ | $1.1262 \times 10^{-30}$ |
| F6 | Best | $\mathbf{6.2313 \times 10^{-7}}$ | $5.1252 \times 10^{-5}$ | $1.4029 \times 10^{-0}$ | 0.00013678 | $4.3456 \times 10^{-5}$ | $6.7265 \times 10^{-0}$ | $7.2831 \times 10^{-6}$ |
| | Mean | $7.2422 \times 10^{-5}$ | 0.00077635 | 0.00077635 | 0.00075583 | 0.0015206 | 0.00084333 | $\mathbf{6.5934 \times 10^{-5}}$ |
| | Std | $5.6658 \times 10^{-5}$ | 0.00066449 | 0.00066449 | 0.00044961 | 0.0016606 | 0.00073516 | $\mathbf{4.0882 \times 10^{-5}}$ |

**Table 5.** Experimental results for four high-latitude multi-objective benchmark test functions.

| F | Index | IDEINFO | INFO | SSA | GWO | WOA | INFO1 | INFO2 |
|---|---|---|---|---|---|---|---|---|
| F7 | Best | $-11205.917$ | $-11365.255$ | **$-12569.485$** | $-7610.723$ | $-12569.482$ | $-10928.05$ | $-12094.216$ |
| | Mean | $-10218.713$ | $-8662.457$ | $-9140.772$ | $-5864.470$ | **$-11201.082$** | $-9034.592$ | $-10451.007$ |
| | Std | 592.216 | 731.942 | 2642.386 | 717.885 | 1541.727 | 641.344 | 617.033 |
| F8 | Best | $\mathbf{4.4409 \times 10^{-16}}$ | $\mathbf{4.4409 \times 10^{-16}}$ | $\mathbf{4.4409 \times 10^{-16}}$ | $1.1102 \times 10^{-14}$ | $\mathbf{4.4409 \times 10^{-16}}$ | $\mathbf{4.4409 \times 10^{-16}}$ | $\mathbf{4.4409 \times 10^{-16}}$ |
| | Mean | $\mathbf{4.4409 \times 10^{-16}}$ | $\mathbf{4.4409 \times 10^{-16}}$ | $\mathbf{4.4409 \times 10^{-16}}$ | $1.5881 \times 10^{-14}$ | $3.6948 \times 10^{-15}$ | $\mathbf{4.4409 \times 10^{-16}}$ | $\mathbf{4.4409 \times 10^{-16}}$ |
| | Std | **0** | **0** | **0** | $2.8846 \times 10^{-15}$ | $2.4875 \times 10^{-15}$ | **0** | **0** |
| F9 | Best | $1.7077 \times 10^{-32}$ | $3.4691 \times 10^{-21}$ | $7.7964 \times 10^{-14}$ | 0.0062327 | 0.00062453 | $4.7647 \times 10^{-21}$ | $\mathbf{1.6593 \times 10^{-32}}$ |
| | Mean | $1.1726 \times 10^{-30}$ | 0.0062119 | $8.5844 \times 10^{-9}$ | 0.037611 | 0.0074199 | 0.0031101 | $\mathbf{1.3445 \times 10^{-31}}$ |
| | Std | $2.7026 \times 10^{-30}$ | 0.028725 | $2.0196 \times 10^{-8}$ | 0.020538 | 0.0087554 | 0.017729 | $\mathbf{5.8543 \times 10^{-31}}$ |
| F10 | Best | 0.043939 | $1.0754 \times 10^{-18}$ | $7.0359 \times 10^{-12}$ | $4.5913 \times 10^{-5}$ | 0.012006 | $\mathbf{4.5503 \times 10^{-19}}$ | 0.043939 |
| | Mean | 1.2208 | 0.050698 | $\mathbf{1.417 \times 10^{-7}}$ | 0.51788 | 0.21943 | 0.057427 | 1.2345 |
| | Std | 1.1314 | 0.07418 | $\mathbf{3.8101 \times 10^{-7}}$ | 0.22404 | 0.15429 | 0.06979 | 1.1609 |

In the low-dimensional test functions, as shown in Table 6, the IDEINFO algorithm proposed in this paper significantly outperforms the other six algorithms, and the optimal solution is found for each of the compared algorithms in the optimization experiment for the F12 function. Therefore, it is verified that the IDEINFO algorithm is highly robust in solving low- and high-dimensional problems, demonstrating that the IDEINFO algorithm has some competitive advantage in solving function optimization problems.

### 4.2. Wilcoxon Rank Sum Test

In the above simulation experiment, the mean and standard deviation alone cannot fully verify the superiority of the IDEINFO algorithm. To ensure the fairness and validity of the algorithm, it is necessary to conduct a statistical test. In this paper, we use the Wilcoxon rank sum test to verify whether the results of each IDEINFO experiment are statistically significantly different from other algorithms. The rank sum test was performed at the 5% significance level, and at $p < 5\%$, it can be considered as rejecting the H0 hypothesis, indicating that there is a significant difference between the two algorithms; for $p > 5\%$, the H0 hypothesis is accepted, indicating that the two algorithms have the same overall performance. Table 7 compares the IDEINFO, INFO1, INFO2,INFO, SSA, WOA, and GWO algorithms for 14 benchmark tests. The Wilcoxon rank sum test on 14 benchmark functions is presented in Table 7. It indicates the comparable performance between the two, where "Na" is "not applicable"; i.e., no significant test can be performed. For the results of the significance tests, "+", "−", and "=", indicate that the performance of IDEINFO is better or worse than that of the compared algorithm.

**Table 6.** Experimental results for four low-latitude benchmark test functions.

| F | Index | IDEINFO | INFO | SSA | GWO | WOA | INFO1 | INFO2 |
|---|---|---|---|---|---|---|---|---|
| F11 | Best | 0.00030749 | 0.00030749 | 0.00030749 | 0.00030749 | **0.00030752** | 0.0030749 | 0.00030749 |
| | Mean | 0.0011592 | 0.00033496 | 0.00031349 | 0.0010126 | 0.00058225 | **0.0043524** | 0.0041142 |
| | Std | 0.0003646 | 0.0001566 | **$6.6315 \times 10^{-5}$** | 0.0037102 | 0.00040975 | 0.0014248 | 0.00032207 |
| F12 | Best | −1.0316 | −1.0316 | −1.0316 | −1.0316 | −1.0316 | −1.0316 | −1.0316 |
| | Mean | −1.0316 | −1.0316 | −1.0316 | −1.0316 | −1.0316 | −1.0316 | −1.0316 |
| | Std | $6.7752 \times 10^{-16}$ | $6.7752 \times 10^{-16}$ | $3.905 \times 10^{-9}$ | $4.2802 \times 10^{-9}$ | $1.3367 \times 10^{-10}$ | **$6.7122 \times 10^{-16}$** | **$6.7122 \times 10^{-16}$** |
| F13 | Best | −10.4029 | −10.4029 | −10.4029 | −10.4029 | −10.4027 | −10.4029 | −10.4029 |
| | Mean | −10.4029 | −8.6849 | −10.4029 | −10.4028 | −8.1503 | −8.6849 | −9.7031 |
| | Std | **$1.8067 \times 10^{-15}$** | 3.1747 | $1.2781 \times 10^{-5}$ | $9.0224 \times 10^{-5}$ | 3.1151 | 3.1747 | 2.1403 |
| F14 | Best | **−10.5364** | −10.5364 | −10.5364 | −10.5364 | −10.5364 | −10.5364 | −10.5364 |
| | Mean | **−10.5364** | −9.8192 | −10.5364 | −10.5363 | −9.4995 | −10.0462 | −10.313 |
| | Std | **$1.6493 \times 10^{-15}$** | 2.1989 | $1.8401 \times 10^{-5}$ | $8.0884 \times 10^{-5}$ | 2.3942 | 1.8886 | 1.2234 |

**Table 7.** Wilcoxon rank sum test results.

| F | Index | INFO | SSA | GWO | WOA | INFO1 | INFO2 |
|---|---|---|---|---|---|---|---|
| F1 | P | $6.3864 \times 10^{-5}$ | 0.00023125 | $6.3864 \times 10^{-5}$ | $6.3864 \times 10^{-5}$ | $6.3864 \times 10^{-5}$ | NaN |
| | R | + | + | + | + | + | = |
| F2 | P | $6.3864 \times 10^{-5}$ | 0.00075118 | $6.3864 \times 10^{-5}$ | $6.3864 \times 10^{-5}$ | $6.3864 \times 10^{-5}$ | NaN |
| | R | + | + | + | + | + | = |
| F3 | P | $6.3864 \times 10^{-5}$ | $6.3864 \times 10^{-5}$ | $6.3864 \times 10^{-5}$ | $6.3864 \times 10^{-5}$ | $6.3864 \times 10^{-5}$ | NaN |
| | R | + | + | + | + | + | = |
| F4 | P | 0.00018267 | 0.00018267 | 0.00018267 | 0.00018267 | 0.00018267 | 0.73373 |
| | R | + | + | + | + | + | - |
| F5 | P | 0.00018165 | 0.00018165 | 0.00018165 | 0.00018165 | 0.00018165 | 1 |
| | R | + | + | + | + | + | - |
| F6 | P | 0.00018267 | 0.001008 | 0.00018267 | 0.00018267 | 0.00018267 | 0.57075 |
| | R | + | + | + | + | + | - |
| F7 | P | 0.00018165 | 0.14047 | 0.00018267 | 0.79134 | 0.00018165 | 0.47268 |
| | R | + | - | + | - | + | - |
| F8 | P | 0.00018165 | NaN | $4.0402 \times 10^{-5}$ | 0.00018923 | 0.00018165 | NaN |
| | R | + | = | + | + | + | = |
| F9 | P | 0.00018267 | 0.00018267 | 0.00018267 | 0.00018267 | 0.00018267 | 0.42736 |
| | R | + | + | + | + | + | - |

**Table 7.** *Cont.*

| F | Index | INFO | SSA | GWO | WOA | INFO1 | INFO2 |
|---|---|---|---|---|---|---|---|
| F10 | P | 0.00018267 | 0.021134 | 0.14047 | 0.00018267 | 0.00018267 | 0.8796 |
|  | R | + | + | - | + | + | - |
| F11 | P | 0.00017562 | 0.00017562 | 0.00017562 | 0.00017562 | 0.00017562 | 0.010336 |
|  | R | + | + | + | + | + | + |
| F12 | P | $6.3864 \times 10^{-5}$ | $6.3864 \times 10^{-5}$ | $6.3864 \times 10^{-5}$ | $6.3864 \times 10^{-5}$ | $6.3864 \times 10^{-5}$ | NaN |
|  | R | + | + | + | + | + | = |
| F13 | P | $6.3864 \times 10^{-5}$ | $6.3864 \times 10^{-5}$ | $6.3864 \times 10^{-5}$ | $6.3864 \times 10^{-5}$ | $6.3864 \times 10^{-5}$ | NaN |
|  | R | + | + | + | + | + | = |
| F14 | P | 0.00010997 | 0.00010997 | 0.00010997 | 0.00010997 | 0.00010997 | 0.58278 |
|  | R | + | + | + | + | + | - |

As shown in Table 7, most of the $p$-values are less than 5%, and the overall performance of IDEINFO is statistically significantly different from the other six algorithms, thus indicating that IDEINFO has better performance than the other algorithms.

## 5. Conclusions

This study proposed an improved INFO algorithm that overcomes the shortcomings of the traditional version. The new version initializes candidate solutions using a two-stage backward learning strategy, which improves the uniformity of their distribution and enhances the search capability of the algorithm. The new INFO algorithm is augmented by combining it with a greedy search algorithm, which results in improved individual vectors per iteration and an improved search capacity. During the iterative search process, a DE strategy is applied to perturb the vectors and generate genetically superior candidate solutions. Furthermore, the search range is expanded probabilistically and combined with a $t$-distribution strategy, which helps avoid local optima traps and improves the global search capability. Using fourteen standard test functions, the improved INFO outperformed the baseline INFO, SSA, GWO, and WOA models. To further verify the efficacy of the improvement points, comparisons were made with INFO1, which performs only two-stage backward learning and a greedy mechanism, and INFO2, which only performs the combined DE and adaptive $t$-distribution actions. The ablative results show that the proposed improved INFO algorithm has better generality, whereas the others present limitations. In the future, we plan to determine how to balance the time and optimization capabilities of the new algorithm, improve its stability, and find practical applications. The convergence rate of IDEINFO proposed in this paper is very impressive because the position of the vectors always tends to move toward the region where a better solution is available. In addition, the IDEINFO algorithm can solve practically complex and challenging optimization problems with constrained and unknown search domains.

For future research, we suggest the following corrections and considerations. Moreover, we suggest enhancing INFO using different types of local search operators in the original INFO; this may further improve the algorithm's optimality seeking ability and its capability to address the challenging nature of complex scenarios. The IDEINFO algorithm proposed in this paper can also be enriched in terms of exploratory and exploitative trends. For example, using different concepts, such as chaotic mapping, could enrich the exploratory and exploitative trends of the proposed IDEINFO. The traditional INFO, or its improved variants, can be applied to applications such as parameter tuning in neural network models, effect enhancement of model prediction methods, and deep learning.

**Author Contributions:** Conceptualization, H.J.; methodology, L.Z.; software, L.Z. All authors have read and agreed to the published version of the manuscript.

**Funding:** This research received no external funding.

**Informed Consent Statement:** Not applicable.

**Data Availability Statement:** The data supporting the findings of this study are available from the corresponding author upon reasonable request.

**Acknowledgments:** The author would like to thank the editor, the academic editor, and anonymous referees who kindly reviewed the earlier version of this manuscript and provided valuable suggestions, comments, and Refs.

**Conflicts of Interest:** The authors declare no conflict of interest.

# References

1. Su, Y.; Han, L.; Wang, H.; Wang, J. The workshop scheduling problems based on data mining and particle swarm optimization algorithm in machine learning areas. *Enterp. Inf. Syst.* **2022**, *2*, 363–378. [CrossRef]
2. Mou, J.; Duan, P.; Gao, L.; Liu, X.; Li, J. An effective hybrid collaborative algorithm forenergy-efficient distributed permutation flow-shop inverse scheduling. *Future Gener. Comput. Syst.* **2022**, *128*, 521–537. [CrossRef]
3. Berkan, A.I. A hybrid firefly and particle swarm optimization algorithm for computationally expensive numerical problems. *App. Soft Comput.* **2018**, *66*, 232–249.
4. Zhang, J.; Kewen, X.; Ziping, H.; Shurui, F. Dynamic multi-swarm differential learning quantum bird swarm algorithm and its application in random forest classification model. *Comput. Intell. Neurosci.* **2020**, *2020*, 6858541. [CrossRef]
5. Zhang, M.; Yu, C.; Jingqiang, L. A privacy-preserving optimization of neighborhood-based recommendation for medical-aided diagnosis and treatment. *IEEE Internet Things J.* **2021**, *8*, 10830–10842. [CrossRef]
6. Pallonetto, F.; De Rosa, M.; Finn, D.P. Impact of intelligent control algorithms on demand response flexibility and thermal comfort in a smart grid ready residential building. *Smart Energy* **2021**, *2*, 100017. [CrossRef]
7. Yongbin, Y.; Yang, C.; Deng, Q.; Nyima, T.; Liang, S.; Zhou, C. Memristive network-basedgenetic algorithm and its application to image edge detection. *J.Syst. Eng. Electron.* **2021**, *32*, 1062–1070. [CrossRef]
8. Ai, N.; Bin, W.; Boyu, L.; Zhipeng, Z. 5G heterogeneous network selection and resource allocation optimization based on cuckoo search algorithm. *Comput. Comm.* **2021**, *168*, 170–177. [CrossRef]
9. Passino, K.M. Biomimicry of bacterial foraging for distributed optimization and control. *IEEE Control Sys.* **2002**, *22*, 52–67.
10. Holland John, H. *Adaptation in Natural and Artificial Systems:An Introductory Analysis with Applications to Biology, Control, and Artificial Intelligence*; The MIT Press: Cambridge, MA, USA, 1992.
11. Holland, J.H. Genetic algorithms. *Sci. Am.* **1992**, *267*, 66–73. [CrossRef]
12. Knowles, J.; Corne, D. The Pareto archived evolution strategy: A new baseline algorithm for Pareto multiobjective optimization. In Proceedings of the 1999 Congress on Evolutionary Computation-CEC99 (Cat. No. 99TH8406), Washington, DC, USA, 6–9 July 1999.
13. Kirkpatrick, S.; Gelatt, C.D.; Vecchi, M.P. Optimization by simulated annealing. *Science* **1983**, *220*, 671–680. [CrossRef]
14. Eker, E.; Kayri, M.; Ekinci, S.; Izci, D. A New Fusion of ASO with SA Algorithm and Its Applications to MLP Training and DC Motor Speed Control. *Arab. J. Sci. Eng.* **2021**, *46*, 3889–3911. [CrossRef]
15. Mirjalili, S.; Mirjalili, S.M.; Lewis, A. Grey wolf optimizer. *Adv. Eng. Softw.* **2014**, *69*, 46–61. [CrossRef]
16. Saxena, A.; Kumar, R.; Das, S. β-chaotic map-enabled grey wolf optimizer. *Appl. Soft Comput.* **2019**, *75*, 84–105. [CrossRef]
17. Ahmadianfar, I.; Bozorg-Haddad, O.; Chu, X. Gradient-based optimizer: A new metaheuristic optimization algorithm. *Information Sciences* **2020**, *540*, 131–159. [CrossRef]
18. Li, S.P.; Sun, Y.Z.; Xiaohui, W. Speech Emotion Recognition Using Hybrid GBO Algorithm-based ANFIS Classifier. *Indian J. Pharm. Sci.* **2019**, *81*, S52–S53.
19. Kadkhodazadeh, M.; Farzin, S. A Novel LSSVM Model Integrated with GBO Algorithm to Assessment of Water Quality Parameters. *Water Resources Management* **2021**, *35*, 3939–3968. [CrossRef]
20. Zhangfang, H. Improved particle swarm optimization algorithm for mobile robot path planning. *Comput. Appl. Res.* **2021**, *38*, 3089–3092.
21. Yuxiang, H.; Huanbing, G.; Zijian, W.; Chuansheng, D. Improved Grey Wolf Optimization Algorithm and Application. *Sensors* **2022**, *22*.
22. Khan, M.K.; Zafar, M.H.; Rashid, S.; Mansoor, M.; Moosavi, S.K.R.; Sanfilippo, F. Improved Reptile Search Optimization Algorithm: Application on Regression and Classification Problems. *App. Sci.* **2023**, *13*, 945. [CrossRef]
23. Ayyarao, T.S.; Ramakrishna, N.; Elavarasan, R.M.; Polumahanthi, N.; Rambabu, M.; Saini, G.; Khan, B.; Alatas, B. War Strategy Optimization Algorithm: A New Effective Metaheuristic Algorithm for Global Optimization. *IEEE Access* **2022**, *10*, 25073–25105. [CrossRef]
24. Wolpert, D.H.; Macready, W.G. No free lunch theorems for optimization. *IEEE Trans. Evol.Comput.* **1997**, *1*, 67–82. [CrossRef]
25. Service, T.C. A No Free Lunch theorem for multi-objective optimization. *Inf. Process. Lett.* **2010**, *110*, 917–923. [CrossRef]
26. Iman, A.; Heidari, A.A.; Noshadian, S.; Chen, H.; Gandomi, A.H. INFO: An efficient optimization algorithm based on weighted mean of vectors. *Expert Syst. Appl.* **2022**, *195*, 116516.
27. Akyol, S.; Alatas, B. Plant intelligence based metaheuristic optimization algorithms. *Artif. Intell. Rev.* **2017**, *47*, 417–462. [CrossRef]

28. Dorigo, M.; Maniezzo, V.; Colorni, A. Ant system: Optimization by a colony of cooperating agents. IEEE transactions on systems, man, and cybernetics. *Part B Cybern. A Publ. IEEE Syst. Man Cybern. Soc.* **1996**, *26*, 29–41. [CrossRef]

29. Karaboga, D.; Basturk, B. A powerful and efficient algorithm for numerical function optimization: Artificial bee colony (ABC) algorithm. *J. Glob. Optim.* **2007**, *39*, 459–471. [CrossRef]

30. Heidari, A.A.; Abbaspour, R.A.; Chen, H. Efficient boosted grey wolf optimizers for global search and kernel extreme learning machine training. *J. App. Soft Comput.* **2019**, *81*, 105521. [CrossRef]

31. Yang, Y.; Chen, H.; Heidari, A.A.; Gandomi, A.H. Hunger games search: Visions, conception, implementation, deep analysis, perspectives, and towards performance shifts. *Expert Syst. Appl.* **2021**, *177*, 114864. [CrossRef]

32. Ahmadianfar, I.; Heidari, A.A.; Gandomi, A.H.; Chu, X.; Chen, H. RUN beyond the metaphor: An efficient optimization algorithm based on Runge Kutta method. *Expert Syst. Appl.* **2021**, *181*, 115079. [CrossRef]

33. Li, S.; Chen, H.; Wang, M.; Heidari, A.A.; Mirjalili, S. Slime mould algorithm: A new method for stochastic optimization. *Future Gener. Comput. Syst.* **2020**, *111*, 300–323. [CrossRef]

34. Mirjalili, S.; Lewis, A. The Whale Optimization Algorithm. *Adv. Eng. Softw.* **2016**, *95*, 51–67. [CrossRef]

35. Bonyadi, M.R.; Michalewicz, Z. Particle Swarm Optimization for Single Objective Continuous Space Problems: A Review. *Evol. Comput.* **2017**, *25*, 1–54. [CrossRef] [PubMed]

36. Jwad, K.I.; Yuegang, T.; Layth, Q. Integration of DE Algorithm with PDC-APF for Enhancement of Contour Path Planning of a Universal Robot. *App. Sci.* **2021**, *11*, 6532.

37. Ghiasi, M.; Niknam, T.; Dehghani, M.; Siano, P.; Haes, A.H.; AlHinai, A. Optimal Multi-Operation Energy Management in Smart Microgrids in the Presence of RESs Based on Multi-Objective Improved DE Algorithm: Cost-Emission Based Optimization. *Appl. Sci.* **2021**, *11*, 3661. [CrossRef]

38. Shihong, Y.; Qifang, L.; Yanlian, D.; Yongquan, Z. DTSMA: Dominant Swarm with Adaptive T-distribution Mutation-based Slime Mould Algorithm. *Math. Biosci. Eng.* **2022**, *19*, 2240–2285.

39. Coufal, P.; Hubálovský, Š.; Hubálovská, M.; Balogh, Z. Snow Leopard Optimization Algorithm: A New Nature-Based Optimization Algorithm for Solving Optimization Problems. *Mathematics* **2021**, *9*, 2832. [CrossRef]

*Article*

# Innovative Forward Fusion Feature Selection Algorithm for Sentiment Analysis Using Supervised Classification

Ayman Mohamed Mostafa *, Meeaad Aljasir, Meshrif Alruily, Ahmed Alsayat and Mohamed Ezz

College of Computer and Information Sciences, Jouf University, Sakaka 72388, Aljouf, Saudi Arabia
* Correspondence: amhassane@ju.edu.sa

**Abstract:** Sentiment analysis is considered one of the significant trends of the recent few years. Due to the high importance and increasing use of social media and electronic services, the need for reviewing and enhancing the provided services has become crucial. Revising the user services is based mainly on sentiment analysis methodologies for analyzing users' polarities to different products and applications. Sentiment analysis for Arabic reviews is a major concern due to high morphological linguistics and complex polarity terms expressed in the reviews. In addition, the users can present their orientation towards a service or a product by using a hybrid or mix of polarity terms related to slang and standard terminologies. This paper provides a comprehensive review of recent sentiment analysis methods based on lexicon or machine learning (ML). The comparison provides a clear vision of the number of classes, the used dialect, the annotated algorithms, and their performance. The proposed methodology is based on cross-validation of Arabic data using a $k$-fold mechanism that splits the dataset into training and testing folds; subsequently, the data preprocessing is executed to clean sentiments from unwanted terms that can affect data analysis. A vectorization of the dataset is then applied using TF–IDF for counting word and polarity terms. Furthermore, a feature selection stage is processed using Pearson, Chi$^2$, and Random Forest (RF) methods for mapping the compatibility between input and target features. This paper also proposed an algorithm called the forward fusion feature for sentiment analysis (FFF-SA) to provide a feature selection that applied different machine learning (ML) classification models for each chunk of $k$ features and accumulative features on the Arabic dataset. The experimental results measured and scored all accuracies between the feature importance method and ML models. The best accuracy is recorded with the Naïve Bayes (NB) model with the RF method.

**Keywords:** sentiment analysis; machine learning; cross-validation; vectorization; feature importance

**Citation:** Mostafa, A.M.; Aljasir, M.; Alruily, M.; Alsayat, A.; Ezz, M. Innovative Forward Fusion Feature Selection Algorithm for Sentiment Analysis Using Supervised Classification. *Appl. Sci.* **2023**, *13*, 2074. https://doi.org/10.3390/app13042074

Academic Editors: Yue Wu, Xinglong Zhang and Pengfei Jia

Received: 2 January 2023
Revised: 2 February 2023
Accepted: 3 February 2023
Published: 5 February 2023

## 1. Introduction

Sentiment analysis is considered a natural language processing (NLP) method for analyzing users' orientations toward services and topics under consideration. The goal of sentiment analysis mechanisms is to differentiate between subjective and objective sentiments. Objective sentiment is used to express general facts, while subjective sentiment is based on polarity terms that express user reviews or opinions. Objective sentences are excluded during the analysis of sentiments, whereas subjective sentiments can be classified into positive, negative, or neutral polarities.

Most peoples' and users' feelings towards different topics are reflected on social media reviews and sites [1]. Social media allow users to share their views, opinions, and emotions to classify the main service and enhance its specifications in the future. The Arabic language is widely applied on most social media platforms, such as Twitter and Facebook. The Arabic language is considered the official language of Middle East countries and North Africa, comprising 27 countries in addition to the other countries that consider the Arabic language one of its popularly used dialects. It has recently attracted more attention due to the increasing use of Arabic in social media platforms [2].

In addition, the Arabic language contains different dialects with high morphological meanings that can be categorized as standard Arabic and colloquial. Most Twitter users, especially those writing and speaking in English, can express their opinions using traditional or colloquial sentences. They can also use mixed terms that make the preprocessing and analysis processes more complex [3]. Since the Arabic language contains a vast number of linguistics and terminology that are challenging to clean and analyze, sentiment analysis has recently attracted a lot of attention. In addition, sentiment analysis and prediction become more difficult with free writing on social media, particularly in the Arabic language [4].

Few research methodologies are conducted to analyze Arabic sentiments on social media due to the high morphological linguistics in each Arabic sentence that is difficult to classify and analyze. Arabic Social Media Analysis for Arabic (ASAD) aims to fill an important gap in analyzing social media in the Arabic language [5]. Marketing analysis of services or products and public responses to events, persons, and pandemics are considered major examples of the dire need to analyze Arabic sentiments efficiently and accurately, especially when the dataset is new and has not been trained earlier [6]. Recent research methodologies of sentiment analysis depend mainly on collecting datasets from social media such as Facebook and Twitter that provide expressive sentiments from several domains. Twitter users create huge volumes of text to convey their views [7] with a wide range of terms, fields, services, and products [8–10]. Twitter datasets are collected from different sources and categories for classifying public events, pandemics, and product marketing [11–13]. The analysis of sentiments on the Twitter dataset has gained more interest as large companies and institutions depend mainly on user reviews to enhance and upgrade their business services. In addition, the simplicity of the Twitter platform makes it one of the most powerful social networks in the world, with a high volume of daily-generated sentiments [14]. Sentiment analysis methodologies concentrate on mining texts and sentences that can explore deep visions and insights into users' attitudes and opinions.

The main analysis strategy is extracting, classifying, and analyzing the sentiments related to several categories, such as emotional, cognition, social, and theoretical, and analyzing complex texts. Furthermore, most retrieved texts from the Twitter dataset contain unstructured texts that need more preprocessing steps to become more concise and clearer. This can increase the complexity of the selected analysis methodology [15]. In addition to the extracted user text, user-generated data is another additional direction for retrieving data. These data can reduce the users' uncertainty towards business or E-commerce products, which helps analyze user opinions and polarity sentiments for applying decision-making strategies [16]. Whenever a machine learning (ML) technique is applied to analyze polarity sentiments, there must be a set baseline and accuracy parameters to follow. The first step in analyzing sentiments is to remove stop words, elongation terms, symbols, and irony terms that can affect the accuracy and performance of the analysis process [17]. The training phase performs a feature extraction applied to the ML technique. In contrast, the testing or prediction phase applies the features to the classifier models to determine the term polarity. The contribution of the paper is presented as follows:

1.  Presenting a current perspective of the primary strategies and algorithms for Arabic sentiment analysis with a thorough examination of applicable dialects, binary and multi-classification, and annotation algorithms.
2.  Providing a multi-stage methodology for feature generation and selection of Arabic terms using TF–IDF.
3.  Proposing a model using the FFF-SA algorithm that adopts a forward filter for the feature selection method for scoring and registering the accuracy of each $k$-chunk feature of Arabic terms and the subsequent accumulative features.
4.  Measuring and scoring the accuracy for each conducted result, proving the high performance of the NB model with the RF method.

The paper sections are organized as follows. Section 2 explains the main sentiment analysis mechanisms that are conducted to predict user opinions. Section 3 highlights the comparative analysis of recent annotation algorithms with their applied dialects and performance. Section 4 explores the proposed methodology for cross-validation, feature generation, and feature selection of data. Section 5 provides the proposed FFF-SA algorithm for sorting, selecting, and filtering polarity term features. The experimental results are explained in Section 6, and the conclusion and future works conclude the paper in Section 7.

## 2. Sentiment Analysis Mechanisms

The sentiment analysis mechanism is based on text analysis and natural language processing (NLP), which aims to identify, extract, and analyze the polarity of sentiments from different sources and languages. The first process for identifying the sentiments is to discriminate between subjective and objective sentences. Subjective sentiments contain polarity terms that reflect the users' attitudes in social media reviews. In contrast, objective sentiments are based on general facts or information that do not reflect the user's orientation.

The sentiment polarity can be verified based on several weights and scales. Most research methods depend mainly on analyzing the sentiments, whether they are positive, negative, or neutral. Positive polarities express the positive orientation of the users towards a service or a topic under consideration. In contrast, negative polarities denote the opposite meaning to express the user's negative orientation towards the service. Neutral orientations mean that the detected positive and negative polarities are equal; therefore, the user orientation towards a topic or a service is fair. In addition, the sentiment classified as neutral may contain neither positive nor negative terms to be detected. In Figure 1, the overall sentiment analysis mechanism is explained. The main process for analyzing user reviews and tweets centrally depends on lexicon-based and machine-learning (ML) approaches. Each approach has its advantage, methodology for implementation, and methods for dealing with input data and user reviews. In addition, each conducted approach must be measured based on data processing, accuracy, and performance. The methodology's performance can also be changed according to the level of analysis of the polarity terms. There are three main analysis levels: aspect, sentence, and document. In addition, analyzing Arabic sentiments with highly complex terms with different linguistics is considered a challenging process.

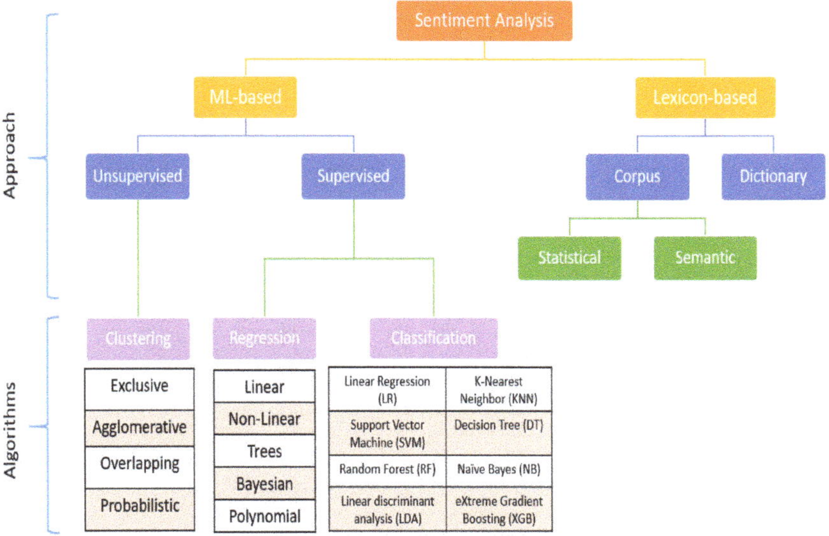


**Figure 1.** Sentiment analysis approaches and algorithms.

### 2.1. Lexicon-Based Approach

The lexicon-based approach aims to score every extracted polarity term from the sentiment sentences, and then the overall polarity of the sentence is calculated. To explain the mechanism of the lexicon-based approach, the polarity weight is first defined to distribute and score each detected polarity term according to its meaning and orientation. Second, the number of detected positive and negative terms is calculated according to their predefined weight score in the sentence. Therefore, the main sentence polarity is defined, and the whole document polarity is analyzed accordingly.

The lexicon-based approach has two main categories: dictionary-based and corpus-based. The dictionary-based stores initial word terms from different sources, and then the dictionary is extended by incorporating additional synonym terms using automated and manual annotations. Therefore, the performance of the dictionary method is constantly changing according to the size of the stored word and polarity terms. The corpus-based method is based on building a corpus that can store different Arabic dialects along with their meaning and orientations. Building a corpus is considered very time-consuming, especially for the Arabic language, which requires adding each term with its corresponding meanings from different dialects.

Arabic is a rich language with additional linguistic terms and several meanings in different dialects. For example, the term "حلو" which means "Good" has several synonyms in different Arabic dialects. The Egyptian and Sudanese dialects use "كويس", the Saudi dialect uses "زين", while some North African dialects use "باهي" or "طيب". All terms reflect the same meaning of "Good" but with different Arabic dialects.

The creation of the corpus depends mainly on statistical or semantic methods. The statistical method measures the behavior of the detected polarity terms in each sentence. If the orientation of the terms is positive, then the sentence polarity will be positive and vice versa. On the other hand, the semantic method assigns a score value to each word term. Words with similar or closer intensified meaning to the word term will have the same score value.

One of the recent methods for managing sentiment sentences was presented in [18], where a mechanism was proposed to embed words to reduce the length of the sentiment. By performing word embedding, each word was converted to its embedded word to reduce its dimension. The dimension reduction can help increase the prediction of sentiment orientation. Therefore, the mapping between predicted and actual polarity scores showed high results. Another enhanced sentiment analysis framework for normalizing the morphological terms of the Arabic language was proposed in [19]. The authors considered two main methods based on the aspect level of sentiments. The two methods were based on the orientation of both category and term polarities, where the normalization of text was executed after the classification. In addition, the authors built a word encoder and decoder to match the word term and polarity term for the given sentiment with their corresponding target meaning in another Arabic dialect.

The process of handling the Arabic language based on its dialects and idioms is considered another major concern. Many social media users and followers use different expressions and idioms based on aphorisms, wisdom, and popular proverbs that can highly affect the analysis performance. The authors of [20] proposed an algorithm for handling this issue. The algorithm's objective was to store the root of the polarity word with the emotions in the sentence that can guide the possible orientation of the sentence.

The authors of [17] built a corpus for measuring the sample percentage with its accuracy and error rate to explore the major concerns that affect the analysis of the Arabic dataset. The authors performed manual, mixed, double-check, and non-check experiments and compared the efficiency of the analysis process. As proposed in [21], a lexicon-based mechanism was presented for analyzing Arabic polarity terms from a Twitter dataset. The proposed method applied a mechanism for distributing different multi-weight polarities based on the number of detected polarity terms in the same sentiment. Therefore, if the number of detected terms increases, the weight polarity will increase due to the diversity and orientation of polarity terms. As presented in [22,23], a set of positive and negative sentences were assigned based on Arabic tweets, where a hybrid strategy was proposed to combine different machine learning approaches. The Lexical-based classifier applied this method to label the training data.

As presented in [24], an automatic sentiment analysis based on supervised classification on the Arabic dataset was proposed. Most sentiment analysis methods verify the sentiments based on their polarity. When the negation term in the sentiment is detected immediately before the polarity term, the sentiment polarity is converted to its reverse polarity. The authors in this paper proposed a methodology for detecting negation terms even if they do not precede the polarity term. Another enhanced sentiment analysis mechanism for analyzing Arabic reviews was proposed in [25], where a lexicon-based analyzer was constructed to analyze polarity terms with different weights to increase efficiency.

Aspect-based sentiment analysis was proposed in [26], where a model was provided for estimating the polarity of user reviews. A lexicon was constructed for acquiring tweets from Twitter, and the dataset was preprocessed to remove any stop words and non-English words. The subjective sentences were processed using Senti-Word-Net to apply a score for each sentence. Based on the provided score, the aspect of each sentence was verified. Another recent research for applying aspect and content analysis of sentiments is presented in [27]. The authors provided a framework for collecting marketing and customer service information from reviews of different universities. The published or posted content of each university was classified into links, photos, videos, and statuses and then the frequency of each content was defined and scored.

The authors of [28] provided a framework for analyzing sentiment opinions during the COVID-19 outbreak. The dataset was collected, classified based on seven clusters, and categorized based on five annotators. The positive, negative, and neutral polarities were determined in each cluster, and the polarity score for each cluster was defined. Another methodology for analyzing sentiments during COVID-19 was proposed in [29]. The authors aimed to examine and measure the influencing factors that affect the orientation of people during the epidemic. The dataset was collected from different social media forums. Based on the proposed influencing factors, the dataset was classified, and the sentiments were analyzed to explore their polarities. The authors of [30] proposed another method of NLP for analyzing user sentiments during the COVID-19 epidemic. A framework was proposed to measure the performance of a set of relevant word terms from different aspects such as economy, social, and health to view the major orientation of the reviews that explained a neutral polarity orientation.

### 2.2. Machine Learning-Based Approach

Machine learning-based (ML) approaches provide powerful methods for analyzing polarity sentiments from different domains. The ML methods are adapted to learn from the input dataset and then provide the prediction from the hidden patterns. Learning and prediction are considered the two major steps in all ML algorithms. As proposed in Figure 1, ML has two major approaches for handling input data for training. These approaches are supervised and unsupervised. In the supervised approach, the input data must be trained first before applying the testing dataset, as the ML algorithms can perform the prediction based on previous experience. Therefore, different classification and regression algorithms are applied to measure the relationship between input and target features [31], and then the accuracy of the prediction is explored. The unsupervised approach depends on an unlabeled dataset where the patterns are discovered by performing complex tasks using clustering algorithms that can group the dataset and learn from unknown patterns. In addition to supervised and unsupervised approaches, the semi-supervised methods can perform the training and prediction with a less labeled dataset that is linked with the unlabeled dataset to produce the result.

Recent research methodologies for applying machine learning (ML) algorithms on sentiment analysis from social media datasets have been proposed. As presented in [32], a sentiment analysis mechanism based on machine learning was applied to Russian and Kazakh languages. The applied dataset was categorized into positive, negative, and neutral. To increase their efficiency, different resampling techniques were used to resample the unbalanced datasets. Another enhanced mechanism for analyzing sentiments from the Twitter dataset was proposed in [33], where different classifiers of machine learning were adopted using the TF–IDF algorithm for extracting features. The sentiments were collected from multi-classes emotion data, and the accuracy was tested. Another sentiment analysis approach was presented in [34] for classifying tweets based on learning models. The research aimed to analyze a large number of social media tweets from Twitter using the Apache Spark model. The experiments were conducted to measure the time consumed using Apache Spark compared with other classifier models. As presented in [35], a machine learning-based approach was applied to analyzing Arabic sentiments. Different ML classifiers were used on the cleaned dataset to remove stop words and word elongation.

As shown in [36], ML classification algorithms were applied to multi-language datasets based on predefined Key Performance Indicators (KPIs). The idea was to group a set of opinions from the leaders of a company and then analyze the comments about the company and distribute these comments over the predefined KPIs. Analyzing and enhancing the accuracy of sentiment analysis can be proposed based on the semantic knowledge and content analysis of the sentiment polarities. This method is called aspect-based sentiment analysis. As presented in [37], aspect-based sentiment analysis was provided to identify sentiment polarities based on different attributes or aspects. The authors applied a framework for extracting aspects from Twitter datasets, and sentiment analysis was applied based on machine learning methods. One of the recent research methodologies for analyzing Arabic user opinions was presented in [38]. The research focused on analyzing Saudi citizens' and residents' opinions about the downloaded programs from Google Play and App Store.

Different machine learning classifiers were provided to measure the dataset's accuracy, classified into negative, positive, neutral, unique, and stem words. Another interactive methodology for analyzing Arabic Twitter sentiments was proposed in [39]. The authors of this research focused on a new topic related to detecting depression terms in Arabic sentiments from Twitter. The authors created three lexicons for storing depression terms and counted the number of tweets for each symptom. Different machine-learning algorithms were deployed to measure the accuracy of the results. Another interactive method for classifying sentiments is shown in [40], where teaching-learning-based optimizers were applied to a Twitter dataset. Different preprocessing steps were executed to remove stop words and symbols before applying four text-processing models. A feature selection algorithm was proposed to classify polarity features into positive and negative polarities, and finally, the results of the models were listed.

As proposed in [41], machine-learning-based algorithms have been applied to different polarity languages. The dataset was collected from Twitter, where each tweet was converted and counted to an integer. The integer was converted to its TF–IDF score value, and finally, the score was applied to ML classifiers for prediction. The authors of [42] proposed another aspect-based sentiment analysis methodology by providing an automatic annotation of datasets from YouTube songs. These songs were extracted and applied based on the number of views and reviews and then an aspect filtration was provided based on five aspect categories. The overall reviews were optimized again based on the predefined aspects to determine the most important aspect category.

Sentiment analysis of users' orientations is also conducted in health sectors to improve the users' feedback and provide deep insight into the provided services. As presented in [43], a sentiment analysis tool was provided for measuring people's attitudes in smart cities during the COVID-19 epidemic. The dataset was tested five times to measure the average accuracy of the classified instances. As shown in [44], sentiment analysis of user polarities based on natural language processing (NLP) was proposed to explore the attitudes of Gulf countries during COVID-19. The goal of the paper was to check whether there were mixed emotions among people or not. The dataset was extracted from Twitter API, and the polarities were classified in different countries. The authors stated that most Gulf people's attitude toward the epidemic was neutral. In addition to previous research, the authors of [45] provided an approach based on semi-supervised machine learning that measured the analyses of datasets collected from different social networks about several epidemics. The first analysis clustered the data from Word-2-Vec and Fast-Text and then explored the sentiment orientation based on the applied classification algorithms. The authors of [46] measured the sentiment analysis of public health but from the financial aspect. The dataset was collected from financial news and then the reviews were grouped based on four polarity attributes to view the overall orientation. Furthermore, different classification algorithms were applied to learn from the polarity attributes and explore sentiment accuracy.

In addition to lexicon-based and ML-based methodologies, different deep learning and transfer learning methods can be applied to Arabic sentiment analysis. As presented in [47], a deep learning model for multitasking was applied for classifying Arabic sentiments. The research aimed to enhance the performance as the Arabic language has low resources and contains high morphological and linguistic terms. The authors proposed a long short-term memory (LSTM) deep learning model that explored the relationship between three and five sentiment polarities into a private layer. This layer contained an encoder for words to add flexibility to the features. The authors of [48] proposed a deep transfer-learning model for manipulating Arabic text. The Convolutional Neural Network (CNN) training was conducted to classify the sentiments based on a pre-annotated dataset. The authors collected different classes of the dataset and performed augmentation of data to enhance the accuracy. Another deep learning model for manipulating Arabic sentiments was presented in [49], where the data was collected based on three classes and then classified using the LSTM model to explore the results. One of the recent methods for extracting and detecting Arabic polarity text was proposed in [50], where transfer learning (TL) techniques were applied to aspect-based Arabic text. The authors proposed an architecture using the BERT model on the HAAD dataset to measure the approach's effectiveness, which showed high results.

Arabic is considered among the richest languages worldwide with many linguistic terms. Arabic sentiment analysis has not been studied as highly as other languages. The Arabic language suffers from a lack of high-quality terms and large-scale training data with the difficulty of manipulating ironic and slang expressions [47]. The Arabic language is considered an unstructured language with many inconsistencies in the spelling of terms and difficulties in identifying key features. Furthermore, using Arabic tweets in social networks cause many word elongations and repetition of terms to convey the user's feelings. This paper proposes a novel algorithm based on a forward fusion feature selection for sentiment analysis using different models such as Pearson, $Chi^2$, and RF, and then different ML classification models for each chunk of k features and accumulative features on the Arabic dataset to explore the winning model with high accuracy.

## 3. Comparative Analysis with Annotation Algorithms of Sentiment Analysis

The analysis of sentiments depends mainly on the extraction of subjective sentences from different social media streams. The customers and users reflect and express their attitudes and opinions towards different services and products using several languages and linguistics with a different number of classes. As presented in Tables 1 and 2, a comprehensive review of recent sentiment analysis researches is conducted. As explained, only positive and negative polarities will be predicted if two classes are applied for classification. The use of three classes for classification adds neutral sentiments where the number of positive and negative sentiments are equal or the overall orientation of the sentiment is fair. Arabic is considered one of the most languages in the world in the context of linguistics and terms that are difficult to identify and analyze. Two main annotation methods are applied for analyzing and predicting sentiments: machine learning (ML) and lexicon-based. The ML-based method applies different algorithms for measuring the accuracy of predicting the sentiments' orientations. In contrast, the lexicon method is based on building a corpus for storing Arabic dialects with their polarities or weight scores.

**Table 1.** Comparison of lexicon-based methods with their performance.

| Ref | Dataset Size | No. of Classes | Dialect | Language | No. of Annotators | Performance |
|---|---|---|---|---|---|---|
| [17] | 30K / 8K | Pos.–Neg. | Single | Arabic | 2 | 77% / 86% |
| [19] | 16.9K / 18.9K | Pos.–Neg. | Single | Arabic | 2 | 76.89% / 76.48% |
| [21] | 5k | Pos.–Neg. / Pos.–Neut.–Neg. | Single | Arabic | 2 / 3 | Senti. Score / Senti. Score |
| [22] | 100K | Pos.–Neut.–Neg. | Single | Arabic | 3 | 66% |
| [23] | 1.1K | Pos.–Neg. | Single | Arabic | 2 | 84% |
| [24] | 3.4K | Pos.–Neut.–Neg. | Single | Arabic | 3 | Senti. Score |
| [25] | 6.3K | High Pos.–Pos. Neut. Neg.–High Neg. | Single | Arabic | 5 | Senti. Score |
| [26] | 6K | Pos.–Neut.–Neg. | Single | English | 3 | Senti. Score |
| [27] | 3.9K | Pos.–Neut.–Neg. | Single | English | 3 | Senti. Score |
| [28] | 12.5K | High Pos.–Pos. Neut. Neg.–High Neg. | Single | English | 5 | 46.4% |
| [29] | 22.5K | Pos.–Neut.–Neg. | Single | English | 3 | 86.2% |

**Table 2.** Comparison of ML-based models with their performance.

| Ref | Dataset Size | No. of Classes | Dialect | Language | No. of Annotators | Performance |
|---|---|---|---|---|---|---|
| [1] | 20K | Pos.–Neut.–Neg. | Single | Arabic | 3 | 79% |
| [5] | 21K | Pos.–Neut.–Neg. | Single | Arabic | 3 | 75.1% |
| [31] | 24K | Pos.–Neut.–Neg. | Single | Turkish | 3 | 91.57% |
| [32] | 80.8K / 15.9K | Pos.–Neut.–Neg. | Multi | Russian / Kazakh | 3 | 77% |
| [33] | 39.8K | Pos.–Neut.–Neg. | Single | English | 3 | 89.92% |
| [34] | 10K | Pos.–Neut.–Neg. | Single | Arabic | 3 | 83% |
| [36] | 685 | Pos.–Neg. | Multi | Eng. Spa. Ita. Ger. Fre. | 2 | 88.17% |
| [37] | 6.7K | Pos.–Neut.–Neg. | Single | English | 3 | 78.76% |
| [38] | 8K | Pos.–Neut.–Neg. | Single | Arabic | 3 | 78.46% |
| [39] | 4.5K | Pos.–Neut.–Neg. | Single | Arabic | 3 | 82.39% |
| [40] | 14.6K | Pos.–Neg. | Single | English | 2 | 76.9% |
| [41] | 11K | Pos.–Neg. | Multi | Urdu / Roman Urdu / English | 2 | 84% / 85% / 77% |
| [43] | 0.5K | Pos.–Neut.–Neg. | Single | English | 3 | 89.4% |

## 4. Proposed Sentiment Analysis Methodology

As discussed in the previous sections, recent sentiment analysis mechanisms depend mainly on different aspects. Firstly, the applied language or dialect contains users' reviews and comments. Secondly, the data preprocessing mechanism for cleaning and adapting datasets and corpus documents for sentiment analysis. Thirdly, the implemented lexicon-based or ML-based methods for manipulating datasets. Finally, the proposed methodology and algorithm for exploring and enhancing the analysis of user polarity is based on word terms and polarity terms.

The methodology of this paper is based on applying ML-based sentiment analysis algorithms on Arabic datasets from [1,51]. The Arabic language is a difficult language in the ramifications of its terminology as it contains many dialects and huge linguistic and morphological terms, which makes analyzing user sentiments a great challenge. The Arabic language has low resources and contains high morphological and linguistic terms [47]. Therefore, the analysis and classification of polarity terms is considered a challenging process. In addition, the Arabic sentiment analysis has not been studied at a level as high as other languages, such as English, Chinese, and French.

Furthermore, the user reviews that use the Arabic language to express their views and orientations can use multi-dialect in the same comment, which causes an additional overhead during the data preprocessing and analysis. As presented in Figure 2, the proposed mechanism depends on interactive multi-level processes for splitting the dataset into training and testing and then performing a 10-fold cross-validation for interchanging the *k* folds. The next step is data preprocessing, where several stages are applied to clean the polarity sentiments from unwanted terms and particles to increase the analysis efficiency. The next feature generation stage transforms the tweets into a set of feature vectors with an encoder that will be applied to the training and testing dataset. The vectorization of features is executed using the Term Frequency—Inverse Document Frequency (TF–IDF) mechanism that generates the features from the overall corpus of documents and sentiment analysis documents with their word and polarity terms. The generated features are selected using a filter method that divides the input features into different feature vectors related to the target feature. The correlation is executed using three methods: Person, Chi$^2$, and Random Forest (RF). The correlated features are sorted according to their relevance to the target features where the best *k* features are ranked and selected. The next *k* features are added until all the features are processed. The final stage is based on modeling the selected features using differing ML algorithms, where the best accuracy for each correlation method with the ML modeling algorithm is recorded.

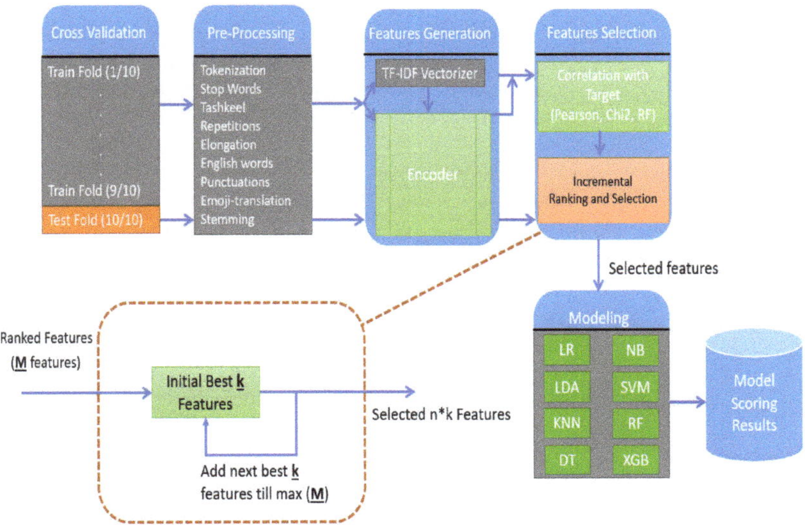


**Figure 2.** Proposed methodology of multi-stage feature generation and selection mechanism.

### 4.1. Cross Validation

The preprocessing stage should be in the correct position during the cross-validation process. Preprocessing steps are intended to be created using training data folds (preprocessing adaptors), and then the procedure is repeated using test data folds (using preprocessing adaptors for transformation). This ensures that the model is only exposed to preprocessed training data during the training phase. This approach aids in avoiding "data leakage," which occurs when the model is exposed to information from the test set during training, resulting in overfitting.

The cross-validation process is applied based on the $k$-fold mechanism. The main objective of $k$-fold cross-validation is to divide or split the dataset into a set of K groups [52]. Each group is treated as a validation unit to evaluate the overall model. In this paper, a 10-fold cross-validation on the dataset is executed randomly where $k = 10$. A number of $k - 1$ folds are used for dataset training while the remaining $i$ fold is used for testing. On the next splitting process, another $i$ fold is used for testing while the remaining $k - 1$ folds are used for dataset training. The 10-fold cross-validation continues until the $i$ testing fold is applied on all splitting stages, where the final model is validated on each $i$ testing fold. Equation (1) summarizes the overall 10-fold process as follows:

$$\forall i \subseteq k \ni i = 1 \ \& \ k - 1 < 10 \tag{1}$$

### 4.2. Data Preprocessing

After performing cross-validation on the sentiment analysis dataset, data preprocessing is executed to eliminate and clean the sentiments and their polarity terms from incorrect or unwanted terms that may affect the accuracy of the analysis process. As presented in Figure 3, data preprocessing depends on a set of sequential steps for removing inconsistent terms from the sentences. The stemming process is applied to the overall dataset to reduce the polarity term length so that the polarity term returns to its root. Stop words and tokenization are eliminated from the sentiment analysis dataset, where the stop words are terms that do not affect the sentence's overall meaning. For example, the stop word terms such as "بعد–قبل–و–لا–على–في–أو" that mean "or–in–on–not–and–before–after", respectively, are removed from the preprocessed data. Tokenization is separating polarity terms using a space or a unique character to be analyzed more efficiently during the machine learning mechanism. For example, the sentence "لم يقدم التطبيق أي خدمة إضافية!" that means "The application does not provide any additional service!" is separated into individual terms to increase the machine learning ability to understand the whole sentiment. Due to the use of the Arabic dataset in our paper, some words that contain "Tashkeel" are also eliminated. The term "Tashkeel" means a set of special characters added to the formation of the words to change the word pronunciation. For example, the sentence "التطبيقُ جَيِد لَكِنَ الخِدْمَاتِ بَطيئَةٌ جِدًّا" that means "The application is good but the services are very slow" contains many special characters to set the word terms. After removing the "Tashkeel" characters, the new sentence becomes "التطبيق جيد لكن الخدمات بطيئة جدا". The English words, punctuation, and repetition of the terms are also removed during data preprocessing to reduce the sentence length during the sentiment analysis process. Emojis are also removed from the sentences as they can contain different ironical or emotional terms that can affect the orientation of the overall sentences from positive to negative and vice versa. Finally, the word elongation is also removed from the sentiments to eliminate any repetition of letters. For example, the polarity term "جمييييل" is processed to be "جميل" that means "nice" or "beautiful".

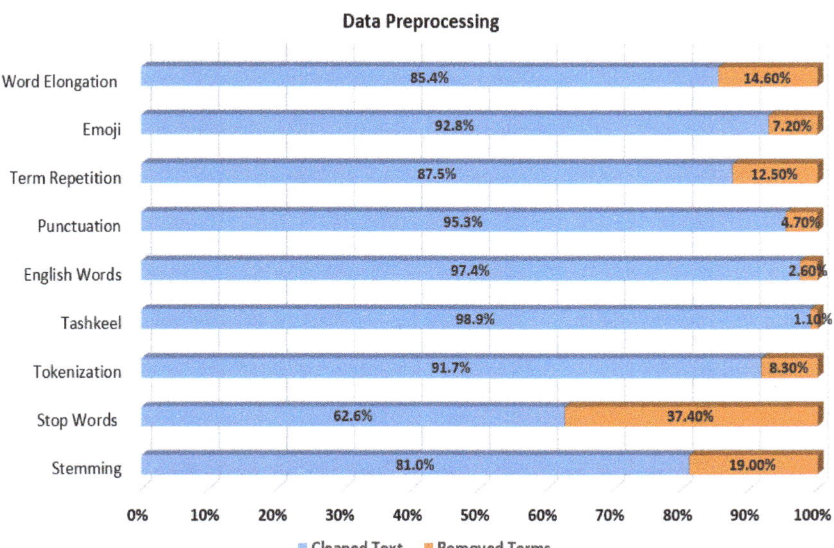

**Figure 3.** Data preprocessing levels on sentiment analysis documents.

The foundation of the classification models is based on data segmentation. The imbalanced dataset uses one chunk of data that contains a large portion, which is called the majority class, while the remaining chunk of data represents the minority class. As a result, an imbalanced dataset is one in which one class has a higher number of occurrences or sentences than the other class. Equal or almost equal numbers of occurrences or sentences from each class are presented in the balanced dataset.

As presented in Table 3, the imbalanced dataset contains 16.7K sentiments that has approximately positive sentiments as 17%, negative sentiments as 16.5%, and neutral sentiments as 66.5%. The dataset was balanced into which two classes of dataset, positive and negative, are selected with 5451 sentiments.

**Table 3.** Balanced dataset with binary classification.

| Dataset / Polarity | Positive | Negative | Neutral | Total |
|---|---|---|---|---|
| Imbalanced | 2843 | 2751 | 11,129 | 16,723 |
| % | 17% | 16.5% | 66.5% | 100% |
| Balanced with 2 Classes | 2725 | 2726 | - | 5451 |
| % | 50% | 50% | - | 100% |

### 4.3. Feature Generation

In this research, the tweets are manually annotated using a unigram model that offers a reasonable coverage degree for the dataset. In order to extract the most important features from the training dataset, data preprocessing and feature generation are executed to convert the tweets into a feature vector. An encoder is generated from the vectorization process of the training dataset to be applied to the testing dataset. The generated encoder converts the tweets into a set of feature vectors that are applied on the training and testing datasets. Text and tweets are applied as vectors using the TF–IDF technique that converts the given text into finite feature vectors. The term frequency (TF) computes the number of times the selected term is repeated in a given document. In contrast, the inverse document frequency (IDF) computes the number of times the selected term is repeated in the overall dataset or corpus.

The TF–IDF method has a linear computational complexity regarding the number of text lines and words per line. In contrast, the RF feature selection algorithm has a computational complexity of $O(n\ estimators \times m \times \log(n))$, where $n$ is the number of samples, $m$ is the number of features, and log is the base-2 logarithm. Finally, the NB classifier is a simple and computationally efficient algorithm with $O(n \times d)$ computational complexity, where $n$ is the number of samples and $d$ is the number of features.

The corpus data contains different sentiment analysis documents with different word and polarity terms. The documents are collected from Twitter to analyze users' orientation and attitudes based on their positive or negative polarity. As presented in Equations (2)–(4), the overall vectorization of the dataset is explained:

$$TF\ (p_t,\ s_d) = \frac{\sum_{i=1}^{n} p_{ti}\ \in\ w_t}{\sum_{j=1}^{m} w_{tj}\ \in\ s_d} \tag{2}$$

$$DF\ (p_t,\ c_d) = log\left(\frac{N}{Count\ (s_d \in c_d\ :\ p_t \in s_d)}\right) \tag{3}$$

$$TF - IDF\ (p_t,\ s_d,\ c_d) = TF\ (p_t,\ s_d) \times IDF\ (p_t,\ c_d) \tag{4}$$

where:
$w_t$ : word terms.
$p_t$ : polarity terms.
$s_d$ : sentiment analysis documents.
$c_d$ : corpus data for all documents.

The corpus data $c_d$ contain different documents of sentiment analysis $s_d$ from different domains and sources that reflect the users' opinions about different services. Each sentiment analysis document $s_d$ contains a large number of sentences with word terms $w_t$ that contain polarity terms $p_t$ that explores users' orientations. As shown in Equation (4), $TF - IDF$ is calculated by measuring the resulting score of the multiplication between $TF$ and $IDF$. The higher the resulting score, the more relevant the polarity term in the sentiment analysis documents $s_d$.

*4.4. Feature Selection*

The feature selection stage aims to reduce the input parameters or features to predict the target values efficiently. In addition, some predictive models contain many variables that can affect the efficiency of the memory or can reduce the system performance due to the incompatibility between the input and the target features. Supervised and unsupervised methods are the main key features for predicting target features. The selection process in unsupervised methods removes redundant features and eliminates the target variable, while supervised methods focus on the target features by removing any insignificant input features.

Other feature selection methods, such as wrapper and filter methods, can be applied by evaluating the model performance on the corpus data $c_d$ and sentiment analysis documents. Regarding the wrapper method for feature selection, the input features are divided into different subsets. The method applies several models on the subsets to select the best model that achieves the highest performance. The filter method for feature selection applies several statistical techniques to estimate the relationship between input and target features. In addition, the filter method scores each resulting value between the input and the output features and then filters the best models based on the recorded scores.

This paper applied the filter method for feature selection by dividing the input features into a different subset of feature vectors and then selecting the subsets that are highly associated or related to the target features. The correlation and selection of the subsets are applied using three feature importance methods: Pearson correlation, Chi$^2$, and Random Forest (RF).

The Pearson method explores the correlation score between the input and target features, where the score ranges from $-1$ to $+1$. If the correlation score is close to $+1$, then the relationship to the target is high and vice versa. Firstly, the covariance between each word term feature $w_t$ and the target feature of the expected polarity term $p_t$ is calculated, and then the Person correlation is measured by dividing the covariance value in Equation (5) by the multiplication of the standard deviation of both word term feature $w_t$ and polarity term features $p_t$ as explained in Equation (6).

$$Cov\ (w_t,\ p_t) = \frac{1}{n} \times \sum\nolimits_{i=1}^{n} ((w_{ti} - \overline{w}_t) \times (p_{ti} - \overline{p}_t)\ ) \tag{5}$$

where:
$w_{ti}$: each input word term feature in the vector.
$\overline{w}_t$: the mean of the overall word term features.
$p_{ti}$: each target polarity term feature.
$\overline{p}_t$: the mean of the overall polarity term features.
$n$: the length of the word terms and polarity terms.

$$PC\ (w_t,\ p_t) = \frac{Cov\ (w_t,\ p_t)}{sd_{wt} \times sd_{pt}} \tag{6}$$

where:
$sd_{wt}$: the standard deviation of word terms.
$sd_{pt}$: the standard deviation of polarity terms.

The Chi$^2$ method for feature selection is used to measure the independence degree between the observed values of the word term features $w_t$ and the expected values of the polarity term features $p_t$. Therefore, the Chi$^2$ method selects the features of both word terms and polarity terms that are highly correlated as shown in Equation (7).

$$Chi^2 = \sum \frac{(w_{ti} - p_{ti}\ )^2}{p_{ti}} \tag{7}$$

The RF is considered a predictive method with high performance and low overfitting value. In addition, the RF combines both filter and wrapper methods and contains several decision trees. Each tree is based on a random extraction of the word term features $w_t$ and a random extraction of the polarity term features $p_t$. Each tree cannot trace all the word term features and polarity term features to reduce the overfitting.

## 5. FFF-SA Algorithm

The filter methods for feature selection measure one input feature at a time with the target feature. Therefore, there is no interaction between the input features. To overcome this issue, this paper proposed an algorithm called the innovative forward fusion-based for feature selection (I-FFF). This algorithm performs a forward chain feature selection by calculating the input feature importance with the target value using Pearson correlation, Chi$^2$, and RF. The Pearson method is applied to numerical variables, while the Chi$^2$ method is applied to categorical variables. The RF method is applied to both numerical and categorical variables. The FFF-SA algorithm is based on three main stages for selecting the best correlation between the input and the target features.

The first stage is based on sorting the feature vectors based on their importance and association to the target. The second stage starts by selecting the best $k$ batch of features and then measures its score during the tuning and validation of data using different modeling approaches of machine learning (ML). The third stage adds a new $k$ batch of features one at a time until all batches of features are added.

After performing the feature selection for all input features of word terms $w_t$ and target features of polarity terms $p_t$, different machine learning models are implemented to predict the labels of the input word terms. In addition, the trained dataset is tuned to maximize the models' performance without overfitting data. The tuning process is based on hyper-parameters that can control the trained dataset. In this modeling stage, eight machine-learning algorithms are applied, where some algorithms have unique hyper-parameters while others have similar hyper-parameters. These algorithms are Logistic Regression (LR) [53], Linear Discriminant Analysis (LDA) [54], K-Nearest Neighbor (KNN) [55], Decision Tree (DT) [56], Naïve Bayes (NB) [57], Support Vector Machine (SVM) [58], Random Forest (RF) [59], and Gradient Boost (XGB) [60]. The overall FFF-SA algorithm is explained as follows:

The proposed FFF-SA algorithm starts by defining the full feature set $FF$, the maximum size of features $M$, the $k$ features for each experiment, and the $n$ number of execution times during the experiment. Each experiment is conducted using the three correlation features, Pearson, $Chi^2$, and RF features, and by dividing the input features to a different subset of feature vectors and then selecting the subsets that are highly associated or related to the target features. The first correlation feature of Pearson is initialized, where the first $k$ features is applied on the experiment, and then the eight ML algorithms are used as models to train the selected $k$ features and then score the accuracy using the testing fold to measure the similarity and correlation between the training and target features. Each time, the first ML algorithm selects the $k$ features and then registers the score. The next $Accum_F + k$ features are added to the previous features as accumulative features to model and score the accuracy on the accumulative features. The process continues until the $M$ features are reached, which represents the maximum size of the features. The second ML algorithm repeats the experiments for the $k$ features and then registers the score at each accumulative feature until all ML algorithms record their accuracy results for the Pearson correlation method. The next experiments are conducted with the $Chi^2$ and RF correlation methods to score and record the accuracy.

| FFF-SA Algorithm | |
|---|---|
| 1 | $Input : FF , M , k , n$ |
| 2 | $Output : Sentiment\ Analysis\ Accuracy$ |
| 3 | $Initialize\ n \leftarrow 1$ |
| 4 | $FOR\ EACH\ CorrF\ in\ \left[ Pearson , Chi^2, RF \right]\ DO$ |
| 5 | $FF_{Cor} = CorrF_i\ (FF)$ |
| 6 | $FOR\ n = 1\ to\ t\ DO$ |
| 7 | $Chunk_F = k$ |
| 8 | $Accum_F = n \times k$ |
| 9 | $WHILE\ (Accum_F < M)\ DO$ |
| 10 | $F_{SELECT} = Extract\ Ranked\ Set\ (FF_{Cor},\ Accum_F)$ |
| 11 | $FOR\ EACH\ Alg_i\ in\ [\ LR,\ LDA,\ KNN,\ DT,\ NB,\ SVM,\ RF,\ XGB\ ]$ |
| 12 | $Model = Train\ (Alg_i ,\ F_{SELECT}\ [\ Train\ k\ fold\ ])$ |
| 13 | $Score = Evaluate\ (\ Model ,\ F_{SELECT}\ [\ Test\ k\ fold\ ]\ )$ |
| 14 | $Register_{Score} = (Alg_i ,\ Accum_F ,\ Score)$ |
| 15 | $Accum_F = Accum_F + k$ |

## 6. Experimental Results

The experimental results based on the proposed FFF-SA algorithm are tested, where the three correlation coefficient methods, Pearson, $Chi^2$, and RF, are used as primary coefficients to measure the accuracy and efficiency of sentiment analysis classification. The hyper-parameters are parameters the user sets rather than learns from the data. Some common hyper-parameters used during the feature selection include the number of selected features, the criteria for selecting features (RF, $Chi^2$, and Person), the feature selection threshold, and finally, the number of iterations for forward feature selection.

As explained before, the collected dataset is based on the Arabic language that contains high and complex morphological sentiments and terms that can affect the performance of the experiments. The Arabic language is rich in many linguistic terms that indicate more than one meaning and orientation. In order to obtain a higher degree of accuracy, the sentiment analysis process depends on many stages to effectively purify and analyze the sentiment analysis documents and terms. In addition, splitting and training the data depends on several steps to find the relationship between the input feature vectors and the target features. The following sections explore the applied feature selection correlation mechanisms with different ML models.

### 6.1. Experiment 1: Feature Importance Using RF

In this experiment, the RF feature selection method is applied to the eight ML algorithms: LR, LDA, KNN, DT, NB, SVM, RF, and XGB. For each conducted ML algorithm, the experiment starts with the first $k = 100$ features where the model is evaluated and the score accuracy is recorded. Based on the FFF-SA algorithm, the experiment is continued by increasing an additional $k + 1 = 100$ to the previous $k$ fold to obtain $k = 200$ as an accumulative feature and then the model is executed again to find the highest accuracy. As explained in Figure 4, the highest accuracy is recorded on the accumulative feature $k = 2400$ with the NB algorithm that achieved an accuracy of 84.4%. The second highest accuracy is recorded on the accumulative feature $k = 2300$ with the same algorithm of NB that achieved an accuracy of 84.17%.

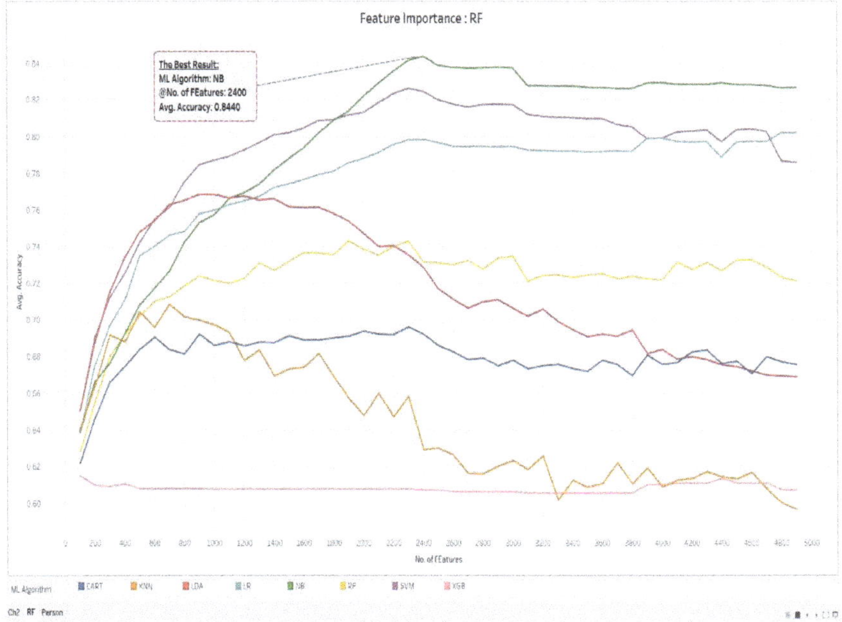


**Figure 4.** Accuracy of RF feature selection method with ML models.

As stated, the four ML algorithms, NB, LR, SVM, and LDA, start the experiments with a linear increase but the results of the LDA algorithm start decreasing at the accumulative feature $k = 1800$. The NB, LR, and SVM score the best results when compared to the remaining ML algorithms where the highest scored accuracy is recorded with NB algorithm.

## 6.2. Experiment 2: Feature Importance Using $Chi^2$

In this experiment, the $Chi^2$ feature selection method is applied to the eight ML algorithms. The experiments are conducted again based on the FFF-SA algorithm where each experiment starts by modeling the ML algorithm with the accumulative set of features $k$. As explained in Figure 5, the experimental results show dispersed results from most ML algorithms except the NB and SVM algorithms that achieved the best results. The NB algorithm recorded the best accuracy of 83.76% with the accumulative feature $k = 2300$. With the accumulative feature $k = 2400$, the accuracy decreased very slightly and recorded 83.62%. The remaining accumulative $k$ features continued in slight decreases, forming a straight line to the end of the accumulative feature with $k = 4900$ that recorded 82.91%. The second-best results are achieved on the SVM algorithm with the accumulative feature $k = 2200$ that recorded an accuracy of 81.49%. On the accumulative feature s $k = 2100$ and $k = 2300$, the SVM recorded accuracies of 81.21% and 80.98%, respectively.

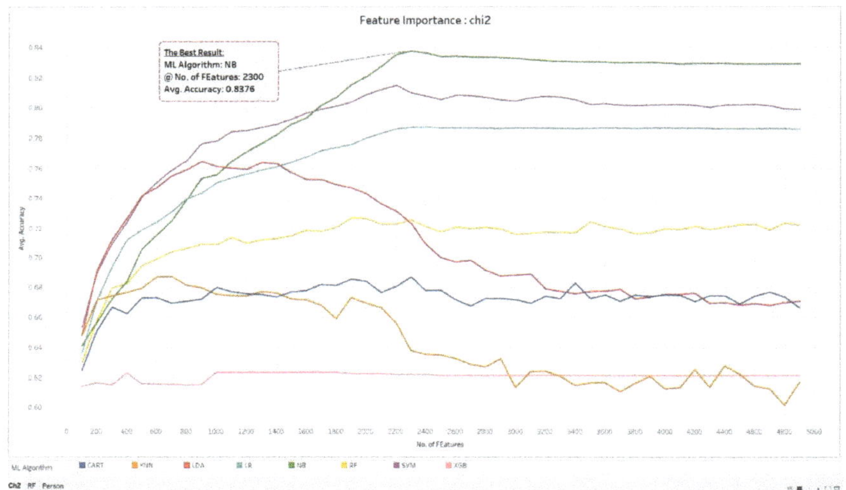


**Figure 5.** Accuracy of $Chi^2$ feature selection method with ML models.

## 6.3. Experiment 3: Feature Importance Using Pearson Correlation

As presented in Figure 6, the experimental results are executed again using the Pearson coefficient for measuring the correlation between the input features and the target features. As explained, all conducted ML algorithms achieved low results, and both LR and SVM algorithms achieved an accuracy of 58.73% on the accumulative feature $k = 700$. Starting from $k = 700$ to the end of the accumulative features where $k = 4900$, the accuracy remains the same score, forming a straight line. The LDA algorithm achieved the second-best results of 58.51% with the accumulative feature $k = 700$, and the stated results continued in a straight line to the end of the accumulative features where $k = 4900$. As noticed from the figure, most ML algorithms start increasing with $k = 100$ until the accumulative feature $k = 700$, where the results remain without change to the end of the features.

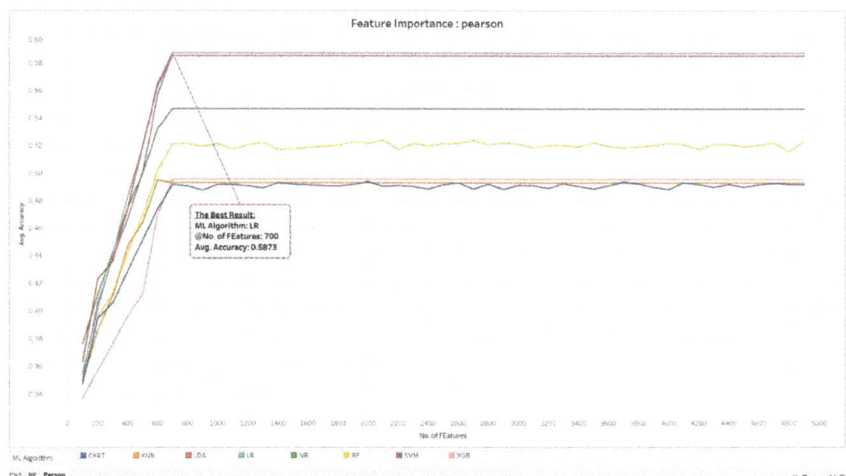



**Figure 6.** Accuracy of Pearson feature selection method with ML models.

## 6.4. Feature Importance Comparison Analysis

Based on the previous experimental results, the FFF-SA algorithm is based on a forward chain of processes for performing a feature selection on the split features. As mentioned before, the sentiment analysis documents and sentiments are divided into $k$ features for training and testing. The correlation methods are applied with different ML models for measuring and scoring each set of $k$ features. On the subsequent stages, the features are accumulated with additional $k$ features, while the ML models perform the scoring at each step. Table 4 presents a summary of the performance analysis for each feature importance method with its corresponding ML models. The table showed that the highest accuracy on all experiments and feature importance stages recorded 84.4% for the RF method with the NB algorithm. The second-best accuracy with the NB algorithm was 83.8% for the Chi$^2$ method. The SVM algorithm also achieved good results, with 82.6% and 81.5% on RF and Chi$^2$, respectively. The provided accuracy provides promising results based on the challenge of pure Arabic sentiment analysis documents and terms collected from different domains [1,51].

**Table 4.** Analysis of accuracy for different feature importance methods with ML models.

| RF | | | | | | | |
|---|---|---|---|---|---|---|---|
| NB | SVM | LR | LDA | RF | KNN | CART | XGB |
| 84.4% | 82.6% | 80.2% | 76.9% | 74.3% | 70.8% | 69.6% | 61.6% |
| **Chi$^2$** | | | | | | | |
| NB | SVM | LR | LDA | RF | KNN | CART | XGB |
| 83.3% | 81.5% | 78.7% | 76.8% | 72.7% | 69% | 68.7% | 62.5% |
| **Pearson** | | | | | | | |
| NB | SVM | LR | LDA | RF | KNN | CART | XGB |
| 54.6% | 58.7% | 58.7% | 58.5% | 52.4% | 49.5% | 49.4% | 49.6% |

In Table 5, the number of features and experiments that are conducted on the dataset with different feature selection methods are explained. The table shows more than 3 million extracted features during the testing of data with about 4423 experiments on all feature selection methods.

**Table 5.** Analysis for the number of features and experiments.

| Experiment<br>Feature Selection | No of Features | No of Experiments |
|---|---|---|
| RF | 1,073,700 | 1455 |
| Chi$^2$ | 1,012,800 | 1596 |
| Pearson | 980,000 | 1372 |
| Total | 3,066,500 | 4423 |

In Table 6, the top 10 accuracies for the experimental results that are executed on all feature importance methods with ML models are provided. The explanation of these results was intended to study the efficiency for feature importance models in all sentiment analysis documents and terms. For each feature $k = 100$, the orientation of polarities were mapped into two classes whether they were positive or negative. The feature importance method was applied to score the accuracy for detecting positive or negative polarities using different ML models. After scoring the first batch of $k$ features, an additional $k + 1$ feature is added to form an accumulative feature where the FFF-SA algorithm is executed again. As stated in the table, the top 10 accuracies are recorded with the NB model with both RF and Chi$^2$ feature importance methods. The best seven results are recorded with the RF method with a different number of accumulative $k$ features. Based on these experiments, it is clear that the NB model with the RF feature importance method scored the best results on Arabic sentiment analysis terms.

**Table 6.** Top 10 accuracies for feature importance methods.

| No. | Vectorization | Feature Importance | No. of Features | Classes | ML Model | Accuracy |
|---|---|---|---|---|---|---|
| 1 | TF–IDF: ngram_range (1,1) | RF | 2400 | Pos.–Neg. | NB | 84.40% |
| 2 | TF–IDF: ngram_range (1,1) | RF | 2300 | Pos.–Neg. | NB | 84.17% |
| 3 | TF–IDF: ngram_range (1,1) | RF | 2500 | Pos.–Neg. | NB | 83.89% |
| 4 | TF–IDF: ngram_range (1,1) | RF | 2900 | Pos.–Neg. | NB | 83.81% |
| 5 | TF–IDF: ngram_range (1,1) | RF | 2600 | Pos.–Neg. | NB | 83.78% |
| 6 | TF–IDF: ngram_range (1,1) | RF | 2800 | Pos.–Neg. | NB | 83.76% |
| 7 | TF–IDF: ngram_range (1,1) | RF | 3000 | Pos.–Neg. | NB | 83.76% |
| 8 | TF–IDF: ngram_range (1,1) | Chi$^2$ | 2300 | Pos.–Neg. | NB | 83.76% |
| 9 | TF–IDF: ngram_range (1,1) | RF | 2700 | Pos.–Neg. | NB | 83.73% |
| 10 | TF–IDF: ngram_range (1,1) | Chi$^2$ | 2400 | Pos.–Neg. | NB | 83.62% |

The conducted results in this paper showed promising results on Arabic sentiment analysis documents and terms that are considered a challenge in data preprocessing, feature extraction, and identification of sentiment polarities. In addition, the Arabic language contains several linguistic terms and colloquial terminologies that are difficult to preprocess and analyze. Figure 7 shows the confusion matrix measured for the last experiments for the final winning model.

In Table 7, a comparison is performed with recent methodologies and frameworks conducted on the Arabic dataset to measure the performance of each model and the annotated algorithm. Furthermore, the research methodologies are applied using ML algorithms or lexicon-based that have different methods for processing and analyzing sentiment polarities with binary classification. As shown, the proposed FFF-SA shows high results compared to related papers.

|  |  | Predicted | |
|---|---|---|---|
|  |  | Pos. | Neg. |
| Actual | Pos. | **578** | **112** |
| | Neg. | **107** | **583** |

**Figure 7.** Confusion Matrix for the final winning model.

**Table 7.** Comparison with recent sentiment analysis performance.

| Ref | Year | No. of Classes | Dialect | Language | Annotation Algorithm | Performance |
|---|---|---|---|---|---|---|
| [17] | 2020 | Pos.–Neg. | Single | Arabic | Lexicon-based | 77% |
| | | Pos.–Neg. | Single | Arabic | Lexicon-based | 86% |
| [19] | 2022 | Pos.–Neg. | Single | Arabic | Lexicon-based | 76.48% |
| [21] | 2023 | Pos.–Neg. | Single | Arabic | Lexicon-based | Senti. Score |
| [23] | 2016 | Pos.–Neg. | Single | Arabic | Lexicon-based | 84% |
| [40] | 2021 | Pos.–Neg. | Single | Arabic | ML-based | 76.9% |
| Proposed FFF-SA Algorithm | | Pos.–Neg. | Single | Arabic | ML-based | 84.4% |

## 7. Conclusions and Future Works

Due to the continuous and increasing use of social networks and E-commerce sites, many platforms depend on the analysis of user opinions to improve the provided services and measure customer satisfaction. One of sentiment analysis's most common problems is the language customers use to express their opinions. The Arabic language is considered one of the most difficult languages in the world because it contains complex linguistic terms and many different dialects that may be used in the same comment or review, making the analysis process more difficult. This paper provides a clear view of recent sentiment analysis approaches and algorithms that depend mainly on ML and lexicon approaches for storing and analyzing a dataset. A comparison of recent research strategies is also provided to compare different methodologies with applied language, including the Arabic language, and their achieved performance. An advanced methodology is proposed using cross-validation that divides the sentiment analysis documents and terms into $k$-folds for training and testing. Data vectorization is applied using the TF–IDF algorithm for counting polarity terms, and an encoder is generated from the training dataset to be applied to the testing dataset. Furthermore, the paper provided an algorithm called FFF-SA based on a forward filter of feature selection that measures and scores the accuracy for each $k$-chuck feature and the following accumulative features. The scoring is processed by executing three feature importance methods, Pearson, Chi$^2$, and RF, with eight ML models where each feature importance is executed with each ML model. Each experiment measures and scores the recorded accuracy for each $k$-chunk feature and then adds accumulative feature to measure the accuracy again. The results proved that the best accuracy is recorded with RF feature importance with the NB model. Future research directions will be directed to apply the same methodology and algorithm on additional Arabic datasets and apply deep learning models to measure the performance and accuracy.

**Author Contributions:** Data curation, A.M.M.; formal analysis, A.M.M. and M.E.; investigation, A.M.M., M.A. (Meshrif Alruily) and A.A.; supervision, M.A. (Meshrif Alruily) and A.A.; writing—original draft, M.E. and M.A. (Meeaad Aljasir); writing—review and editing, A.M.M., M.E., M.A. (Meshrif Alruily), and A.A. All authors have read and agreed to the published version of the manuscript.

**Funding:** This research received no external funding.

**Institutional Review Board Statement:** Not applicable.

**Informed Consent Statement:** Not applicable.

**Data Availability Statement:** Furnished on request.

**Acknowledgments:** The authors acknowledge the Deanship of Scientific research at Jouf University.

**Conflicts of Interest:** The authors declare no conflict of interest.

## References


1. Alamro, H.; Alshehri, M.; Alharbi, B.; Khayyat, Z.; Kalkatawi, M.; Jaber, I.; Zhang, X. Overview of the Arabic Sentiment Analysis 2021 competition at KAUST. *King Abdullah Univ. Sci. Technol.* **2021**, *10754*, 1–9. Available online: http://hdl.handle.net/10754/672089 (accessed on 15 November 2022).
2. Zirikly, A.; Diab, M. Named Entity Recognition for Arabic Social Media. In Proceedings of the1st Workshop on Vector Space Modeling for Natural Language Processing, Denver, CO, USA, 5 June 2015. [CrossRef]
3. Alruily, M. Classification of Arabic Tweets: A Review. *Electronics* **2021**, *10*, 1143. [CrossRef]
4. Oueslati, O.; Cambria, E.; HajHmida, M.; Ounelli, H. A Review of Sentiment Analysis Research in Arabic Language. *Future Gener. Comput. Syst. Elsevier* **2020**, *112*, 408–430. [CrossRef]
5. Hassan, S.; Mubarak, H.; Abdelali, A.; Darwish, K. ASAD: Arabic Social Media Analytics and Understanding. In Proceedings of the 16th Conference of the European Chapter of the Association for Computational Linguistics: System Demonstrations, Kiev, Ukraine, 19–23 April 2021. [CrossRef]
6. Alomari, K.; ElSherif, H.; Shaalan, K. Arabic Tweets Sentimental Analysis Using Machine Learning. In Proceedings of the International Conference on Industrial, Engineering and Other Applications of Applied Intelligent Systems, Arras, France, 27–30 June 2017. [CrossRef]
7. Ansari, M.; Aziz, M.; Siddiqui, M.; Mehra, H.; Singh, K. Analysis of Political Sentiment Orientations on Twitter. *Procedia Comput. Sci. Elsevier* **2020**, *167*, 1821–1828. [CrossRef]
8. Vidya, N.; Fanany, M.; Budi, I. Twitter Sentiment to Analyze Net Brand Reputation of Mobile Phone Providers. *Procedia Comput. Sci. Elsevier* **2015**, *72*, 519–526. [CrossRef]
9. Adilah, M.; Supendar, H.; Ningsih, R.; Muryani, S.; Solecha, K. Sentiment Analysis of Online Transportation Service Using the Naïve Bayes Methods. *J. Phys.* **2020**, *1641*, 012093. [CrossRef]
10. Bakliwal, A.; Foster, J.; van der Puil, J.; O'Brien, R.; Tounsi, L.; Hughes, M. Sentiment Analysis of Political Tweets: Towards an Accurate Classifier. In Proceedings of the Workshop on Language in Social Media, Atlanta, GA, USA, 13 June 2013; Available online: https://aclanthology.org/W13-1106 (accessed on 1 January 2023).
11. Rao, A.; Kanade, V.; Motarwar, C.; Girme, S. Election Result Prediction Using Twitter Analysis. In Proceedings of the International Conference on Inventive Computation Technologies (ICICT), Coimbatore, India, 19 January 2017. [CrossRef]
12. Patel, R.; Passi, K. Sentiment Analysis on Twitter Data of World Cup Soccer Tournament Using Machine Learning. *IoT* **2020**, *1*, 218–239. [CrossRef]
13. Zhang, X.; Yang, Q.; Albaradei, S.; Lyu, X.; Alamro, H.; Salhi, A.; Ma, C.; Alshehri, M.; Jaber, I.; Tifratene, F.; et al. Rise and Fall of the Global Conversation and Shifting Sentiments during the COVID-19 Pandemic. *Humanit. Soc. Sci. Commun. Nat.* **2021**, *8*, 120. [CrossRef]
14. Wang, Y.; Guo, J.; Yuan, C.; Li, B. Sentiment Analysis of Twitter Data. *Appl. Sci.* **2022**, *12*, 1775. [CrossRef]
15. Gutierrez, E.; Karwowski, W.; Fiok, K.; Davahli, M.; Liciaga, T.; Ahram, T. Analysis of Human Behavior by Mining Textual Data: Current Research Topics and Analytical Techniques. *Symmetry* **2021**, *13*, 1276. [CrossRef]
16. Li, S.; Liu, F.; Zhang, Y.; Zhu, B.; Zhu, H.; Yu, Z. Text Mining of User-Generated Content (UGC) for Business Applications in E-Commerce: A Systematic Review. *Mathematics* **2022**, *10*, 3554. [CrossRef]
17. Kwaik, K.; Saad, M.; Chatzikyriakidis, S.; Dobnik, S.; Johansson, R. An Arabic Tweets Sentiment Analysis Dataset (ATSAD) Using Distant Supervision and Self-Training. In Proceedings of the 4th Workshop on Open-Source Arabic Corpora and Processing Tools, Marseille, France, 12 May 2020. Available online: https://aclanthology.org/2020.osact-1.1 (accessed on 22 November 2022).
18. Li, Q.; Li, Z.; Du, Y.; Fan, Y.; Chen, X. A New Sentiment-Enhanced Word Embedding Method for Sentiment Analysis. *Appl. Sci.* **2022**, *12*, 10236. [CrossRef]
19. Chennafi, M.; Bedlaoui, H.; Bedlaoui, A.; Al-qaness, M. Arabic Aspect-Based Sentiment Classification Using Seq2Seq Dialect Normalization and Transformers. *Knowledge* **2022**, *2*, 388–401. [CrossRef]
20. Alwakid, G.; Osman, T.; El Haj, M.; Alanazi, S.; Humayun, M.; Us Sama, N. MULDASA: Multifactor Lexical Sentiment Analysis of Social-Media Content in Nonstandard Arabic Social Media. *Appl. Sci.* **2022**, *12*, 3806. [CrossRef]
21. Mostafa, A. Enhanced Sentiment Analysis Algorithms for Multi-Weight Polarity Selection on Twitter Dataset. *Intell. Autom. Soft Comput.* **2023**, *35*, 1015–1034. [CrossRef]
22. Alharbi, B.; Alamro, H.; Alshehri, M.; Khayyat, Z.; Kalkatawi, M.; Jaber, I.; Zhang, X. ASAD: A Twitter-Based Benchmark Arabic Sentiment Analysis Dataset. *arXiv* **2022**, arXiv:2011.00578. [CrossRef]

23. Aldayel, H.; Azmi, A. Arabic Tweets Sentiment Analysis—A Hybrid Scheme. *J. Inf. Sci.* **2016**, *42*, 782–797. [CrossRef]

24. Mostafa, A. An Automatic Lexicon with Exceptional-Negation Algorithm for Arabic Sentiments Using Supervised Classification. *J. Theor. Appl. Inf. Technol.* **2017**, *95*, 3662–3671. Available online: http://www.jatit.org/volumes/Vol95No15/25Vol95No15.pdf (accessed on 1 January 2023).

25. Mostafa, A. Advanced Automatic Lexicon with Sentiment Analysis Algorithms for Arabic Reviews. *Am. J. Appl. Sci.* **2017**, *14*, 754–765. [CrossRef]

26. Banjar, A.; Ahmed, Z.; Daud, A.; Abbasi, R.; Dawood, H. Aspect-Based Sentiment Analysis for Polarity Estimation of Customer Reviews on Twitter. *Comput. Mater. Contin.* **2021**, *67*, 2203–2225. [CrossRef]

27. Mehmood, S.; Ahmad, I.; Khan, M.; Khan, F.; Whangbo, T. Sentiment Analysis in Social Media for Competitive Environment using Content Analysis. *Comput. Mater. Contin.* **2022**, *71*, 5603–5618. [CrossRef]

28. Ibrahim, A.; Hassaballah, M.; Ali, A.; Nam, Y.; Ibrahim, I. COVID19 Outbreak: A Hierarchical Framework for User Sentiment Analysis. *Comput. Mater. Contin.* **2022**, *70*, 2507–2524. [CrossRef]

29. Oglah, M.; Baniata, A.; Asghar, S. Sentiment Analytics: Extraction of Challenging Influencing Factors from COVID-19 Pandemics. *Intell. Autom. Soft Comput.* **2021**, *30*, 821–836. [CrossRef]

30. Abdukhamidov, E.; Juraev, F.; Abuhamad, M.; El-Sappagh, S.; AbuHmed, T. Sentiment Analysis of Users' Reactions on Social Media during the Pandemic. *Electronics* **2022**, *11*, 1648. [CrossRef]

31. Deniz, E.; Deniz, E.; Cosar, M. Multi-Label Classification of e-Commerce Customer Reviews via Machine Learning. *Axioms* **2022**, *11*, 436. [CrossRef]

32. Mutanov, G.; Karyukin, V.; Mamykova, Z. Multi-Class Sentiment Analysis of Social Media Data with Machine Learning Algorithms. *Comput. Mater. Contin.* **2021**, *69*, 913–930. [CrossRef]

33. Saranya, S.; Usha, G. A Machine Learning-Based Technique with Intelligent Word-Net Lemmatize for Twitter Sentiment Analysis. *Intell. Autom. Soft Comput.* **2023**, *36*, 339–352. [CrossRef]

34. Iqbal, M.; Latha, K. A Parallel Approach for Sentiment Analysis on Social Networks Using Spark. *Intell. Autom. Soft Comput.* **2023**, *35*, 1831–1842. [CrossRef]

35. Hnaif, A.; Kanan, E.; Kanan, T. Sentiment Analysis for Arabic Social Media News Polarity. *Intell. Autom. Soft Comput.* **2021**, *28*, 107–119. [CrossRef]

36. Grande-Ramírez, J.; Roldán-Reyes, E.; Aguilar-Lasserre, A.; Juárez-Martínez, U. Integration of Sentiment Analysis of Social Media in the Strategic Planning Process to Generate the Balanced Scorecard. *Appl. Sci.* **2022**, *12*, 12307. [CrossRef]

37. Al-Absi, A.; Kang, D.; Al-Absi, M. Sentiment Analysis and Classification Using Deep Semantic Information and Contextual Knowledge. *Comput. Mater. Contin.* **2023**, *74*, 671–691. [CrossRef]

38. Hadwan, M.; Al-Hagery, M.; Al-Sarem, M.; Saeed, F. Arabic Sentiment Analysis of Users' Opinions of Governmental Mobile Applications. *Comput. Mater. Contin.* **2022**, *72*, 4675–4689. [CrossRef]

39. Musleh, D.; Alkhales, T.; Almakki, R.; Alnajim, S.; Almarshad, S.; Alhasaniah, R.; Aljameel, S.; Almuqhim, A. Twitter Arabic Sentiment Analysis to Detect Depression Using Machine Learning. *Comput. Mater. Contin.* **2022**, *71*, 3463–3477. [CrossRef]

40. Muhammad, A.; Abdullah, S.; Sani, N. Optimization of Sentiment Analysis Using Teaching-Learning Based Algorithm. *Comput. Mater. Contin.* **2021**, *69*, 1783–1799. [CrossRef]

41. Bhatti, M.; Azhar, S.; Sohail, A.; Hijji, M.; Ayemen, H.; Ramzan, A. Multilingual Sentiment Mining System to Prognosticate Governance. *Comput. Mater. Contin.* **2022**, *71*, 389–406. [CrossRef]

42. Qureshi, M.; Asif, M.; Hassan, M.; Mustafa, G.; Ehsan, M.; Ali, A.; Sajid, U. A Novel Auto-Annotation Technique for Aspect Level Sentiment Analysis. *Comput. Mater. Contin.* **2022**, *7*, 4987–5004. [CrossRef]

43. Hilal, A.; Alfurhood, B.; Al-Wesabi, F.; Hamza, M.; Duhayyim, M.; Iskandar, H. Artificial Intelligence Based Sentiment Analysis for Health Crisis Management in Smart Cities. *Comput. Mater. Contin.* **2022**, *71*, 143–157. [CrossRef]

44. Albahli, S.; Algsham, A.; Aeraj, S.; Alsaeed, M.; Alrashed, M.; Rauf, H.; Arif, M.; Mohammed, M. COVID-19 Public Sentiment Insights: A Text Mining Approach to the Gulf Countries. *Comput. Mater. Contin.* **2021**, *67*, 1613–1627. [CrossRef]

45. Qin, Z.; Ronchieri, E. Exploring Pandemics Events on Twitter by Using Sentiment Analysis and Topic Modelling. *Applied Sciences* **2022**, *12*, 11924. [CrossRef]

46. Alanazi, S.; Khaliq, A.; Ahmad, F.; Alshammari, N.; Hussain, I.; Zia, M.; Alruwaili, M.; Alanazi, R.; Alsayat, A.; Afsar, S. Public's Mental Health Monitoring via Sentimental Analysis of Financial Text Using Machine Learning Techniques. *Environ. Res. Public Health* **2022**, *19*, 9695. [CrossRef]

47. Alali, M.; Sharef, N.; Murad, M.; Hamdan, H.; Husin, N. Multitasking Learning Model Based on Hierarchical Attention Network for Arabic Sentiment Analysis Classification. *Electronics* **2022**, *11*, 1193. [CrossRef]

48. Omara, E.; Mosa, M.; Ismail, N. Emotion Analysis in Arabic Language Applying Transfer Learning. In Proceedings of the IEEE International Conference on Computer Engineering, Cairo, Egypt, 9 March 2020. [CrossRef]

49. Alwehaibi, A.; Roy, K. Comparison of Pre-trained Word Vectors for Arabic Text Classification using Deep Learning Approach. In Proceedings of the IEEE International on Machine Learning and Applications, Orlando, FL, USA, 17 January 2019. [CrossRef]

50. Chouikhi, H.; Alsuhaibani, M.; Jarray, F. BERT-Based Joint Model for Aspect Term Extraction and Aspect Polarity Detection in Arabic Text. *Electronics* **2023**, *12*, 515. [CrossRef]

51. Arabic Sentiment Analysis 2021 @ KAUST. Available online: https://kaggle.com/competitions/arabic-sentiment-analysis-2021-kaust (accessed on 28 December 2022).

52. Zhang, X.; Liu, C. Model Averaging Prediction by *K*-Fold Cross-Validation. *J. Econom.* 2022, in press. [CrossRef]
53. Criminisi, A.; Shotton, J.; Konukoglu, E. Decision Forests: A Unified Framework for Classification, Regression, Density Estimation, Manifold Learning and Semi-Supervised Learning. *Found. Trends Comput. Graph. Vis.* **2012**, *7*, 81–227. [CrossRef]
54. Gupta, R.; Agrawalla, R.; Naik, B.; Evuri, J.; Thapa, A.; Singh, T. Prediction of Research Trends Using LDA Based Topic Modeling. *Glob. Transit. Proc.* **2022**, *3*, 298–304. [CrossRef]
55. Balaji, T.; Annavarapu, C.; Bablani, A. Machine Learning Algorithms for Social Media Analysis: A Survey. *Comput. Sci. Rev.* **2021**, *40*, 100395. [CrossRef]
56. Jordan, M.; Mitchell, T. Machine learning: Trends, perspectives, and prospects, Science. *Science* **2015**, *349*, 255–260. [CrossRef]
57. Saritas, M.; Yasar, A. Performance Analysis of ANN and Naive Bayes Classification Algorithm for Data Classification. *Int. J. Intell. Syst. Appl. Eng.* **2019**, *7*, 88–91. [CrossRef]
58. Istia, S.; Purnomo, H. Sentiment Analysis of Law Enforcement Performance Using Support Vector Machine and K-Nearest Neighbor. In Proceedings of the 3rd IEEE International Conference on Information Technology, Information System and Electrical Engineering, Yogyakarta, Indonesia, 13–14 November 2018. [CrossRef]
59. Chen, J.; Li, K.; Tang, Z.; Bilal, K.; Yu, S.; Weng, C.; Li, K. A Parallel Random Forest Algorithm for Big Data in a Spark Cloud Computing Environment. *IEEE Trans. Parallel Distrib. Syst.* **2016**, *28*, 919–933. [CrossRef]
60. Zhou, J.; Qiu, Y.; Armaghani, D.; Zhang, W.; Li, C.; Zhu, S.; Tarinejad, R. Predicting TBM Penetration Rate in Hard Rock Condition: A Comparative Study among Six XGB-Based Metaheuristic Techniques. *Geosci. Front.* **2021**, *12*, 101091. [CrossRef]

*applied sciences*

*Article*

# RJA-Star Algorithm for UAV Path Planning Based on Improved R5DOS Model

Jian Li [1,2], Weijian Zhang [1], Yating Hu [1,2,*], Shengliang Fu [1], Changyi Liao [1] and Weilin Yu [1]

[1] College of Information Technology, Jilin Agricultural University, Changchun 130118, China
[2] Bioinformatics Research Center of Jilin Province, Changchun 130118, China
* Correspondence: huyating@jlau.edu.cn; Tel.: +86-178-3312-5736

**Abstract:** To improve the obstacle avoidance ability of agricultural unmanned aerial vehicles (UAV) in farmland settings, a three-dimensional space path planning model based on the R5DOS model is proposed in this paper. The direction layer of the R5DOS intersection model is improved, and the RJA-star algorithm is constructed with the improved jump point search A-star algorithm in our paper. The R5DOS model is simulated in MATLAB. The simulation results show that this model can reduce the computational complexity, computation time, the number of corners and the maximum angles of the A-star algorithm. Compared with the traditional algorithm, the model can avoid obstacles effectively and reduce the reaction times of the UAV. The final fitting results show that compared with A-star algorithm, the RJA-star algorithm reduced the total distance by 2.53%, the computation time by 97.65%, the number of nodes by 99.96% and the number of corners by 96.08% with the maximum corners reduced by approximately 63.30%. Compared with the geometric A-star algorithm, the running time of the RJA-star algorithm is reduced by 95.84%, the number of nodes is reduced by 99.95%, and the number of turns is reduced by 67.28%. In general, the experimental results confirm the effectiveness and feasibility of RJA star algorithm in three-dimensional space obstacle avoidance.

**Keywords:** R5DOS intersection matrix; RJA-star algorithm; jump point search algorithm; path planning

**Citation:** Li, J.; Zhang, W.; Hu, Y.; Fu, S.; Liao, C.; Yu, W. RJA-Star Algorithm for UAV Path Planning Based on Improved R5DOS Model. *Appl. Sci.* **2023**, *13*, 1105. https://doi.org/10.3390/app13021105

Academic Editor: Paolo Renna

Received: 5 December 2022
Revised: 9 January 2023
Accepted: 10 January 2023
Published: 13 January 2023

## 1. Introduction

Smart agriculture means that rural workers use advanced technology and experience to carry out agricultural production. The development of social, economic and technological smart agriculture in rural areas is closely related [1]. which is essential in eliminating poverty, helping developed economies "catch up" and forming strategies that promote development in China. Intelligent agriculture includes areas such as: internet technologies, wireless sensor technology [2,3] and remote control technology. As a very important part of intelligent equipment, UAVs are the focus of more attention because of their practical usefulness. UAVs are widely used in agriculture [4,5], forestry [6], disaster relief [7], and geological exploration [8]. The Agricultural UAV [9,10] plays an important role in the fields of crop monitoring, crop yield assessment and plant protection. However, when the UAV sprays precisely at low altitude, farmland obstacles such as plant protection network, residence, electric pole, communication tower, lighting objects and various organisms will pose a serious threat to the UAVs [11]. The main task in agricultural work is to effectively avoid obstacles and achieve the set goals. Significant research has been carried out to solve this problem. Autonomous flight can be achieved to a certain extent by means of sensors, reliable control algorithms and pre-measured obstacle position information [12,13].

A large amount of path planning and obstacle avoidance algorithms have been developed for the path planning of UAVs. For example, A-star algorithm [14], ant colony algorithm [15], artificial potential field [16], DIJKSTRA [17] and so on. Deep learning technology can also be used as an effective method for UAV path planning. A-star algorithm is an effective tool, but the computational complexity of the traditional A-star algorithm space will increase exponentially with the size of the map, resulting in a significant increase in computational complexity and computational time. In order to solve these problems, domestic and foreign scholars have done significant research. Anh et al. [18]., proposed a zigzag global planner based on a-star, which improves the search efficiency of a-star algorithm in narrow space. Hong et al. [19]., proposed an improved A-star algorithm based on terrain data, reducing the calculation time of the algorithm. Zhang et al. [20]., added A-star algorithm to the artificial potential field algorithm, which optimizes the path xu of the UAV and better reflects the actual environment of the UAV. Li et al. [21]., improved the genetic algorithm with A-star algorithm. Ma et al. [22]., adopted the idea of collision ji to improve A-star algorithm. Li et al. [23]., proposed an improved A-star algorithm combined with the jump point search algorithm to reduce the computational cost of A-star algorithm. However, there are still problems in using the nearest neighbor interpolation method to process the path.

To sum up, the a-star algorithm in two-dimensional space has been greatly improved. But there is little improvement regarding the three-dimensional space A-star algorithm [24]. Dilip Mandloi et al. [25], made a detailed comparison on the path planning strategies of star a, Lazy Theta and Theta in a three-dimensional environment. Zhang et al. [26], constructed an improved path planning model using three-dimensional two-way sector expansion method and variable step search strategy. Although the moving distance is reduced by 7.53%, the time cost is increased by 2.66 times. These researches still have the problems of complex computation and large time cost, to solve this problem, this paper presents an algorithm of RJA-star (R5DOS Jump A-star) based on the R5DOS(RCC5-Direction-Octant-Strongly-exists model) model in three-dimensional space. Firstly, the topology of UAV and UAV detection area is expressed, and the A-star algorithm is improved by the JPS(jump point search) algorithm. The model combines the JPS algorithm with the three-dimensional space algorithm to improve A-star algorithm.

The purpose of this study is as follows:

(1) Reduce the complexity of A-star algorithm in three-dimensional space.
(2) Optimize the path selection of A-star algorithm, reduce the corners of the path and make the path more smooth.
(3) Reduce the mobile cost of UAVs, including computing time, path length and access nodes.

## 2. Materials and Methods

### 2.1. Abstract Topological Representation of UAV

Li et al., improved the R5DOS model and proposed a multi UAV formation model, dividing the space into 16 regions, combined with Topology [27]. We modified the R5DOS model, cancelled its formation, and placed the UAV at the center of the model.

According to the R5DOS model, the UAV can be divided into two parts: the body area and the detection area. The detection area is used as the area for UAV to obtain information and perceive the surrounding environment, and is mainly responsible for detecting obstacles and target points, this can be considered a safe area. UAV fuselage area is the area where obstacles need to be avoided during flight.

According to reference [28], we can get the definition of five topological relationships, which are: Discrete (DR), Partial Overlap (PO), Proper Part (PP), Equal (EQ), Proper Part Inverses (PPI), as shown in Figure 1.

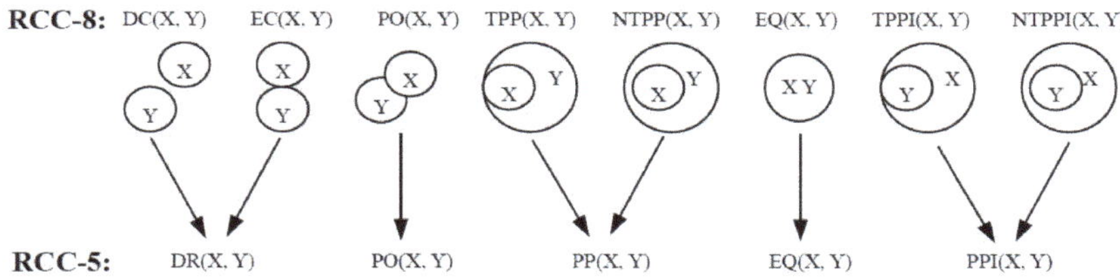

**Figure 1.** Definition and representation of five topological relationships.

The UAV Area C is included in area B, satisfies topological relationship PP (B, C), note that the coordinates of the UAV are $(x_u, y_u, z_u)$, and the coordinates of the target point are $(x_t, y_t, z_t)$, as shown in Figure 2:

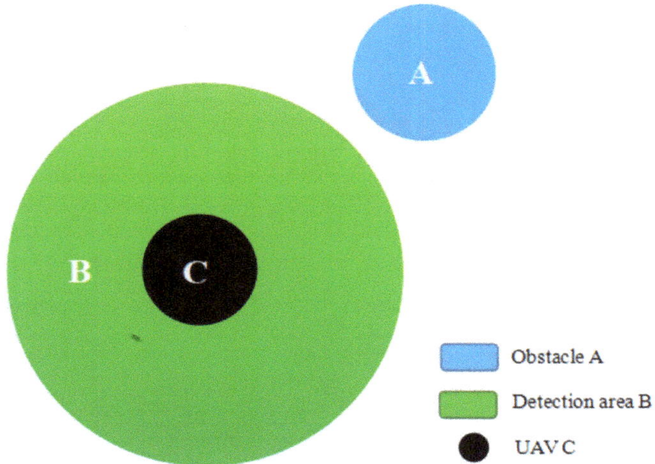


**Figure 2.** Abstract topology representation of UAV area.

## 2.2. Brief Introduction to A-Star Algorithm

The A-star algorithm is a method of finding the shortest path in static scenes. The calculation principle of the A-star algorithm is both simple and fast, but this will slow down with the increase of map size. The increase in computing space will also be exponential, with the larger the map, the greater the computing time and storage footprint. The A-star algorithm will take significant time and space for path calculation. In addition, the computation time will determine the routing efficiency of A star algorithm.

The A-star algorithm is a fast and efficient way to find the right heuristic. Among them, $H(n)$ is the function expression of the shortest path cost of UAV starting from the nth node $(x_n, y_n, z_n)$ to the target $(x_t, y_t, z_t)$. $G(n)$ represents the shortest path between the starting point $(x_s, y_s, z_s)$ and the nth node. Wherein, $F_{cot}(n)$, $H(n)$, $G(n)$ can be expressed by Equation (1).

$$H(n) = \sqrt{(x_t - x_n)^2 + (y_t - y_n)^2 + (z_t - z_n)^2}$$
$$G(n) = \sqrt{(x_n - x_s)^2 + (y_n - y_s)^2 + (z_n - z_s)^2} \tag{1}$$
$$F_{cot}(n) = G(n) + H(n)$$

### 2.3. Jump Point Search Algorithms

A-star requires a lot of unnecessary computation and memory space [29], to improve this an adaptive JPS algorithm was introduced. JPS algorithm is a special search strategy, which only accesses special nodes in the calculation process [30]. This can be thought of as a pre-processing method. After pre-processing, the remaining nodes to be searched are called jump points. As shown in Figure 3, the current node is dotted in blue. From the blue node A5 to the orange node G5, we can ignore the grey node B4, C4, B6, C6 and so on, because the cost of going through B5 to G5 is minimal. When extending to node G5, since node G6 is an obstacle, G5 is the forced neighbor of hop G6.

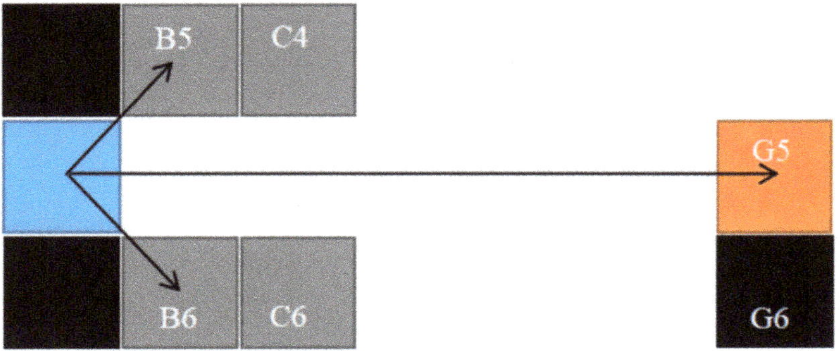


**Figure 3.** The shortest path selection example of JPS algorithm.

This will greatly increase computing time while reducing computing efficiency. In this paper, we will add a preprocessing process to all nodes based on the A-star algorithm to obtain a batch of special hops. In this way, the improved RJA star algorithm can reduce a lot of unnecessary analysis and calculation.

This article defines it according to the actual situation. Take the three-dimensional view of Figure 4a and the cross-section of Figure 4b as an example, assuming that the grey area is an obstacle. The connecting line between UAV and target is defined as $L_{ut}$, the length as $d_{ut} = \sqrt{(x_t - x_u)^2 + (y_t - y_u)^2 + (z_t - z_u)^2}$, and the direction vector as $\vec{L}_{ut} = \left( \frac{(x_t - x_u)}{d_{ut}}, \frac{(y_t - y_u)}{d_{ut}}, \frac{(z_t - z_u)}{d_{ut}} \right)$. Taking the connection line $L_{ut}$ as the central axis of the detection area B, the length of the UAV body is r, and in order to ensure that the UAV can safely avoid obstacles without being affected by other factors, we set the diameter of B as 3r, to determine if there is an obstacle $A_i, (i = 1, 2, \cdots)$ and B intersecting, that is, the topological relationship is $R(A_i, B, C) = \begin{pmatrix} 0 & 1 & 0 & 1 \\ 1 & 1 & 0 & 1 \end{pmatrix}$, where B represents the detection area and C represents the UAV body Area.

If such an obstacle exists, the nearest obstacle to the UAV is considered as the forcing neighbor, the vertices and boundary points with special distance on the obstacle are regarded as the next round of search nodes, namely, that is, the yellow nodes in the Figure 4a,b.

When the algorithm executes the JPS algorithm once, it repeats the above steps until no obstacle intersects with $L_{ut}$ in Figure 5b, and then completes the JPS. Note:

$$\begin{cases} J_0, \text{ the coordinates of the starting point of the UAV} \\ J_i(i = 1, 2, \cdots, n - 1), \text{the coordinates of the jump point} \\ J_n, \text{ the coordinates of the target point} \end{cases}$$

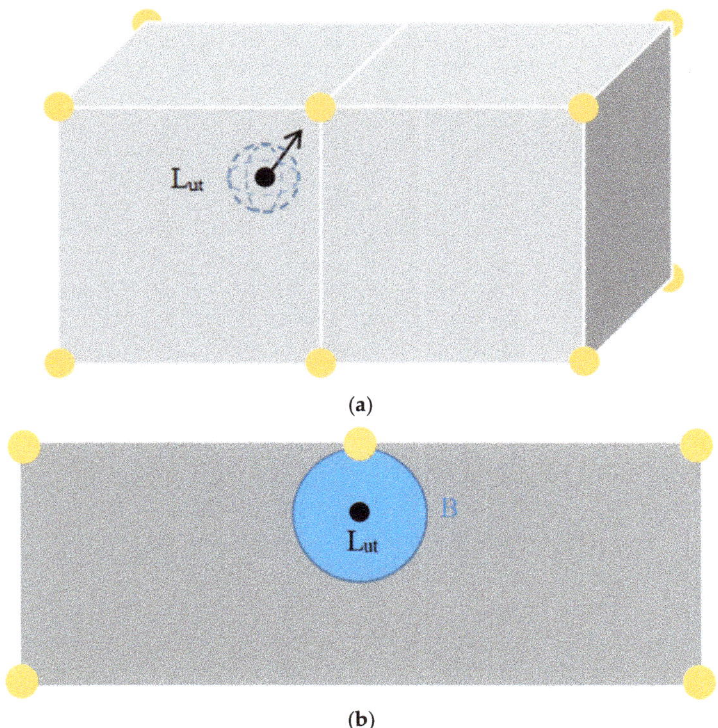



**Figure 4.** (**a**) Schematic diagram of obstacles and UAV. (**b**) Cross sectional sketch of obstacles and UAV.

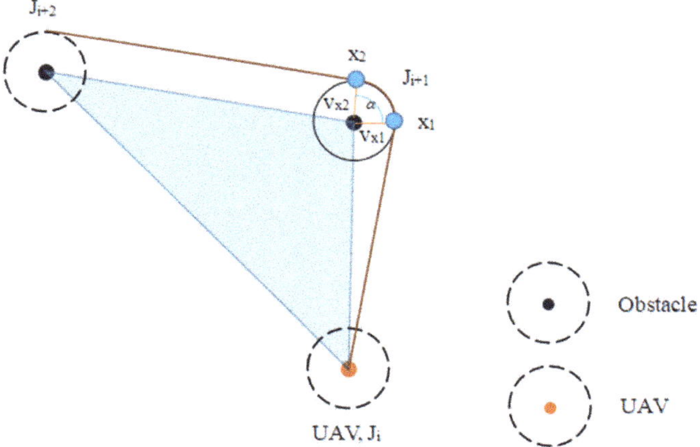



**Figure 5.** Path Selection for UAV Obstacle Avoidance.

During the implementation of the A-star algorithm, the path is not smooth enough because there are many corners, which does not conform to the movement mode of the UAV. To ensure that the UAV can move smoothly and be safe enough, first, we filter out all the jumping points. Then consider expanding a sphere with a diameter of 3r at each node and call it a "jumping body". At the same time, the moving target of UAV is not the selected jumping point, but the jumping body corresponding to each jumping point. Starting from the origin, every time the UAV moves to the next jump body, we will obtain the coordinates of the current node and the next two jump points of the UAV. In the three-dimensional plane the two jumping bodies will intersect on the corresponding spherical tangent plane.

There are two types of UAV paths:

1. When flying from the starting point, move the tangent line from the current position to the circle.
2. Obtain two tangent points $x_1$ and $x_2$ of the same circle on the path, connect them with the center of the circle to obtain the vectors $V_{x1}$ and $V_{x2}$, then the rotation angle is: $\cos(\alpha) = \frac{V_{x1} \bullet V_{x2}}{|V_{x1}| \bullet |V_{x2}|}$.
3. When moving from one body to another, first move along the circle until you leave the body, then move toward the tangent of the next body, as shown in Figure 5.

The current position of the UAV is updated after each movement, and the process is repeated until the movement reaches the target point.

*2.4. Improved Path Planning Algorithm*

According to Sections 2.1–2.4, RJA star path planning Algorithm 1 is as follows:

1. Detects obstacles in the path between the UAV current position and the target point.
2. If there is an obstacle, consider the one closest to the UAV as a coercive neighbor. If there is no obstacle, end the search and fly directly to the target point.
3. Consider the Vertex of the forcing neighbor and its closest boundary point as the jump point.
4. Repeat steps 2–3 until all hops are searched.
5. Compute the cost functions for all hops, compare them, and choose the best path.
6. The UAV chooses the jumping body to move in turn, and the moving path is on the plane formed by the adjacent jumping points.
7. Move in sequence until you reach the target point.

The pseudo-code of RJA-star is:

---

**Algorithm 1: RJA-star**

---

1   algorithm RJA-star (start point, current_node, child node, goal point)
2     $F_{cot}$(current_node) = G(current_node) + H(current_node)
3   if current_node = goal point found the feasible path; break
4   Generate child node that come after current_node
5   for child node of current_node
6     if $R(A_i, B, C) = \begin{pmatrix} 0 & 0 & 0 & 1 \\ 1 & 1 & 0 & 1 \end{pmatrix}$ then Jumpbody.f $\leftarrow \varnothing$
7      if $R(A_i, B, C) = \begin{pmatrix} 0 & 1 & 0 & 1 \\ 1 & 1 & 0 & 1 \end{pmatrix}$ then open $\leftarrow$ Jump bodys
8       else closed $\leftarrow$ Jump bodys
9     end if
10    end if
11     closed $\leftarrow$ current_node
12      if Path(current_node) and RJAPath(current_node) then
13       final $\leftarrow$ closed
14     end if
15 end for
16 end
17 return
18 if (current_node ! = goal point) exit with error (the open list is empty)

---

The algorithm flow is shown in Figure 6.

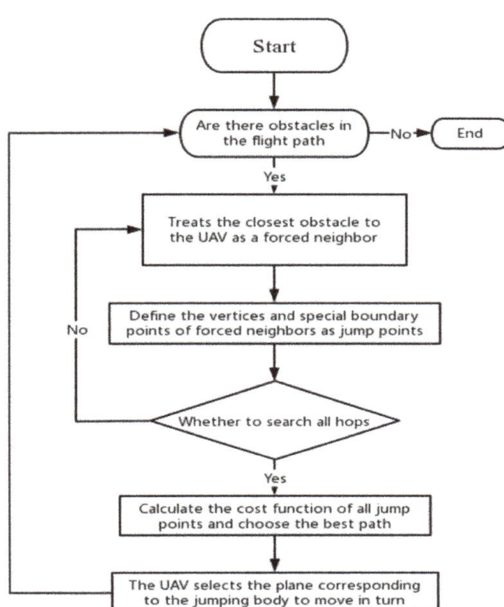


**Figure 6.** Work flow diagram of RJA star algorithm.

## 3. Results

We simulated maps of different map sizes in Matlab. The starting point is (0,0,0), aand the target point is the point farthest from the origin under the current dimension, and generate disjoint obstacles based on the xoy plane. The number of obstacles is 8/10 of the projected area of the current map on the xoy plane, and then a random height is given.

In order to discuss the advantages of RJA star algorithm more intuitively, we simulate A-star algorithm and RJA star algorithm on maps of different sizes. Calculate the running time of the algorithm to evaluate the effectiveness of the RJA star algorithm. The running time starts from the map generation and ends when the UAV reaches the target point

Our simulation environment is as follows:

CPU:Intel® Core™ i7-8750H;

GPU:NVIDIA GTX 1060 Max-Q 6 GB.

We repeated the average value of 100 simulation experiments as the result. We fixed the height of the experimental map at 15 m and set several scenarios, as shown in Table 1.

**Table 1.** Scenario Names of Different Map Sizes.

| Scenario Name | Map Size | Scenario Name | Map Size |
|---|---|---|---|
| scenario 1 | $20 \times 20 \times 15$ | scenario 7 | $80 \times 80 \times 15$ |
| scenario 2 | $30 \times 30 \times 15$ | scenario 8 | $90 \times 90 \times 15$ |
| scenario 3 | $40 \times 40 \times 15$ | scenario 9 | $100 \times 100 \times 15$ |
| scenario 4 | $50 \times 50 \times 15$ | scenario 10 | $110 \times 110 \times 15$ |
| scenario 5 | $60 \times 60 \times 15$ | scenario 11 | $30 \times 90 \times 15$ |
| scenario 6 | $70 \times 70 \times 15$ | | |

We simulate the RJA star algorithm and A-star algorithm in scenario 11, as shown in Figure 7a–d.

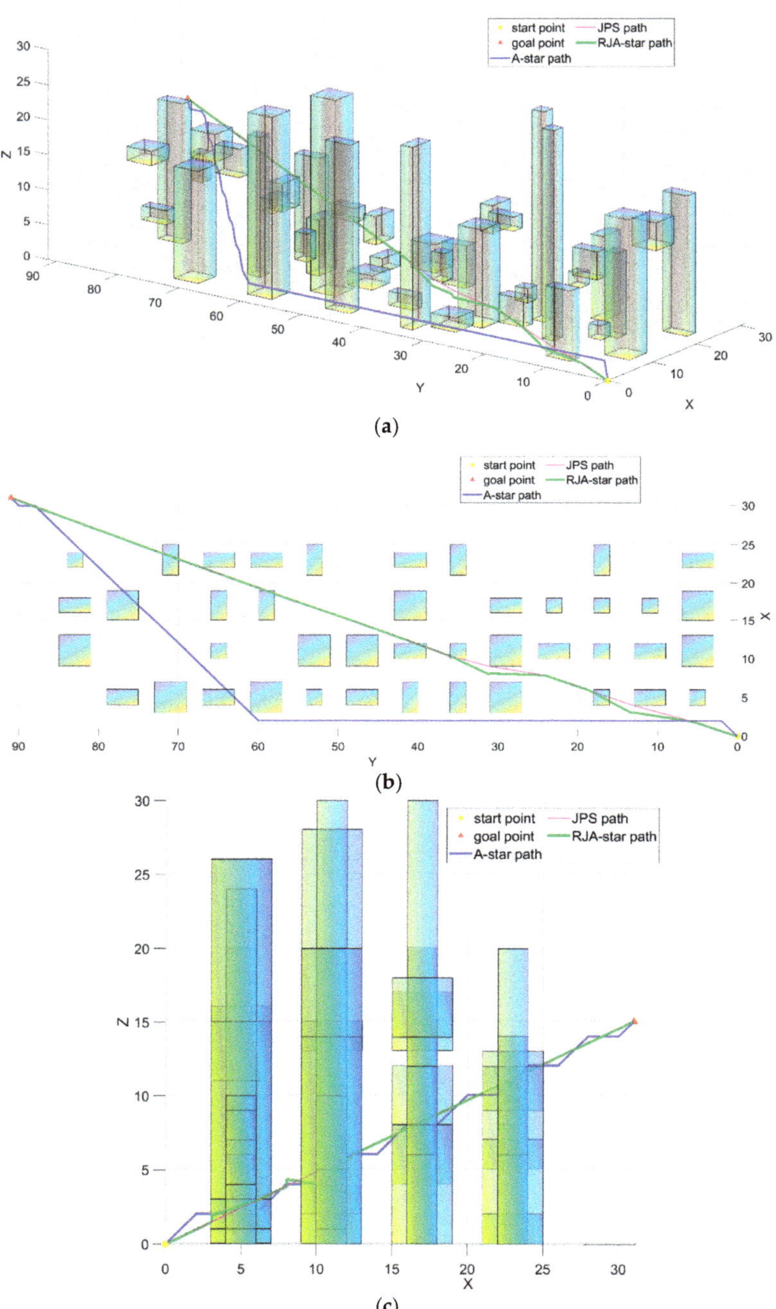

**Figure 7.** *Cont.*

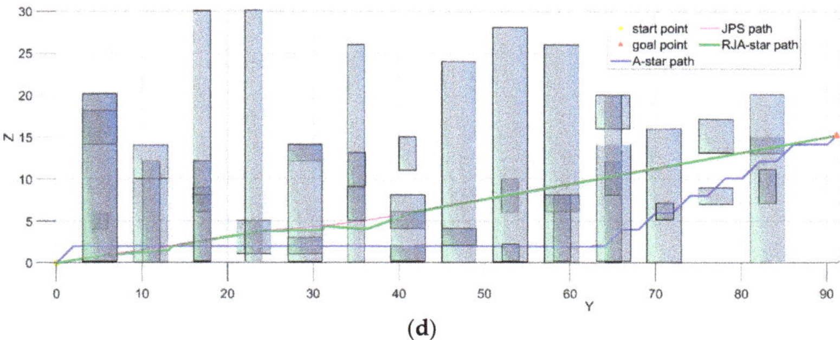



**(d)**

**Figure 7.** (**a**) The fitting results and details of RJA star algorithm on scenario 11. (**b**) Fitting result of XY plane of RJA star algorithm in scenario 11. (**c**) Fitting result of XZ plane of RJA star algorithm in scenario 11. (**d**) Fitting result of YZ plane of RJA star algorithm in scenario 11.

To compare the experimental results, we use Equation (2) to calculate the results of RJA star algorithm and A-star algorithm in computing time, exploration nodes, and path length.

$$\frac{(I_A - I_{RJA})}{I_A} \times 100\%, I_{A/RJA} = \{computing\ time,\ exploration\ nodes,\ path\ length\} \qquad (2)$$

Through Equation (2), we can calculate the reduced computation time, probe node and path length of the RJA star algorithm. Figure 8 is the results of a simulation run under scenario 11. Compared with A-star algorithm, RJA star algorithm reduces 94.4% rotation angle, 99.87% calculation time and 10.05% path length. It can be seen from the results in Figure 9 that the path of the JPS algorithm will stick to the obstacle, which will threaten the safety of the UAV. The RJA star path is smoother and safer than the JPS algorithm, and it can avoid obstacles to reach the destination in many cases. We can see that the RJA-star algorithm avoided obstacles well.

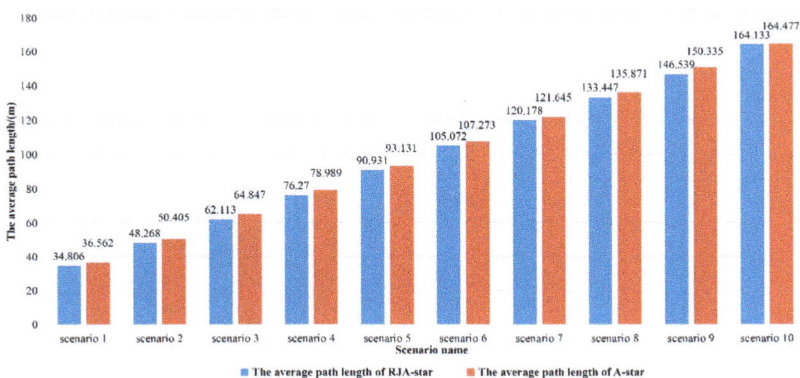



**Figure 8.** Comparison of path length between the two algorithms.

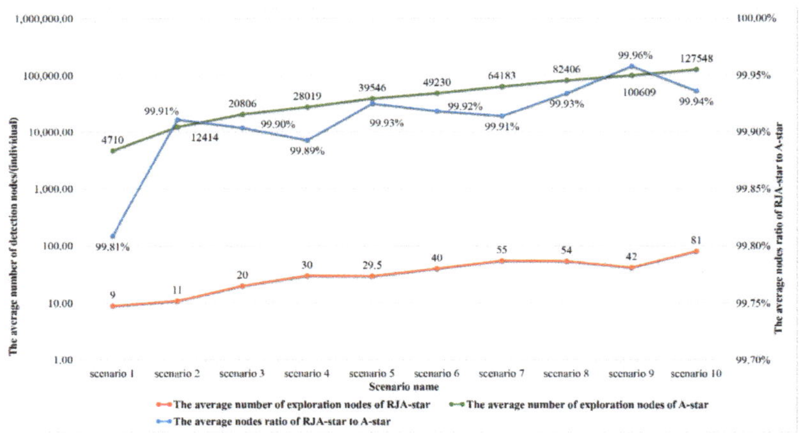

**Figure 9.** Comparison of two algorithms for exploring nodes.

After 100 simulation experiments, we fit the two algorithms into the scenario 1—scenario 10. We limit the random range within the current map, and generate disjoint obstacles based on the xoy plane. The number of obstacles is 8/10 of the projected area of the current map on the xoy plane, and then a random height is given. We compare the two algorithms on a 20 m long map and a 10 m long map. We can obtain the path lengths (as shown in Figure 8), exploration nodes (as shown in Figure 9), computation time (as shown in Figure 10) and the number of corners (as shown in Figure 11) for the A-star algorithm and the RJA-star algorithm. The maximum number of corners are shown in Figure 11.

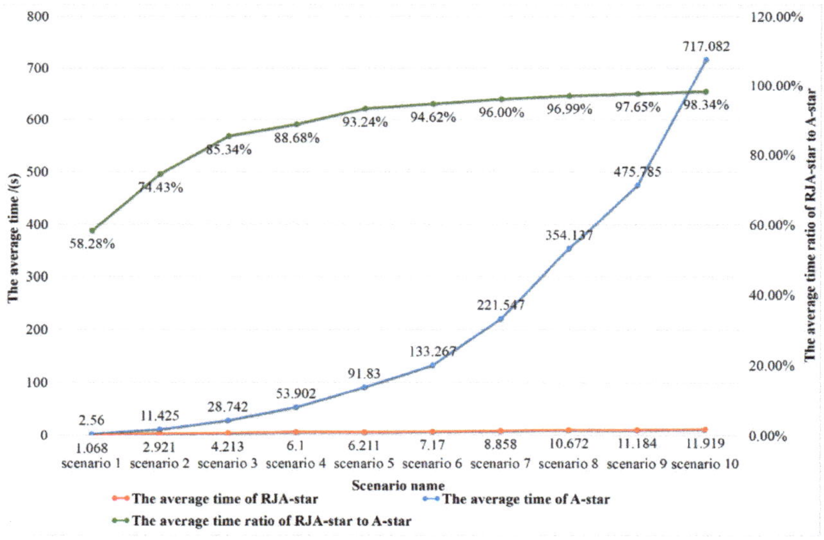

**Figure 10.** Comparison of the average time of the two algorithms.

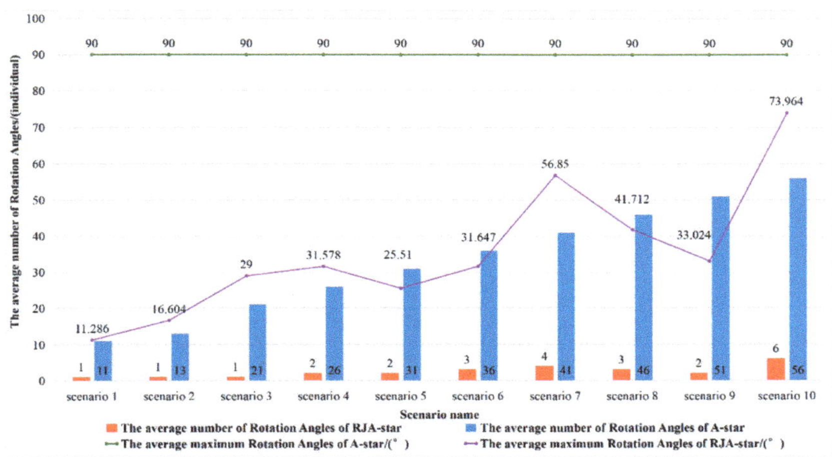

**Figure 11.** Comparison of the number of corners and the maximum angle between the two algorithms.

The result shows that the path length of the RJA star algorithm is not much different from that of the A-star algorithm.

It can be seen from the results in the Figure that the number of nodes of RJA star algorithm is far less than that of A star algorithm.

The y-axis on the left of Figures 9 and 10 respectively represents the number of exploration nodes and average time, and the y-axis on the right represents the percentage reduction of RJA star algorithm compared with A-star algorithm,, which can be calculated by Equation (2). We may know from the Figure that RJA star algorithm has the advantage of short computing time. The five sizes of the map are compared in Table 2.

**Table 2.** Comparison of experimental data of three different specifications of network diagrams.

| Map Size | Path Parameters | Traditional A-Star | RJA-Star | RJA-Star Reduced Proportion |
|---|---|---|---|---|
| | Average running time/s | 2.56 | 1.068 | 58.28% |
| | Average number of nodes | 4710 | 9 | 99.81% |
| 20 × 20 | Total distance/s | 36.806 | 34.806 | 4.80% |
| | Max turning | 90° | 11.286° | 96.59% |
| | Number of turns | 11 | 1 | 87.46% |
| | Average running time/s | 28.742 | 4.213 | 85.34% |
| | Average number of nodes | 20,806 | 20 | 99.90% |
| 40 × 40 | Total distance/s | 64.847 | 62.113 | 4.22% |
| | Max turning | 90° | 29° | 67.78% |
| | Number of turns | 21 | 1 | 95.24% |
| | Average running time/s | 91.83 | 6.211 | 93.24% |
| | Average number of nodes | 39,546 | 29.5 | 99.93% |
| 60 × 60 | Total distance/s | 93.131 | 90.931 | 2.36% |
| | Max turning | 90° | 25.51° | 71.66% |
| | Number of turns | 31 | 2 | 93.55% |

**Table 2.** *Cont.*

| Map Size | Path Parameters | Traditional A-Star | RJA-Star | RJA-Star Reduced Proportion |
|---|---|---|---|---|
| | Average running time/s | 221.547 | 8.858 | 96.00% |
| | Average number of nodes | 64,183 | 55 | 99.91% |
| 80 × 80 | Total distance/s | 121.645 | 120.178 | 1.21% |
| | Max turning | 90° | 56.85° | 36.83% |
| | Number of turns | 41 | 4 | 90.24% |
| | Average running time/s | 475.785 | 11.184 | 97.65% |
| | Average number of nodes | 100,609 | 81 | 99.96% |
| 100 × 100 | Total distance/s | 150.335 | 146.839 | 2.53% |
| | Max turning | 90° | 33.024° | 63.31% |
| | Number of turns | 56 | 2 | 96.08% |

*Discussion*

We select scenario 1, 4, 7, 9. Make the number of obstacles equal to 5/10 of the area of the xoy plane, run the algorithm 20 times on each map, and conduct statistical analysis on the algorithm. From the results, we can get the mean and variance of RJA star algorithm and A-star algorithm under each size map, as shown in Table 3.

**Table 3.** Mean and variance of different indexes of two algorithms.

| Map Size | Map ID | Evaluating Indicator | Estimated Parameters | A-Star | RJA-Star |
|---|---|---|---|---|---|
| 20 × 20 | 1–20 | Average running time/s | mean | 1.194 | 0.501 |
| | | | variance | 0.106 | 0.031 |
| | | Average number of nodes | mean | 1937.950 | 12.300 |
| | | | variance | 3,139,070.892 | 85.484 |
| | | Total distance/s | mean | 34.729 | 34.007 |
| | | | variance | 0.293 | 1.181 |
| 50 × 50 | 21–40 | Average running time/s | mean | 33.634 | 2.900 |
| | | | variance | 619.427 | 0.352 |
| | | Average number of nodes | mean | 23,298.050 | 19.050 |
| | | | variance | 57,624,445.210 | 96.787 |
| | | Total distance/s | mean | 77.010 | 74.192 |
| | | | variance | 0.275 | 0.624 |
| 50 × 50 | 41–60 | Average running time/s | mean | 119.688 | 4.893 |
| | | | variance | 308.581 | 0.844 |
| | | Average number of nodes | mean | 48,848.400 | 37.550 |
| | | | variance | 23,670,949.090 | 350.892 |
| | | Total distance/s | mean | 119.351 | 117.597 |
| | | | variance | 0.020 | 15.461 |
| 100 × 100 | 61–80 | Average running time/s | mean | 323.165 | 2.900 |
| | | | variance | 50,306.815 | 0.352 |
| | | Average number of nodes | mean | 79,975.700 | 40.900 |
| | | | variance | 829,201,538.100 | 526.937 |
| | | Total distance/s | mean | 147.947 | 145.607 |
| | | | variance | 0.454 | 12.160 |

It can be concluded from Table 3 that the RJA star algorithm has smaller variance and mean value in both running time and nodes, which means that the RJA star algorithm is more stable and excellent in running time and nodes performance. The RJA star algorithm considers the safety distance, so the mean value of path is less than A-star algorithm, but the variance is greater than A-star algorithm. Regression analysis was conducted for our data. As shown in Figure 12a–c.

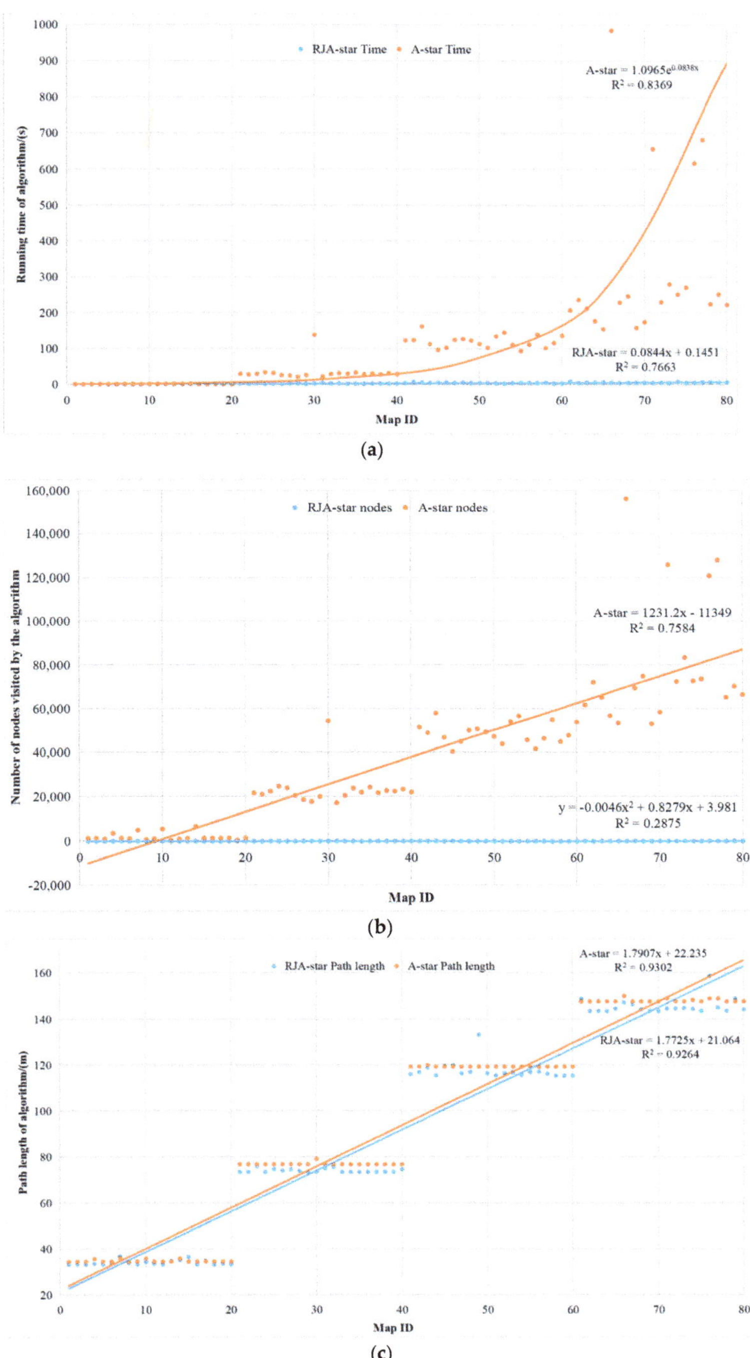

**Figure 12.** (**a**) Regression analysis of computing time of two algorithms. (**b**) Regression analysis of two algorithms' access nodes. (**c**) Regression analysis of path length of two algorithms.

We can see that the number of nodes and path moving length of the two algorithms show a linear regression trend. The regression accuracy of their path lengths can reach 93.02% and 92.64% respectively. The running time of A-star algorithm is exponential regression, with an accuracy of 83.69%. The running time of RJA star algorithm is linear regression, with an accuracy of 76.63%.

Tang et al. proposed the geometric A-star algorithm [31]. To compare the advantages and disadvantages of A-star algorithm, geometric A-star algorithm and improved RJA algorithm, we simulated 100 experiments in scenario 9, and took the average value to Table 4.

**Table 4.** Comparison of indexes of three algorithms.

| Path Parameters | Traditional A-Star | RJA-Star | Geometric A-Star Reduced Proportion | RJA-Star Reduced Proportion |
|---|---|---|---|---|
| Average running time/s | 475.785 | 11.184 | 41.1% | 97.65% |
| Average number of nodes | 100,609 | 81 | 19% | 99.96% |
| Total distance/s | 150.335 | 146.839 | 13% | 2.53% |
| Max turning | 90° | 33.024° | 66.7% | 63.30% |
| Number of turns | 56 | 2 | 84.1% | 96.08% |

Overall, the RJA-star algorithm performs better in terms of computing time and detecting nodes.

## 4. Conclusions

This paper proposes a new path planning algorithm, namely RJA star algorithm. The experimental results show that the RJA star algorithm reduces the mobile path by 2.53%, the computation time by 97.65%, the number of nodes by 99.96% and the number of corners by 96.08% with the maximum corners reduced by approximately 63.30%.smaller angles and smoother paths. In general, our method can effectively reduce the number of turns, calculate the nodes and the angle of turns in the process of UAV motion.

The contributions of this study are as follows:

(1) In order to solve the problem of high computational complexity of three-dimensional A-star algorithm, RJA star path algorithm is proposed.

(2) The pretreatment process is added to screen out a batch of hops and optimize the path between hops. The access and calculation time of nodes are reduced, and the calculation speed is improved.

(3) When obstacles are detected, the algorithm generates hops to optimize the moving path to avoid obstacles.

The organizational structure of this paper is: In the Section 2, we introduce the specific improved form of RJA star algorithm and pseudo code. In Section 3, we simulated in a 3D map to verify the progressiveness of our algorithm. Finally, we draw conclusions in Section 4 and conceive the future work.

The RJA-star algorithm mainly works offline on already known scenarios, without considering the fact that the actual working environment of the UAV should be dynamic and complex. Future work should focus on studying complex dynamic scenarios, including the consideration of flying birds and moving agricultural machinery and other factors.

**Author Contributions:** Conceptualization, J.L. and W.Z.; methodology, J.L. and Y.H.; software, S.F. and C.L.; validation, S.F., C.L. and W.Y.; data curation, C.L.; writing—original draft preparation, W.Z.; writing—review and editing, W.Z.; funding acquisition, J.L. and Y.H. All authors have read and agreed to the published version of the manuscript.

*Appl. Sci.* **2023**, *13*, 1105

**Funding:** This research was funded by Jilin Province Development and Reform Commission China, grant number 2020C037-7; The Education Department of Jilin Province China, grant number JJKH20220332KJ; The Science and Technology Project of Education Department of Jilin Province grant number JJKH20220330KJ; Changchun Science and Technology Development Plan China, grant number 21ZGN26.

**Data Availability Statement:** Not applicable.

**Conflicts of Interest:** The authors declare no conflict of interest.

# References


1. Janc, K.; Konrad, C.; Marcin, W. In the starting blocks for smart agriculture: The internet as a source of knowledge in transitional agriculture. *NJAS-Wagening J. Life Sci.* **2019**, *90*, 100309. [CrossRef]
2. Li, H.P.; Zhang, C.; Zhang, S.Q.; Ding, X.H.; Atkinson, P.M. Iterative Deep Learning (IDL) for agricultural landscape classification using fine spatial resolution remotely sensed imagery. *Int. J. Appl. Earth Obs. Geoinf.* **2021**, *102*, 102437. [CrossRef]
3. Li, H.P.; Zhang, C.; Zhang, Y.; Zhang, S.; Ding, X.; Atkinson, P.M. A Scale Sequence Object-based Convolutional Neural Network (SS-OCNN) for crop classification from fine spatial resolution remotely sensed imagery. *Int. J. Digit. Earth* **2021**, *14*, 1528–1546. [CrossRef]
4. Basudeb, B.; Anusha, V.; Ashok, K.D.; Pascal, L.; Muhammad, K.K. Private blockchain-envisioned drones-assisted authentication scheme in IoT-enabled agricultural environment. *Comput. Stand. Interfaces* **2022**, *80*, 103567.
5. Hovhannisyan, T.; Efendyan, P.; Vardanyan, M. Creation of a digital model of fields with application of DJI phantom 3 drone and the opportunities of its utilization in agriculture. *Ann. Agrar. Sci.* **2018**, *16*, 177–180. [CrossRef]
6. Liang, M.; Delahaye, D. Drone fleet deployment strategy for large scale agriculture and forestry surveying. In Proceedings of the 2019 IEEE Intelligent Transportation Systems Conference (ITSC), Auckland, New Zealand, 27–30 October 2019; pp. 4495–4500.
7. Al-Naji, A.; Perera, A.G.; Mohammed, S.L.; Chahl, J. Life signs detector using a drone in disaster zones. *Remote Sens.* **2019**, *11*, 2441. [CrossRef]
8. Park, S.; Choi, Y. Applications of unmanned aerial vehicles in mining from exploration to reclamation: A Review. *Minerals* **2020**, *10*, 663. [CrossRef]
9. Chen, C.J.; Huang, Y.Y.; Li, Y.S.; Chen, Y.C.; Chang, C.Y.; Huang, Y.M. Identification of fruit tree pests with deep learning on embedded drone to achieve accurate pesticide spraying. *IEEE Access* **2021**, *9*, 21986–21997. [CrossRef]
10. Colloredo-Mansfeld, M.; Laso, F.J.; Arce-Nazario, J. Drone-Based participatory mapping: Examining local agricultural knowledge in the Galapagos. *Drones* **2020**, *4*, 62. [CrossRef]
11. Wang, L.; Lan, Y.; Zhang, Y.; Zhang, H.; Tahir, M.N.; Ou, S.; Liu, X.; Chen, P. Applications and prospects of agricultural unmanned aerial vehicle obstacle avoidance technology in China. *Sensors* **2019**, *19*, 642. [CrossRef]
12. Hosseinpoor, H.R.; Samadzadegan, F.; DadrasJavan, F. Pricise target geolocation and tracking based on UAV video imagery. *ISPRS-Int. Arch. Photogramm. Remote Sens. Spat. Inf. Sci.* **2016**, *XLI-B6*, 243–249. [CrossRef]
13. Huang, A.S.; Bachrach, A.; Henry, P.; Krainin, M.; Maturana, D.; Fox, D.; Roy, N. *Visual Odometry and Mapping for Autonomous Flight Using an RGB-D Camera*; Springer International Publishing: Cham, Switzerland, 2016; pp. 235–252.
14. Liu, Z.; Liu, H.; Lu, Z. A dynamic fusion pathfinding algorithm using delaunay triangulation and improved a-star for mobile robots. *IEEE Access* **2021**, *9*, 20602–20621. [CrossRef]
15. Miao, C.; Chen, G.; Yan, C. Path planning optimization of indoor mobile robot based on adaptive ant colony algorithm. *Comput. Ind. Eng.* **2021**, *156*, 107230. [CrossRef]
16. Orozco-Rosas, U.; Montiel, O.; Sepúlveda, R. Mobile robot path planning using membrane evolutionary artificial potential field. *Appl. Soft Comput.* **2019**, *77*, 236–251. [CrossRef]
17. Luo, M.; Hou, X.; Yang, J. Surface optimal path planning using an extended Dijkstra algorithm. *IEEE Access* **2020**, *8*, 147827–147838. [CrossRef]
18. Le, A.V.; Prabakaran, V.; Sivanantham, V.; Mohan, R.E. Modified A-Star algorithm for efficient coverage path planning in tetris inspired self-reconFigurable robot with integrated laser sensor. *Sensors* **2018**, *18*, 2585. [CrossRef] [PubMed]
19. Hong, Z.; Sun, P.; Tong, X.; Pan, H.; Zhou, R.; Zhang, Y.; Han, Y.; Wang, J.; Yang, S.; Xu, L. Improved A-Star algorithm for long-distance off-road path planning using terrain data map. *ISPRS Int. J. Geo Inf.* **2021**, *10*, 785. [CrossRef]
20. Zhang, J.; Wu, J.; Shen, X.; Li, Y. Autonomous land vehicle path planning algorithm based on improved heuristic function of A-Star. *Int. J. Adv. Robot. Syst.* **2021**, *18*, 17298814211042730. [CrossRef]
21. Li, Y.; Dong, D.; Guo, D. Mobile robot path planning based on improved genetic algorithm with A-star heuristic method. In Proceedings of the 2020 IEEE 9th Joint International Information Technology and Artificial Intelligence Conference (ITAIC), Chongqing, China, 11–13 December 2020; pp. 1306–1311.
22. Ma, L.; Zhang, H.T.; Meng, S.J.; Liu, J.Y. Volcanic ash region path planning based on improved A-star algorithm. *J. Adv. Transp.* **2022**, *2022*, 9938975. [CrossRef]
23. Li, J.; Liao, C.; Zhang, W.; Fu, H.; Fu, S. UAV Path Planning Model Based on R5DOS Model Improved A-Star Algorithm. *Appl. Sci.* **2022**, *12*, 11338. [CrossRef]

24. Zhao, X.; Qin, N.; Huang, J.; Miao, Y. The Application of Adaptive A-star Algorithm in Layout of Spatial Pipeline. In Proceedings of the IEEE 9th Data Driven Control and Learning Systems Conference (DDCLS), Liuzhou, China, 19–21 June 2020; pp. 272–277.
25. Mandloi, D.; Arya, R.; Verma, A.K. Unmanned aerial vehicle path planning based on A* algorithm and its variants in 3d environment. *Int. J. Syst. Assur. Eng. Manag.* **2021**, *12*, 990–1000. [CrossRef]
26. Zhang, Z.; Wu, J.; Dai, J.; He, C. Optimal path planning with modified A-Star algorithm for stealth unmanned aerial vehicles in 3D network radar environment. *Proc. Inst. Mech. Eng. Part G J. Aerosp. Eng.* **2022**, *236*, 72–81. [CrossRef]
27. Li, J.; Zhang, W.; Hu, Y.; Li, X.; Wang, Z. Leader-Follower UAV formation model based on R5DOS-Intersection model. *CMC-Comput. Mater. Contin.* **2021**, *69*, 2493–2511. [CrossRef]
28. Gao, X.; Li, G. A KNN model based on manhattan distance to identify the SNARE proteins. *IEEE Access* **2020**, *8*, 112922–112931. [CrossRef]
29. Liu, H.; Shan, T.; Wang, W. Automatic routing study of spacecraft cable based on A-star algorithm. In Proceedings of the IEEE 5th Information Technology and Mechatronics Engineering Conference (ITOEC), Chongqing, China, 12–14 June 2020; pp. 716–719.
30. Xiang, D.; Lin, H.; Ouyang, J. Combined improved A-star and greedy algorithm for path planning of multi-objective mobile robot. *Sci. Rep.* **2022**, *12*, 13273. [CrossRef]
31. Tang, G.; Tang, C.; Claramunt, C.; Hu, X.; Zhou, P. Geometric A-Star algorithm: An improved A-star algorithm for AGV path planning in a port environment. *IEEE Access* **2021**, *9*, 59196–59210. [CrossRef]

*Article*

# High Performing Facial Skin Problem Diagnosis with Enhanced Mask R-CNN and Super Resolution GAN

**Mira Kim [1,*] and Myeong Ho Song [2]**

[1] IBM Corporation, Costa Mesa, CA 92626, USA
[2] Department of Software, Soongsil University, Seoul 06978, Republic of Korea
* Correspondence: mirakim1012@gmail.com; Tel.: +1-213-308-4596

**Abstract:** Facial skin condition is perceived as a vital indicator of the person's apparent age, perceived beauty, and degree of health. Machine-learning-based software analytics on facial skin conditions can be a time- and cost-efficient alternative to the conventional approach of visiting facial skin care shops or dermatologist's offices. However, the conventional CNN-based approach is shown to be limited in the diagnosis performance due to the intrinsic characteristics of facial skin problems. In this paper, the technical challenges in facial skin problem diagnosis are first addressed, and a set of 5 effective tactics are proposed to overcome the technical challenges. A total of 31 segmentation models are trained and applied to the experiments of validating the proposed tactics. Through the experiments, the proposed approach provides 83.38% of the diagnosis performance, which is 32.58% higher than the performance of conventional CNN approach.

**Keywords:** facial skin problem; mask R-CNN; super resolution; Generative Adversarial Network (GAN); tactics for high performance

## 1. Introduction

Facial skin condition is perceived as a vital indicator of the person's apparent age, perceived beauty, and degree of health. A face with shining, silky, bright, hydrated, and trouble-free skin indicates a high degree of beauty and thus attractiveness, which creates an initial impression of the person. As people get older, their facial skin also ages, revealing symptoms of aging such as wrinkles, age spots, and visible pores. The biological age from his or her facial skin condition can intuitively be predicted. For this reason, people wish to maintain youthful facial skin without aging symptoms.

The conventional way of assessing our facial skin condition is to visit facial skin care shops or dermatologists' offices. However, this requires a burden of locating the right facial skin clinic, making appointments, and visiting the clinics. In addition, the cost incurred for the visit can be substantial.

Machine learning-based software analytics on facial skin conditions can be a time- and cost-efficient alternative to the conventional approach of visiting the clinics. In recent years, researchers have applied Convolutional Neural Network (CNN) based deep learning models to diagnose facial skin problems [1–12]. However, current CNN-based approaches have been shown to be limited in delivering a diagnosis with high performance and, hence, limited in their applicability in clinics. This is mainly due to the following technical challenges in diagnosing facial skin problems with CNN models.

- Detecting small-sized skin problems such as pores, moles, and acne
- Handling the high complexity of detecting about 20 different facial skin problem types
- Handling appearance variability of the same facial skin problem type among people
- Handling appearance similarity of different facial skin problem types
- Handling false segmentations on non-facial areas

**Citation:** Kim, M.; Song, M.H. High Performing Facial Skin Problem Diagnosis with Enhanced Mask R-CNN and Super Resolution GAN. *Appl. Sci.* **2023**, *13*, 989. https://doi.org/10.3390/app13020989

Academic Editors: Yue Wu, Xinglong Zhang and Pengfei Jia

Received: 24 December 2022
Revised: 5 January 2023
Accepted: 8 January 2023
Published: 11 January 2023

The goal of this study is to devise effective software methods to overcome the technical challenges that can be provided as a clinic-level high performance of facial skin diagnosis. This paper is to propose a set of five software tactics that can effectively remedy the challenges and provide a high level of performance in diagnosing facial skin problems. The paper also presents a technical assessment of the proposed methods through experiments and comparison to other approaches.

The paper is organized as the following. Section 2 is to summarize related works and their contributions. Section 3 is to present the intrinsic limitations of conventional CNN approach to diagnosing facial skin problems. Section 4 is to elaborate the set of five tactics that can effectively overcome the technical limitations and provide a high performance of diagnosis. Section 5 is to present the datasets used for training facial skin diagnosis models and the results of the experiments for evaluating the proposed tactics and comparing to other approaches.

The contribution of this study is twofold: (1) proposing a set of give tactics that overcoming the intrinsic limitations of CNN models in diagnosing face skin problems and (2) seamlessly integrating the tactics into software implementation. Through a proof-of-concept implementation of the system and experiments with it, the set of proposed tactics is shown to provide 83.38% of the diagnosis, improve the diagnosis performance by 32.58% compared to the conventional CNN approach, and outperform by an average of 17.89% compared to diagnosis models with MobileNetV2, Xception, ResNet, VGG16, and VGG19.

The proposed facial skin diagnosis system with the tactics can potentially be utilized as a supplementary approach in face skin care clinics and a cost-effective alternative to visiting clinics by individuals.

## 2. Related Works

There have been a number of studies for diagnosing facial skin problems with deep neural networks. Shen's study [1], Quattrini's study [2], and Zhao's study [3] explored the diagnosis of a specific facial skin problem, such as acne or rosacea, using CNN models. VGG-19 models were utilized in Shen's work and Quattrini's work. Liu's study proposed a system for detecting moles [4] using UNet segmentation models.

There have been studies to analyze multiple types of facial skin problems, including Wu [5] and Gerges [6]. Both works utilize a CNN network for diagnosing multiple types of facial skin problems.

There are works to utilize effective methods to improve the performance of analyzing spatial information in the domain of facial skin diagnosis. Yadav's study [7] and Junayed's study [8] proposed a pre-processing method for improving the performance of diagnosing facial skin problems. Yadav applied a method of changing image color space from RGB to HSV to emphasize the acne area. Junayed applied a method for generating multiple images by changing color spaces to reduce noise and emphasize the acne scar areas. Bekmirzaev [9] proposed a segmentation model structure for multiple facial skin problems, which consists of Long Short-Term Memory (LSTM) layers and convolutional layers. Gessert [10] proposed a deep neural network structure whose input is multiple divided sections from high-resolution skin photos and consists of convolutional layers and recurrent layers. Gulzar [11] proposed a segmentation model for skin lesions, which combines a vision transformer and U-Net structure.

There exist works to effectively detect small-sized objects with CNN models [13–21]. Cui [13] proposed a CNN structure for detecting small-size objects by revising the Single Shot Detector (SSD) structure by applying fusion layers for multiple convolutional layers with deconvolutional layers. Liu [14] proposed a modified mask R-CNN model to detect cracks in asphalt pavement using ground-penetrating radar images by adding a feature pyramid network to the original backbone network of the mask R-CNN.

The studies applying the small-sized object detection model in specific domains are summarized as shown in Table 1.

**Table 1.** Representative Studies for Detecting Small Size Objects.

| Study | Input Image | Functionality | Enhanced Section |
| --- | --- | --- | --- |
| Liu [14] | Ground-Penetrating Radar Image | To detect signals for cracks in asphalt pavement | Applied two feature pyramid networks as backbone networks in Mask R-CNN. |
| Nie [15] | Remote Sensing Images | To detect and segment ships | Added bottom-up structure to the feature pyramid networks in Mask R-CNN. |
| Liu [16] | Captured images by Unmanned aerial vehicle (UAV) | To detect objects in UAV perspective | Modified backbone network in YOLO to increase receptive fields. |
| Seo [17] | Medical Image | To segment liver and liver tumor | Added a residual path with deconvolution and activation operations to the skip connection of the U-Net. |
| Peng [18] | 3D Magnetic Resonance Imaging Image | To segment breast tumor in 3D image | Proposed a segmentation structure consisting of localization and segmentation by two Pseudo-Siamese networks (PSN). |
| Tian [19] | Captured images by UAV | To detect objects in UAV perspective | Proposed a detection model consisting of two detection networks to compute low-dimensional features |

The small-sized object detections and segmentations are required in multiple domains. The proposed studies enhanced the original CNN models to add or modify networks.

There exist works to enhance the quality of input images, such as face photos. They proposed to enlarge images and improve the quality of the image. Various approaches for enhancing resolutions have been proposed in [22–27].

There exist works to analyze directional images including [28–30]. Wei [28] proposed a CNN model for segmenting brain areas from MRI by segmenting the brain MRI image into coronal, sagittal, and transverse axials. The feature maps for each direction are applied to generate the 3D brain segmentation result.

There exist works to handle false segmentation with machine learning including [31–35]. Sander proposed a process to discard the failure detection area using the CNN model and knowledge-based filtering [31].

There exist works to distinguish the target class images among similar images including [36,37]. Khan's study proposed a medical image classification by the appearance similarity of each organ in different medical images using a method fused scale-invariant feature transform (SIFT) descriptor and Harris corner algorithm [36].

The current related works provide various deep-learning approaches to diagnose facial skin problems, but they do not address the technical hardship in diagnosing facial skin problems. Our study is distinct on identifying the specific challenges of the diagnosis problems and proposing a set of 5 practical tactics to handle the challenges. As a result, the diagnosis performance is high enough to be utilized in clinics.

### 3. Technical Challenges in Diagnosing Facial Skin Problems

Due to the intrinsic characteristics of facial skin problems, the diagnosis of skin problems—even with advanced machine learning algorithms—presents the following technical challenges.

*3.1. Challenge #1: Detecting Small-Sized Skin Problems Such as Pores, Moles, and Acne*

The CNN algorithm can effectively analyze spatial information. CNN is based on the shared-weight architecture of the convolution kernels and filters that slide along input features and provide translation-equivariant responses known as feature maps [38].

However, the performance of CNN models drops when the size of the target object in an image is considerably small. Some studies have proposed the problems of detecting small-sized objects by CNN models and proceeded experiments to find better algorithms for detecting the objects [39–41]. This is due to the limited spatial features exposed in the small-sized object image, and consequently, the limited extraction of spatial features with filters.

Using our collection of 2225 facial photos (at a resolution of 576 × 576 pixels), the average occupation ratios of facial skin problem areas on the photos are measured, as shown in Table 2.

**Table 2.** Comparing occupation ratios of face, face section, and facial skin problem areas.

| Instances on Facial Photo | Average Number of Pixels | Occupation Ratio on Photo |
| --- | --- | --- |
| *Whole Face* | 101,000 pixels | 33.06% |
| *Sections of Face*, i.e., Eye, Nose, Mouth, and Ear | 3690 pixels | 3.36% |
| *Facial Skin Problems* such as Pore, Mole, and Acne | 25 pixels | 0.02% |

The whole face occupies an average of 33.06% of the photo. A face section such as eye, nose, mouth, or ear occupies an average of 3.36% of the photo. A facial skin problem, such as pore, mole, or acne, occupies an average of 0.02% of the photo. The average radius of a pore is 0.02~0.05 mm [42], and the average radius of a mole is about 6 mm [43].

The hypothesis from this observation is that the detection of such small-sized facial skin problems with CNN models results in significantly low performance. To validate our hypothesis, CNN models using the Mask R-CNN algorithm [44] were trained to detect and visually segment 3 different types of objects. The model was trained with ResNet as the backbone, 0.001 for the learning rate, and (16, 32, 64, 128, 256) as the RPN anchor size. The performance of detection results using the Dice Similarity Coefficient (DSC) metric is shown in Table 3.

**Table 3.** Performance measurements of detecting objects in different sizes.

| Target Objects in Different Sizes | Average of DSC Measurements |
| --- | --- |
| Whole Face | 95.6% |
| Mouth as a *Section of the Face* | 90.9% |
| Mole *as a Facial Skin Problem* | 31.5% |

The average DSC of a mole is shown to be only 31.5%, which is considerably lower than the average DSC of 95.6% for the entire facial area and the DSC of 90.9% for the mouth area.

This is due to the size of the mole, which is too small for the filters in a CNN model to detect the spatial characteristics. If a mole is represented with 25 pixels and the size of a filter is (5 × 5 (i.e., 25 pixels)), then the visual features of the mole are not captured enough by the filter; rather, the features are even simplified and lost through the process of convolution. The resulting feature map could not represent the mole with sufficient information.

Moreover, the mask placed around such a small object like a mole by a Mask R-CNN model cannot present its boundary with a high distinction.

*3.2. Challenge #2: Detecting about 20 Different Types of Facial Skin Problems*

There are about 20 different types of commonly known facial skin problems, including acne, hyperpigmentation, scars, birthmarks, spider veins, white spots, rosacea, ingrown hair, moles, wrinkles, dark circles, eye bags, dry skin, oily skin, dull skin, large pores, and black heads. Some of the common skin problems are shown in Figure 1.

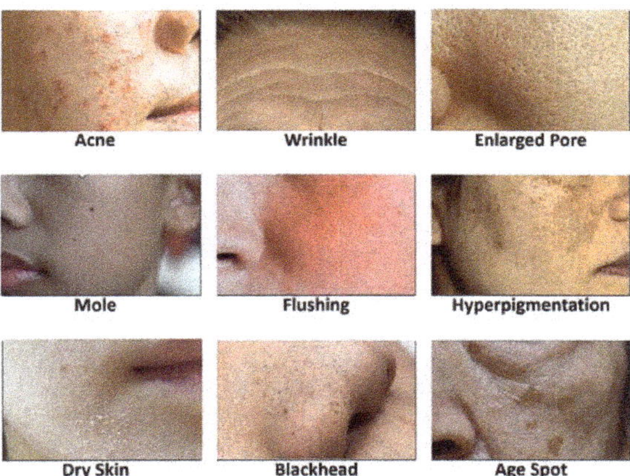

Figure 1. Different types of facial skin problems.

It is challenging to train a CNN model that can detect all the different types of facial skin problems with a high level of performance. This is because the 20 facial skin problem types are not distinct in their appearances; rather, they have a high similarity. Consequently, training a CNN model from a training set of instances from different classes but a high similarity would result in a low performance of classification.

Moreover, because facial skin problem areas are quite small in size, training a CNN model for detecting all the skin types with a high performance becomes infeasible.

### 3.3. Challenge #3: Variability on Appearances of Same Facial Skin Problem Type

There also exists a high variability on the appearances of a facial skin problem type among people. Figure 2 shows four different appearances for acne.

Figure 2. Different appearances of acnes.

For a given facial skin problem type, there can be tens of different appearances, varying in the shape, size, depth, darkness, borderline vividness, and direction. When considering about 20 different facial skin problem types and an average of 'm' different appearances for each facial skin problem type, there exist (20 × m) spatial patterns to be recognized by a CNN model. When 'm' is 50, there are 1000 spatial patterns to handle.

It is challenging to train a CNN model that can detect that many different spatial patterns with a high level of performance. This is because the variability in appearances for the same facial skin problem type expands the heterogeneity of spatial features within a facial skin problem type and the complexity of the spatial features to handle for all 20 facial skin problem types.

### 3.4. Challenge #4: Similarity on Appearances of Different Facial Skin Problem Types

As discussed earlier, some facial skin problem types are not highly distinguishable; rather, they show some similarities. As an example, consider the instances of a mole, blackhead, hyperpigmentation, and age spot as shown in Figure 3.

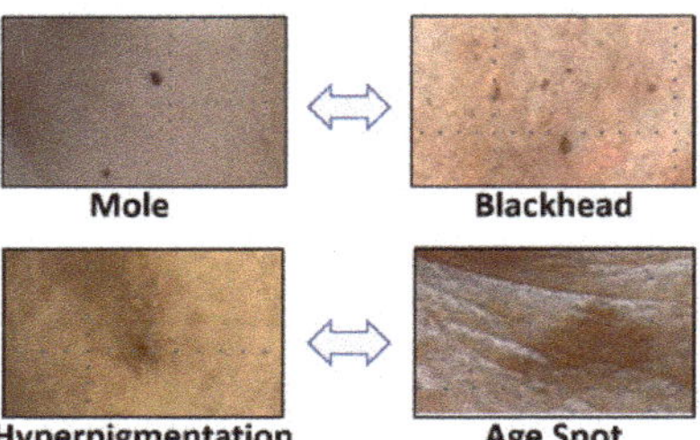

**Figure 3.** Appearance similarity among 4 different facial skin problem types.

They are not highly distinguishable, but exhibit a considerable level of similarities. In the figure, the mole is similar to the blackhead and the concentral part of hyperpigmentation, which is similar to the age spot.

It is challenging to train a CNN model that can distinguish among different facial skin problem types with a high appearance similarity. This is because this appearance similarity between different facial skin problem types should be learned by a CNN model and training the model requires a sufficiently large training set that is configured to represent all appearance variants. Moreover, this similarity adds the complexity of the spatial features to recognize.

*3.5. Challenge #5: False Segmentations on Non-Facial Areas*

Facial skin problems occur only on the facial area, and hence detection of the skin problems should occur on the facial area. A photo or image for a face typically includes images of eyebrows and hairs on the head. Consequently, a trained CNN model for the purpose of detecting facial skin problems could falsely-detect skin problem instances on non-facial areas.

Figure 4 shows examples of the false-detection of facial skin problems on non-facial areas using a Mask R-CNN model.

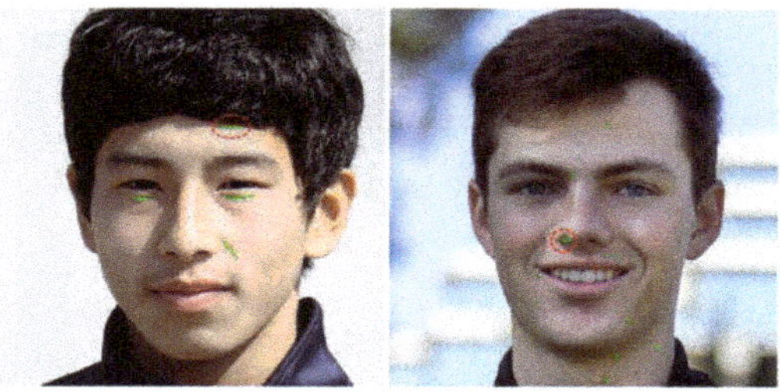

**Figure 4.** Examples of false detection on non-facial areas.

The left image shows a false detection of a winkle around hairs on the forehead and the right image shows a false detection of acne on the nose. This type of false detection could occur whenever a non-facial area contains shapes that are similar to the facial skin problem types.

## 4. Design of Tactics for Remedying the Technical Challenges

To remedy the technical challenges presented earlier and yield a high performance of detecting and segmenting facial skin problems, a set of 5 effective technical tactics are presented in this section. Each tactic is used to handle one or more technical challenges as shown in Figure 5.

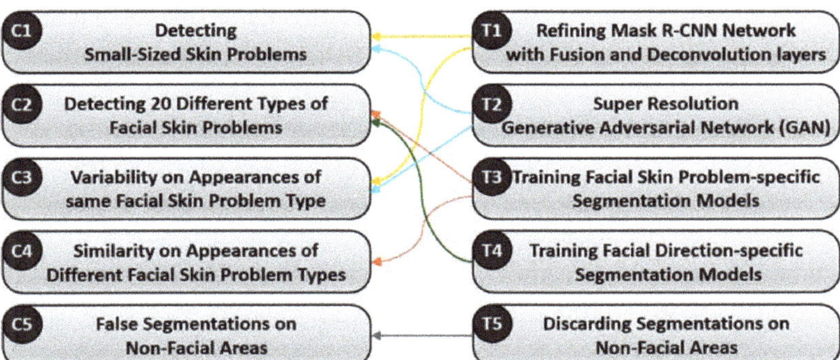

**Figure 5.** Effectiveness of the tactics on remedying the technical challenges.

*4.1. Design of Tactic #1: Refining Mask R-CNN Network with Fusion and Deconvolution Layers*

This tactic is to devise a refined version of Mask R-CNN network structure that is suitable for detecting small-sized objects such as facial skin problems. To overcome the limitations of CNN algorithms in detecting small-sized objects, the Mask R-CNN structure is refined with two elements: *Fusion Layers* and *Deconvolution Layers*.

The structure of our refined Mask R-CNN structure is shown in Figure 6.

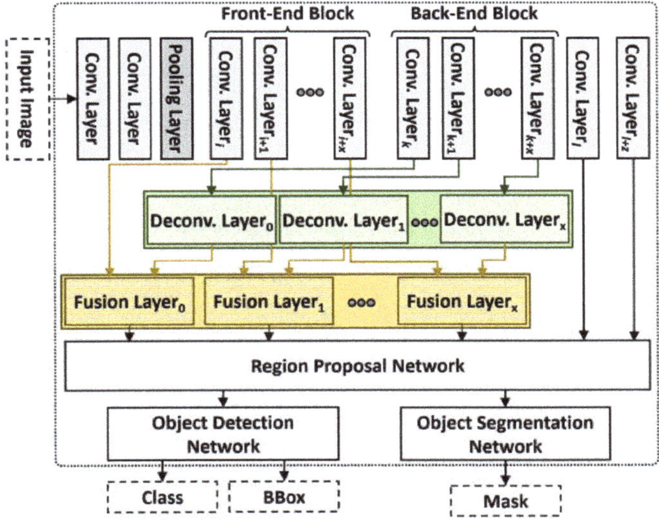

**Figure 6.** Structure of refined mask R-CNN for small-sized objects.

The network structure of a CNN model is shown on the top of the figure, consisting of convolution layers and pooling layers. The CNN structure is refined by performing the following six steps.

Step 1 is to identify the *Front-end Block* that captures finer-grained features of the input image. The block consists of 'x' number of layers that perform convolution and pooling operations to extract features from the input image. The size of this block is determined by the kernel size of each layer in this block and the average size of annotated facial skin problem instances. *Front-end Block* is defined by the layers that have a smaller kernel size rather than the average size of facial skin problem areas captured on its immediately preceding feature map.

Step 2 is to identify the *Back-end Block* that captures coarser-grained features of the input image. The block consists of the same 'x' number of layers that extract the features of larger-sized objects. The size of this block is same as the same size of the *Front-end Block* because the *Fusion Block* requires pairs of a layer in *Front-end Block* layer and a layer in *Back-end Block* layer as shown in Figure 6.

Step 3 is to generate *Deconvolution Block* that consists of 'x' deconvolution layers. This block is used to enlarge the size of input feature maps from *Back-end Block*, which are fed into a *Fusion Block*. A deconvolution layer performs the reverse operation of convolution, i.e., enlarging the size of the feature map created by a convolution layer. That is, each vector in a feature map is padded with a value of zero.

Step 4 is to generate a *Fusion Block* that consists of 'x' fusion layers. This block is used to fuse two feature maps from two sources: *Front-end Block* and *Back-end Block*. That is, each fusion layer receives a feature map from $(i + t)$th layer in *Front-end Block* and a feature map from $(i + x + t)$th layer in *Back-end Block*, sums up the two feature maps and returns a feature map.

Step 5 is to refine the structure of the *Region Proposal Network* of the Mask R-CNN model by entering the feature maps from the *Fusion Block* as inputs to the *Regional Proposal Network*. Note that the basic structure of the *Regional Proposal Network* in the Mask R-CNN model is constructed from the feature maps from layers appearing later in the network. In contrast, our refined *Regional Proposal Network* is enhanced with features maps from *Front-end Block* that capture the spatial features of small-sized objects.

Step 6 is to apply the *Object Detection Network* of the Mask R-CNN model to detect facial skin problem instances and the *Object Segmentation Network* of Mask R-CNN to segment the detected problem areas.

By applying this process of six steps, the enhanced version of the Mask R-CNN model can detect target objects of a small size, and consequently, the performance of facial skin problem diagnosis can significantly be increased.

Hyperparameters of the proposed network are set to 0.001 as the learning rate, 0.3 as the detection non-maximum suppression (NMS) threshold, 0.5 for the region of interest (ROI) positive ratio, 0.7 as a threshold for RPN and NMS, (0.5, 1, 2) as RPN anchor ratio, (8, 16, 32, 64, 128) as the RPN anchor size, localization loss (smooth L1) and confidence loss (Softmax) for loss functions in proposed front-end and back-end blocks, localization, and average binary cross-entropy loss. In addition, the loss function for the refined mask R-CNN is computed by the sum of loss functions for classification, bounding box detection, and segmentation.

*4.2. Design of Tactic #2: Super Resolution Generative Adversarial Network (GAN) for Small-sized and Blurry Images*

This tactic is to enhance the quality of small-sized object images, i.e., facial skin problem instances, by applying a Super-Resolution Generative Adversarial Network (SR-GAN) [45]. Generative Adversarial Network (GAN) consists of a generator network and a discriminator network to compete with each other to generate accurate predictions. GR-GAN is a GAN model that upscales and improves the quality of low-resolution images. That is, the structure of the Generator in GAN is enhanced with *Sub-Pixel Convolution Layers* as shown in Figure 7.

A *Sub-Pixel Convolutional Layer* is to enlarge the size of the feature map by combining the vectors in feature maps into a single feature map. Then, the resulting feature map consists of a larger number of vectors than the input feature map. Accordingly, an original image is enhanced with more detailed image features.

The effect of applying SR-GAN on facial skin problem diagnosis is to enlarge the small-sized facial skin problem instances and to result in a more accurate problem diagnosis with Mask R-CNN models.

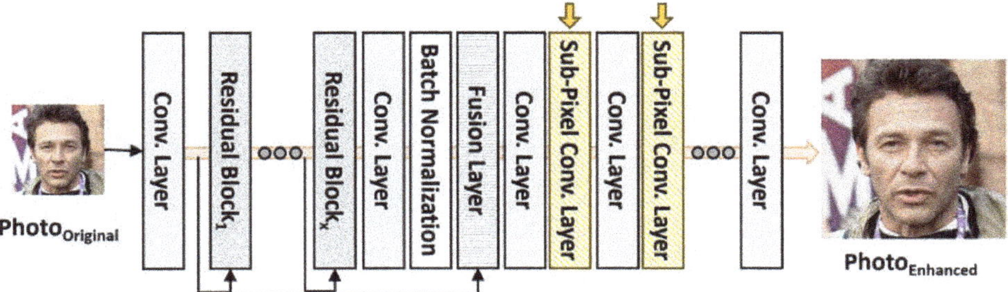


**Figure 7.** Generator in SR-GAN with Sub-Pixel convolution layers.

*4.3. Design of Tactic #3: Training Facial Skin Problem-Specific Segmentation Models*

This tactic is to train a segmentation model for each type of facial skin problems. This is based on the observation on the feature extraction schemes of CNN algorithm. CNN network consists of convolution layers to learn spatial characteristics of objects, pooling layers to reduce the dimensions of the feature maps, flattening layers to convert the resultant 2-dimensional arrays into a single long continuous linear vector, and fully connected layers to connect every input neuron to every output neuron [46,47].

However, CNN models provide lower performance for detecting multi-class objects due to the learning scheme of spatial features with convolution layers and the dimension reduction scheme with pooling layers [46,47]. Generating this phenomenon, detecting multi-class objects is harder than detecting single-class objects [48,49]. For example, a CNN model detecting people would perform better than a CNN model detecting the people with gender information, i.e., male and female.

Another example is to detect animals in a zoo. Detecting a single-class object, such as detecting 'dog', should outperform comparing to detecting 100 different animal types in a zoo. Then, the CNN model for 100-class objects must handle the appearance features of all 100 types of animals. Through the repetition of applying convolution and pooling in CNN, the spatial features of 100 animal types are abstracted by cancelling some of the acquired features through activation functions, such as *ReLU*.

Another cause of the lower performance of the multi-class CNN model is the technical hardship in distinguishing objects of different types but having some degree of appearance similarity [49–52]. For example, domestic dogs, wolves, coyotes, foxes, jackals, and dingoes belong to different animal classes, but there exist a number of similar appearance features among different types of animals, such as between domestic dogs and wolves and between coyotes and foxes.

Hence, this tactic is to train $k$ segmentation models for $k$ types of facial skin problems, rather than training a single segmentation model to recognize all $k$ types of facial skin problems. That is, for a given facial image, the $k$ segmentation models are individually applied to detect and segment its specific facial skin problem type. Then, the results of applying $k$ segmentation models are integrated into a single output as shown in Figure 8.

In the figure, the facial image is fed into $k$ segmentation models, which will detect and segment its specific facial skin problem type. The results are integrated into a single output.

By specializing segmentation models by their facial skin problem types, the performance of facial skin diagnosis is increased over employing a single integrated segmentation model.

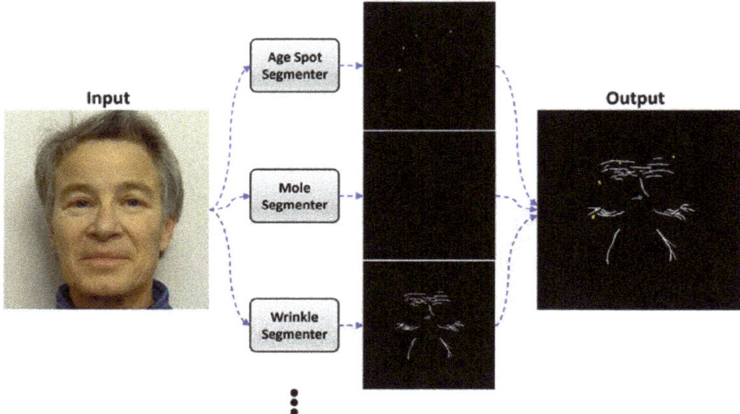


**Figure 8.** Applying *k* segmentation models and integrating the results of all segmentations.

### 4.4. Design of Tactic #4: Training Face Direction-Specific Segmentation Models

This tactic is to train a segmentation model for each direction of a face, i.e., left-side face, frontal face, and right-side face. A face photo is taken from one direction, and the photo cannot capture the facial skin problem instances on other directions, such as instances near ears or side-cheeks. As a result, a single segmentation model to detect facial skin problem instances on all different areas cannot correctly detect all the skin problem instances.

This tactic is to handle this problem by applying face direction-specific segmentation models. To utilize this tactic, a set of 3 facial photos is required as input for the facial skin problem diagnosis system. In order to determine the direction of the face photo, a *Face Direction Identifier* is designed by applying Facial Landmark Detection model [53]. Using the face contours and locations of nose, mouth, and the eyes, its direction can automatically be identified.

An example of applying face direction-specific segmentation models is shown in Figure 9.

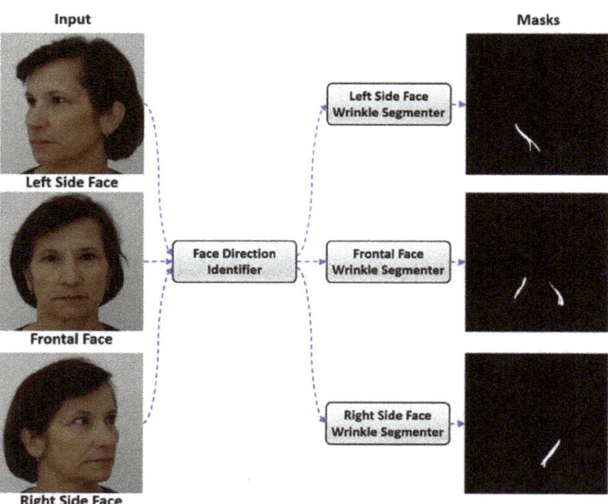


**Figure 9.** Applying face direction-specific segmentation models.

As shown in the figure, *Face Direction Identifier* determines the direction of each input face photo. Then, its specific segmentation model is applied to diagnose the facial skin problem instances, and their results are integrated.

By specializing segmentation models by the directions of a face photo, the performance of facial skin diagnosis is increased over employing a single integrated segmentation model.

### 4.5. Design of Tactic #5: Discarding Segmentations on Non-Facial Areas Using Facial Landmark Model

This tactic is to discard the false segmentations made on non-facial areas by applying a Facial Landmark Detection model. The false segmentations can effectively be discarded with the following steps.

Step 1 is to detect facial skin problem instances and facial landmarks for each facial image. The Facial Landmark Detection model is used to detect the landmarks on the face image.

Step 2 is to generate a mask around only the skin area of the facial image. That is, the eyes, eyebrows, mouth, nostril, and hairs are excluded in this mark.

Step 3 is to remove the facial skin problem instances in non-facial areas. This is done by overlaying the two types of masks, masks of facial skin problem instances and the mask of the facial skin area, and discarding segmentations made on the outside of the mask of the facial skin area.

An example of discarding false segmentations is shown in Figure 10.

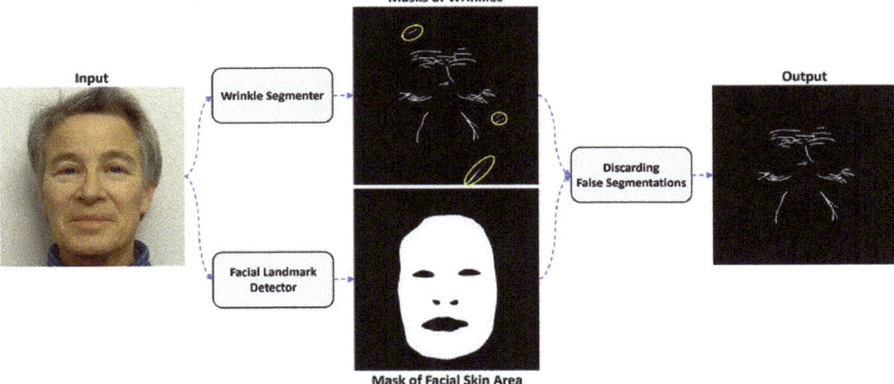


**Figure 10.** Example of discarding false segmentations.

In the figure, the *Wrinkle Segmenter* model produces marks of detected wrinkles. Three of the wrinkle segmentations are made on a non-facial area. The *Facial Landmark Detector* produces a mask of facial skin area. By overlaying two types of masks, the false segmented wrinkles are discarded.

### 4.6. Design of the Main Control Flow

The main control flow of the facial skin diagnosis system is to invoke the functional components that implement the proposed 5 tactics. The control flow is shown in the following algorithm as shown in Algorithm 1.

As shown in the algorithm, the main control of the diagnosis system reads facial photos as the input and invokes the functional components that implement the 5 tactics.

---

**Algorithm 1.** Main control flow of 'Facial Skin Problem Diagnosis' system.

---

**Input**: *photos*: A list of 3 face photos (per person)
**Output**: *FSPResults*: A list of detected facial skin problem instances
**1:**  Main() {
**2:**      FSPResults = [];
**3:**      SRGAN = // SR-GAN Model for upscaling and improving quality of Images
**4:**      FaceLandmarkDetector = // Model for Face Landmark Detector
**5:**      upscalingRatio = // Ratio of upscaling image by SRGAN
**6:**      segmenters = // set of segmentation models for face direction and face skin problems
**7:**
**8:**      **for** (photo **in** photos){
**9:**          // **Step 1. Identify Face Directions (regarding the tactic #4)**
**10:**         landmarks = FaceLandmarkDetector.identify(photo);
**11:**         locMouth = // Location of Mouth from detected Landmarks
**12:**         locNose = // Location of Nose from Detected Landmarks
**13:**         locRt = // Location of right side of face in detected landmarks
**14:**         locLt = // Location of Left side of face in detected landmarks
**15:**         **if** ((|locMouth-locLT| < |locMouth-locRT|) & (|locNose-locLT| < |locNose-locRT|))
**16:**             curDirection = LEFT;
**17:**         **else if** ((|locMouth-locLT| > |locMouth-locRT|) & (|locNose-locLT| > |locNose-
**18:** locRT|))
**19:**             curDirection = RIGHT;
**20:**         **else** curDirection = FRONTAL;
**21:**
**22:**         // **Step 2. Invoke Facial Skin Problem-specific Segmenters (regarding the tactic #3)**
**23:**         listFSPs = [];
**24:**         **for** (segmenter_type **in** segmenters[curDirection]){
**25:**           SEGRef_type = // Segmenter based on Refined mask R-CNN from segmenter_type
**26:**           SEGOrg_type = // Segmenter based on Original mask R-CNN from segmenter_type
**27:**           // **Step 3. Applying Refined mask R-CNN (regarding the tactic #1)**
**28:**           resultRef = SegRef_type.segment(photo);
**29:**
**30:**           // **Step 4. Enhance the Quality of Facial Images with SR-GAN model (regarding**
**31:** the tactic #2)
**32:**             sections = {SECi | SEC in photo, ∀SEC = photo}; // Divide Photos
**33:**             resultOrg = [];
**34:**             **for** (SEC ∈ sections){
**35:**               enlargedSEC = SRGAN.enlarge(SEC);
**36:**               result = SEGOrg_type.segment(enlargedSEC);
**37:**               resultOrg ← ( result // After Decreasing size of result to (1/upscalingRatio)
**38:**             }
**39:**             // Determine the facial skin problem
**40:**             // (1) Select a result from step 3 and Step 4
**41:**             **if**(*size*(resultOrg) < thInstanceSize)
**42:**                 result = resultRef;
**43:**             **else**
**44:**                 result = resultOrg;
**45:**             // (2) Check whether the segmented instances classified by different FSP
**46:**             **for** (fsp **in** listFSPs){
**47:**                 **if** (*size*(fsp∧result)/*max*(*size*(fsp), *size*(result)) > thSize){
**48:**                     **if** ((confidence score of fsp) > (confidence score of result))
**49:**                         // Remain fsp
**50:**                     **else**{
**51:**                         // discard fsp from listFSPs and add result
**52:**                     }

---

```
53:              }
54:            }
55:          }
56:          // Step 5. Discard false segmentations on non-facial skin area (regarding the tactic
57: #5)
58:          faceArea = // Mask for face skin area excepting eyes, nostrils, and mouth.
59:          segResult = FSPArea ∧ faceArea; // Overlay both segmented area
60:          FSPResults ← ( (curDirection, segResult);
        }
      return FSPResults;
    }
```

## 5. Experiments and Assessment

This section is to present the results of experiments by applying the facial skin segmentation models trained with the proposed tactics.

### 5.1. Datasets for Training Models

The data collection of face photos used for training, evaluation, and experiments consists of 2225 face photos at a resolution of 576 × 576 pixels. Each photo contains one or more instances of acne, age spots, moles, rosacea, and wrinkles.

In the experiments, only 5 types of facial skin problem types are considered for the following criteria.

- A minimal set of essential facial skin problem types is needed; performing experiments with photos of all 20 different facial skin problem types requires a dataset of more than 10,000 photo images and annotating the facial skin problem areas on each photo manually by researchers would require an effort of more than 30 person-months.
- A dataset including facial skin problem types that are relatively large in size and also small in size is needed. Hence, photos showing wrinkles and rosacea are selected for the large-sized types and acne, mole, and age spots are selected for the small-sized types.
- A dataset including different facial skin problem types but having some similarities in their appearances is needed. Hence, photos showing acne, age spots, and moles are selected.
- A dataset including blurry boundaries between the problem-free skin areas and facial skin problem areas is needed. Hence, photos showing acne, rosacea, and wrinkles are selected.

The face photos were acquired from 3 different sources: Acne 04 dataset [54] and Flickr-Faces-HQ dataset [55] available on GitHub repositories [56,57] and FEI face dataset available on Centro University website [58].

To train CNN models to detect facial skin problems, each photo had to be manually annotated in all the areas of facial skin problem instances. For this task, COCO Annotator was utilized as an annotation software tool [59] and an XP-Pen Artist 15.6 Pro stylus pen was used as a touch-pad device. Since a photo may contain several kin problem instances, the task of manually annotating the 2225 photos demanded a high effort of 8 person-months.

An example of manual annotations for facial skin problem instances is shown in Figure 11.

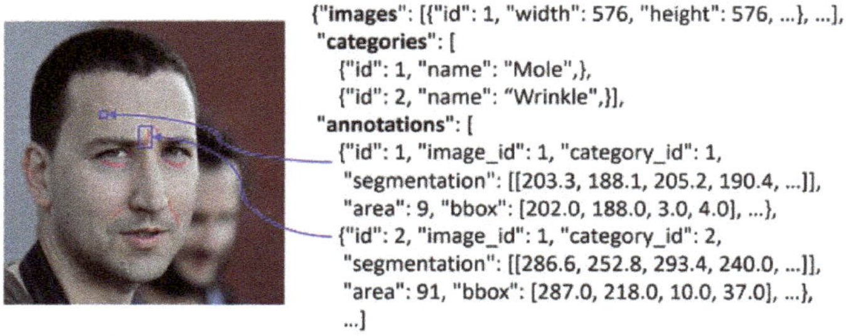

{"images": [{"id": 1, "width": 576, "height": 576, ...}, ...],
 "categories": [
   {"id": 1, "name": "Mole",},
   {"id": 2, "name": "Wrinkle",}],
 "annotations": [
   {"id": 1, "image_id": 1, "category_id": 1,
    "segmentation": [[203.3, 188.1, 205.2, 190.4, ...]],
    "area": 9, "bbox": [202.0, 188.0, 3.0, 4.0], ...},
   {"id": 2, "image_id": 1, "category_id": 2,
    "segmentation": [[286.6, 252.8, 293.4, 240.0, ...]],
    "area": 91, "bbox": [287.0, 218.0, 10.0, 37.0], ...},
   ...]

**Figure 11.** Face photo with annotations and its JSON representation.

The left-side of figure shows annotations for two facial skin problems. Moles are annotated in Blue, and wrinkles are annotated in Red. The right-side of the figure shows the JSON representation of the annotations, which is required to train the model.

By using our collection of 2225 face photos with annotations on facial skin problem instances, a total of 31 Mask R-CNN models were trained. The data collection is utilized as a training set, validation set, and test set, as shown in Table 4.

**Table 4.** Distribution of Data Collection.

| Subset Type | Training Set | Validation Set | Test Set | Total |
|---|---|---|---|---|
| **Ratio (%)** | 69.98 | 10.02 | 20.0 | 100 |
| **# of Photos** | 1557 | 223 | 445 | 2225 |

The training set consists of 1557 photos that are 70% of the data collection. The validation set consists of 223 photos that are 10% of the data collection. The test set consists of 445 photos that are 30% of the data collection.

Hyperparameters for training models are set to 0.001 as the learning rate, 1000 as epochs, and 5 as batch size. The early stopping is applied to prohibit overfitting the model to the training set.

*5.2. Proof-of-Concept Implementation*

A web-based system of facial skin diagnosis has been implemented in Python using the following libraries: TensorFlow for developing CNN models, NumPy for processing operations for mask data, OpenCV for processing facial photos, MySQL for managing database, and Django framework for building the web site.

Figure 12 shows the web user interface of this system.

The original image, masks generated around the facial skin problems, and overlay of the mask on the original image are shown on the left-side of the figure. The right-side of the figure shows the results of identifying facial skin problem instances.

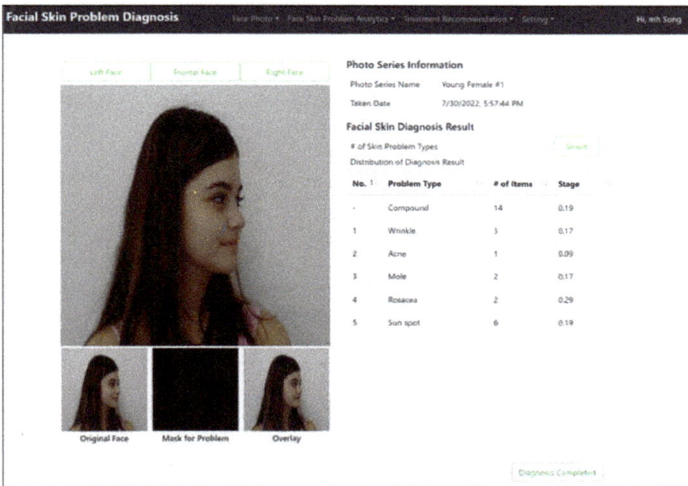


**Figure 12.** User Interface of Facial Skin Problem Diagnosis System.

*5.3. Performance Metric for Facial Skin Problem Diagnosis*

Dice Similarity Coefficient (DSC) is an appropriate measure in evaluating the performance of segmenting the models. DSC is to measure the ratio of matching the segmented areas on the labeled areas on images. Its metric is given below.

$$DSC = \frac{2 * |LB \cap SEG|}{|LB| + |SEG|}$$

Let *LB* be a mask for the label data of a photo, and Let *SEG* be a mask of the segmentation results from the photo. Let $|X|$ be the area of masked instances in an input mask *X*. *DSC* is measured by the sum of masked areas for *LB* and *SEG* over the area containing the intersection area in *LB* and *SEG*. The range of *DSC* is between 0 and 1. The more accurate the segmented results, a *DSC* value close to 1 is returned.

*5.4. Experiment Scenarios and Results*

A set of experiments were conducted to evaluate the effectiveness of the 5 tactics, an experiment to evaluate the integration of all the tactics, and an experiment to compare our approach to other known approaches.

5.4.1. Experiment for Tactic #1: Refined Mask R-CNN Segmentation Models

This experiment is to evaluate the effectiveness of the Refined Mask R-CNN model. This is done by training two segmentation models: a model with Mask R-CNN structure and a model with the refined Mask R-CNN structure with fusion and deconvolution layers.

The dataset of 410 face photos was used and each photo contains one or more facial skin problems of acne, age spots, and moles. These facial skin problem types are quite small in size, and the bounding box of the largest problem instance in the dataset is sized to (12 × 12) pixels.

The performances of the two models are compared in Figure 13.

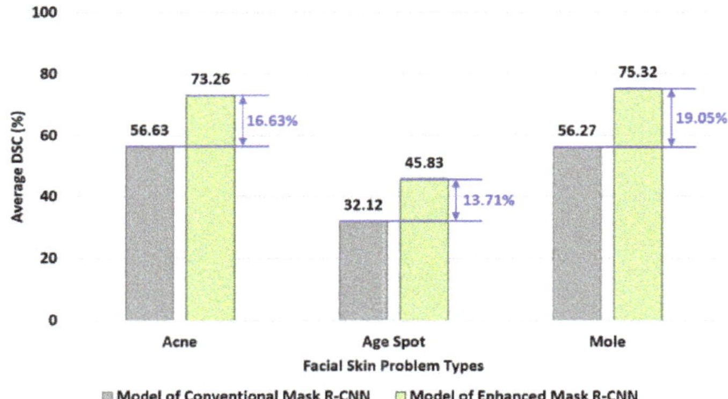

**Figure 13.** Comparing performances of conventional Mask R-CNN and enhanced Mask R-CNN models.

In this experiment, a segmentation model of conventional Mask R-CNN was applied to detect facial skin problems on all the face photos, the performances were measured with DSC, and computed the average of all DSC measurements. Then, a segmentation model trained with the enhanced Mask R-CNN was trained and applied to measure its performance in DSC.

For the facial photos with acne problems, the conventional Mask R-CNN model yielded 56.63% of DSC where the Enhanced Mask R-CNN model yielded 73.26% as shown in the figure. A significant degree of performance has been gained.

As the summary of the experiment, the performances of segmenting acne, age spots, and mole problems were increased by 16.63%, 13.7%, and 19.05%, respectively, and the average of performance gains for all 3 types of facial skin problems is 16.46%.

5.4.2. Experiment for Tactic #2: Super Resolution GAN Model

This experiment is to evaluate the effectiveness of applying Super Resolution GAN model in enhancing the quality of facial images. A dataset of 3 facial skin problem types was made: acne, rosacea, and wrinkle. A segmentation model was trained with both the Mask R-CNN and the SR-GAN structure and performed the experiments. The results of applying the super resolution tactic with SR-GAN are shown in Figure 14.

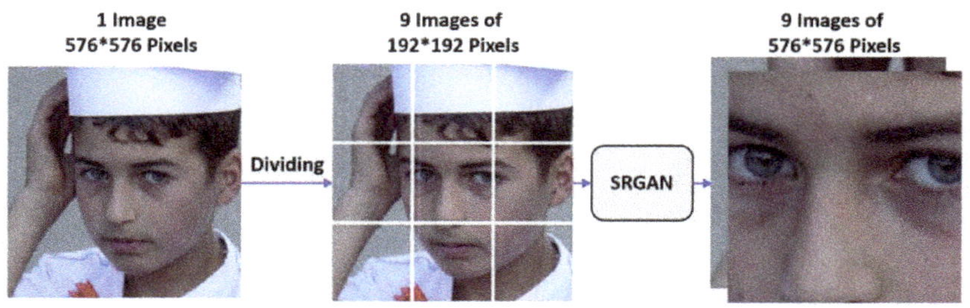


**Figure 14.** Result of applying super resolution with SR-GAN.

In the figure, the original face image is partitioned into 9 images of $(192 \times 192)$ size, which is required by the trained Mask R-CNN model. Then, each image is fed into the SR-GAN model that will enhance the quality of images as shown in the figure.

To compare the performance of super resolution, a segmentation model was trained with conventional Mask R-CNN and another segmentation model with both Mask R-CNN and SR-GAN. The comparison of the performances of the two models is shown in Figure 15.

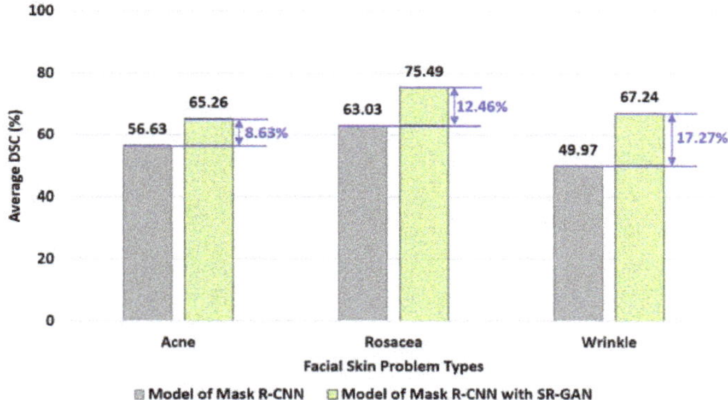

**Figure 15.** Comparing performances of segmentation with and without SR-GAN.

The Mask R-CNN segmentation model was applied to detect all the face photos, the performance was measured in DSC, and the average of all the DSC measurements was computed. Then, a segmentation model with both Mask R-CNN and SR-GAN was trained and applied to perform the same operations.

For wrinkle problems, the DSC with the Mask R-CNN model yielded 49.97% of DSC where the Enhanced Mask R-CNN model yielded 67.24% of DSC, as shown in the figure. A significant degree of performance has been gained.

As the summary of the experiment, the performances of segmenting acne, rosacea, and wrinkle problems were increased by 8.63%, 12.46%, and 17.27%, respectively, and the average of performance gains for all 3 types of facial skin problems is 12.79%.

### 5.4.3. Experiment for Tactic #3: Facial Skin Problem-Specific Models

This experiment is to evaluate the effectiveness of applying facial skin problem-specific models instead of using a single integrated model. A dataset of face photos for 5 types of skin problems was used: acne, age spots, moles, rosacea, and wrinkles.

An integrated Mask R-CNN model for all 5 types of facial skin problem types was trained. Then, a set of 5 individual segmentation models for 5 different facial skin problem types were trained. Then, their performances were measured and compared as shown in Figure 16.

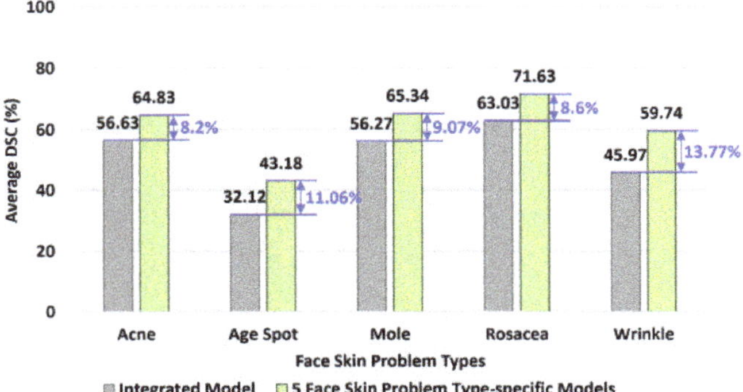

**Figure 16.** Comparing performance of integrated model and facial skin problem type-specific model.

For wrinkle problems, the integrated segmentation model yielded 45.97% of DSC where the wrinkle-specific segmentation model yielded 59.74% of DSC, as shown in the figure. A significant degree of performance has been gained.

The performances of segmenting acne, age spots, moles, rosacea, and wrinkles were increased by 8.2%, 11.06%, 9.07%, 8.6%, and 13.77%, respectively, and the average of performance gains for all 5 types of facial skin problems is 10.14%.

### 5.4.4. Experiment for Tactic #4: Face Direction-Specific Models

This experiment is to evaluate the effectiveness of applying face direction-specific models instead of using a single integrated model. A dataset of face photos taken in 3 different directions was used in this experiment.

A Mask R-CNN model was trained to segment face photos in any direction. Then, a set of 3 individual models for 3 face directions were trained: left-side, frontal, and right-side directions. Then, their performances were measured and compared, as shown in Figure 17.

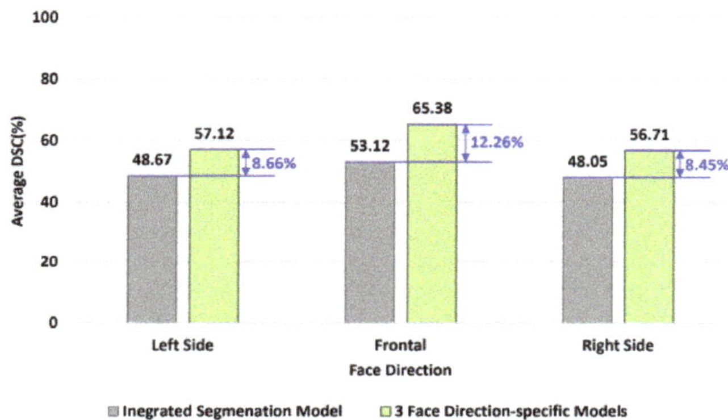


**Figure 17.** Comparing performances of integrated model and face direction-specific models.

For the photos of frontal face, the integrated segmentation model yielded 53.12% of DSC where the frontal direction-specific model yielded 65.38%, as shown in the figure. A significant degree of performance has been gained.

The average performances of segmenting face photos showing the left side face, frontal face, and right side face were increased by 8.45%, 12.26%, and 8.66%, respectively, and the average of performance gains for all 3 face directions is 9.79%

### 5.4.5. Experiment for Tactic #5: Discarding False Segmentations

This experiment is to evaluate the effectiveness of discarding false segmentations made on non-facial areas. A Mask R-CNN model was trained and applied to detect facial skin problems. Then, the software component implementing the tactic of discarding false segmentation was implemented and applied to discard any resulting false segmentations.

The performance of the diagnosis without applying this tactic and the performance of the diagnosis by applying this tactic were measured and compared as shown in Figure 18.

For the wrinkle problems, the segmentation with the Mask R-CNN yielded 45.97% of DSC where the performance measure after discarding false segmentations was 61.23%, as shown in the figure. A significant degree of performance has been gained.

The performances of segmenting acne, age spots, moles, rosacea, and wrinkles were increased by 5.49%, 9.17%, 8.01%, 6.65%, and 15.26%, respectively, and the average of performance gains for all 5 types of facial skin problems is 8.92%.

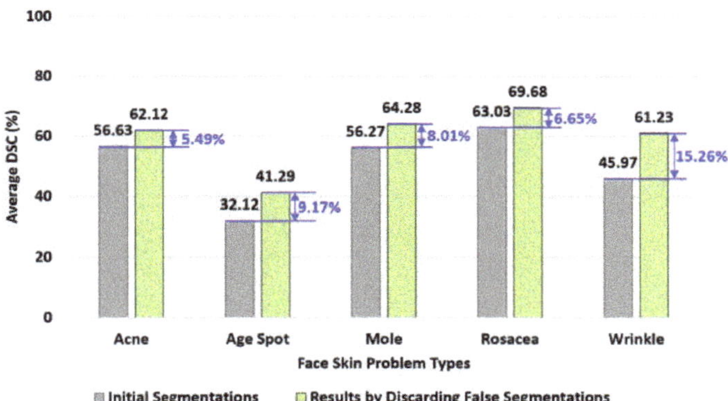



**Figure 18.** Comparing performances of segmentations with- and without discarding false segmentation.

### 5.4.6. Experiment for Integrating all 5 Tactics

This experiment is to evaluate the performance of facial skin problem diagnosis by integrating all 5 tactics. A conventional Mask R-CNN segmentation model was trained for all the skin problem types. Then, a total of 30 individual segmentation models were trained for the 5 different types of tactics and the 3 facial photo directions.

Then, the performances of 3 different approaches were measured: (1) performance of the conventional Mask R-CNN model, (2) the average performances of 5 different tactics, and (3) the average performance of applying all 30 segmentation models.

The performances of the three approaches are compared in Figure 19.

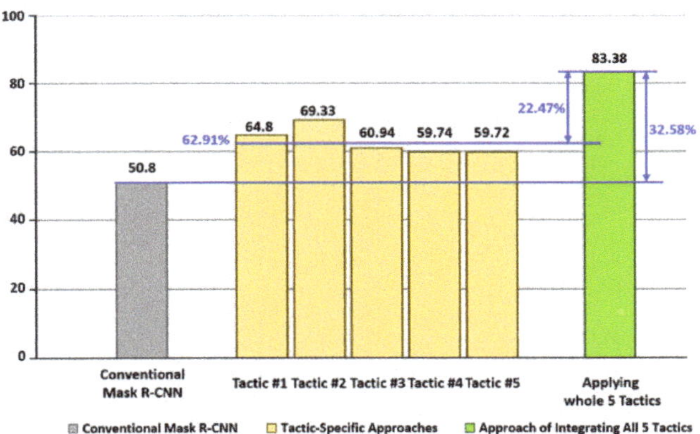



**Figure 19.** Comparing performances of the three approaches.

The conventional Mask R-CNN yielded 50.8% of DSC, the tactic-specific approaches yielded (64.6%, 69.33%, 60.94%, 59.74%, and 59.72%) of DSC, and the approach of integrating all 5 tactics yielded 83.38% of DSC.

The integrated approach outperformed the conventional Mask R-CNN approach by 32.58%, and it outperformed the tactic-specific approach by an average of 22.47%.

### 5.4.7. Experiment for Comparing with Other Backbone Networks

This experiment was to compare the performance of our proposed approach with diagnosis models trained with 6 different backbone networks: MobileNetV2, Xception, VGG16, VGG19, ResNet50, and ResNet101.

Table 5 shows the code segments of training the segmentation models using the 6 different backbone networks.

**Table 5.** Code segment of training segmentation models using the 6 backbone networks.

**Code.** Implementing a Python Class for training 6 segmentation models

```
 1:  import tensorflow as tf
 2:  import tensorflow.keras as keras
 3:  import tensorflow.keras.layers as KL
 4:  from .backbone import build_mobilenet, build_xception, build_vgg16, build_vgg19,
      build_resnet50,
 5:  build_resnet101, build_fusion_deconv
 6:  from .mrcnn import build_rpn_and_mrcnn
 7:
 8:  class FSPSegmenter:
 9:  def __init__(self, backbone_type, config, p_model):
10:  self.config = config
11:      self.path_model = p_model
12:      self.model = self.build(backbone_type)
13:
14:  def build(self, backbone_type):
15:  input = KL.Input(shape= self.config.img_shape)
16:      # To build backbone
17:      feature_maps_MobileNetV2 = self.build_mobilenet(input_img)
18:  feature_maps_xception = build_xception(input_img)
19:  feature_maps_vgg16 = build_vgg16(input_img)
20:  feature_maps_vgg19 = build_vgg19(input_img)
21:  feature_maps_resnet50 = build_resnet50(input_img)
22:  feature_maps_resnet101 = build_resnet101(input_img)
23:  feature_maps_fusion_deconv = build_fusion_deconv(input_img)
24:
25:  # To initialize structures for region proposal networks and networks for segmentation and
      detection
26:  self.model_mobileNetV2 = build_rpn_and_mrcnn(feature_maps_MobileNetV2)
27:  self.model_xception = build_rpn_and_mrcnn(feature_maps_xception)
28:  self.model_vgg16 = build_rpn_and_mrcnn(feature_maps_vgg16)
29:  self.model_vgg19 = build_rpn_and_mrcnn(feature_maps_vgg19)
30:  self.model_resnet50 = build_rpn_and_mrcnn(feature_maps_resnet50)
31:  self.model_resnet101 = build_rpn_and_mrcnn(feature_maps_resnet101)
32:  self.model_fusion_deconv = build_rpn_and_mrcnn(feature_maps_fusion_deconv)
```

The code segment is to configure (6 + 1) segmentation backbone structures inside of the *build* method (lines 17 to 23) and to configure the remaining network structures of Mask R-CNN inside in *build_rpn_and_mrcnn* methods (lines 26–32).

Once the network structures are configured, then all 7 segmentation models are trained and applied to detecting facial skin problems using the test set of 445 photos.

The comparison of their average performances is shown in Figure 20.

As shown in the figure, each of the segmentation models with MobileNetV2, Xception, VGG16, VGG19, ResNet50, and ResNet101 shows an average 46.91% of detection performance. It is 17.89% lower than the performance of our proposed model.

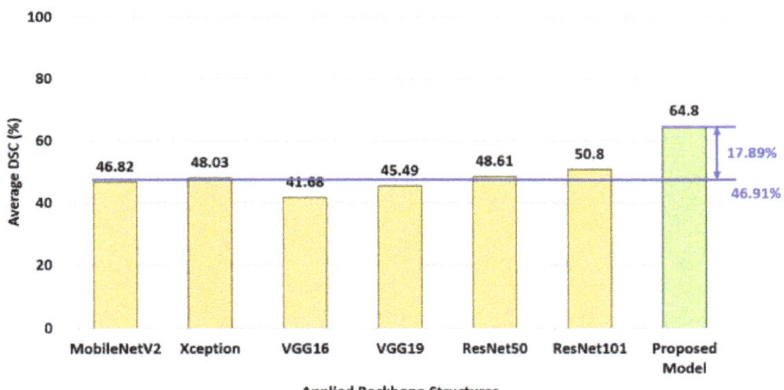



**Figure 20.** Comparing the proposed approach with 6 other backbone networks.

## 6. Concluding Remarks

The condition of the facial skin is perceived as a vital indicator of the person's apparent age, perceived beauty, and degree of health. For this reason, people wish to maintain youthful facial skin without aging symptoms.

Machine-learning-based software analytics on facial skin conditions can be a time- and cost-efficient alternative to the conventional approach of visiting the clinics. However, the current CNN-based approaches have been shown to be limited in the diagnosis performance and, hence, limited in their applicability in clinics.

In this paper, the set of 5 technical challenges in diagnosing facial skin problems were addressed. Then, a set of 5 effective design tactics to overcome the technical challenges in diagnosing facial skin problems were presented. Each proposed tactic is devised to resolve one or more technical challenges.

Using a data collection of 2225 photos, a total of 30 segmentation models were trained and applied to the experiments. The experiments showed 83.38% of the diagnosis performance when applying all 5 tactics, which outperforms conventional CNN approaches by 32.58%. The diagnosis system presented in this study can potentially be utilized in developing clinical diagnosis systems.

**Author Contributions:** Conceptualization, M.K.; Methodology, M.K. and M.H.S.; Software, M.K. and M.H.S.; Investigation, M.H.S.; Writing—original draft, M.K. and M.H.S.; Supervision, M.K. All authors have read and agreed to the published version of the manuscript.

**Funding:** This research received no external funding.

**Institutional Review Board Statement:** Not applicable.

**Informed Consent Statement:** Not applicable.

**Data Availability Statement:** Not applicable.

**Conflicts of Interest:** The authors declare no conflict of interest.

## References

1. Shen, X.; Zhang, J.; Yan, C.; Zhou, H. An Automatic Diagnosis Method of Facial Acne Vulgaris Based on Convolutional Neural Network. *Sci. Rep.* **2018**, *8*, 5839. [CrossRef]
2. Quattrini, A.; Boër, C.; Leidi, T.; Paydar, R. A Deep Learning-Based Facial Acne Classification System. *Clin. Cosmet. Investig. Dermatol.* **2022**, *15*, 851–857. [CrossRef] [PubMed]
3. Zhao, Z.; Wu, C.-M.; Zhang, S.; He, F.; Liu, F.; Wang, B.; Huang, Y.; Shi, W.; Jian, D.; Xie, H.; et al. A Novel Convolutional Neural Network for the Diagnosis and Classification of Rosacea: Usability Study. *JMIR Public Health Surveill.* **2021**, *9*, e23415. [CrossRef]
4. Liu, S.; Chen, Z.; Zhou, H.; He, K.; Duan, M.; Zheng, Q.; Xiong, P.; Huang, L.; Yu, Q.; Su, G.; et al. DiaMole: Mole Detection and Segmentation Software for Mobile Phone Skin Images. *J. Health Eng.* **2021**, *2021*, 6698176. [CrossRef] [PubMed]

5.    Wu, H.; Yin, H.; Chen, H.; Sun, M.; Liu, X.; Yu, Y.; Tang, Y.; Long, H.; Zhang, B.; Zhang, J.; et al. A deep learning, image based approach for automated diagnosis for inflammatory skin diseases. *Ann. Transl. Med.* **2020**, *8*, 581. [CrossRef]

6.    Gerges, F.; Shih, F.; Azar, D. Automated Diagnosis of Acne and Rosacea using Convolution Neural Networks. In Proceedings of the 2021 4th International Conference on Artificial Intelligence and Pattern Recognition (AIPR 2021), Xiamen, China, 24–26 September 2021. [CrossRef]

7.    Yadav, N.; Alfayeed, S.M.; Khamparia, A.; Pandey, B.; Thanh, D.N.H.; Pande, S. HSV model-based segmentation driven facial acne detection using deep learning. *Expert Syst.* **2021**, *39*, e12760. [CrossRef]

8.    Junayed, M.S.; Islam, B.; Jeny, A.A.; Sadeghzadeh, A.; Biswas, T.; Shah, A.F.M.S. ScarNet: Development and Validation of a Novel Deep CNN Model for Acne Scar Classification With a New Dataset. *IEEE Access* **2021**, *10*, 1245–1258. [CrossRef]

9.    Bekmirzaev, S.; Oh, S.; Yo, S. RethNet: Object-by-Object Learning for Detecting Facial Skin Problems. In Proceedings of the 2019 IEEE/CVF International Conference on Computer Vision Workshop (ICCVW), Seoul, Republic of Korea, 27–28 October 2019. [CrossRef]

10.   Gessert, N.; Sentker, T.; Madesta, F.; Schmitz, R.; Kniep, H.; Baltruschat, I.; Werner, R.; Schlaefer, A. Skin Lesion Classification Using CNNs With Patch-Based Attention and Diagnosis-Guided Loss Weighting. *IEEE Trans. Biomed. Eng.* **2019**, *67*, 495–503. [CrossRef]

11.   Gulzar, Y.; Khan, S.A. Skin Lesion Segmentation Based on Vision Transformers and Convolutional Neural Networks—A Comparative Study. *Appl. Sci.* **2022**, *12*, 5990. [CrossRef]

12.   Li, H.; Pan, Y.; Zhao, J.; Zhang, L. Skin disease diagnosis with deep learning: A review. *Neurocomputing* **2021**, *464*, 364–393. [CrossRef]

13.   Cui, L.; Ma, R.; Lv, P.; Jiang, X.; Gao, Z.; Zhou, B.; Xu, M. MDSSD: Multi-scale deconvolutional single shot detector for small objects. *Sci. China Inf. Sci.* **2020**, *63*, 120113. [CrossRef]

14.   Liu, Z.; Yeoh, J.K.; Gu, X.; Dong, Q.; Chen, Y.; Wu, W.; Wang, L.; Wang, D. Automatic pixel-level detection of vertical cracks in asphalt pavement based on GPR investigation and improved mask R-CNN. *Autom. Constr.* **2023**, *146*, 104689. [CrossRef]

15.   Nie, X.; Duan, M.; Ding, H.; Hu, B.; Wong, E.K. Attention Mask R-CNN for Ship Detection and Segmentation From Remote Sensing Images. *IEEE Access* **2020**, *8*, 9325–9334. [CrossRef]

16.   Liu, M.; Wang, X.; Zhou, A.; Fu, X.; Ma, Y.; Piao, C. UAV-YOLO: Small Object Detection on Unmanned Aerial Vehicle Perspective. *Sensors* **2020**, *20*, 2238. [CrossRef] [PubMed]

17.   Seo, H.; Huang, C.; Bassenne, M.; Xiao, R.; Xing, L. Modified U-Net (mU-Net) with Incorporation of Object-Dependent High Level Features for Improved Liver and Liver-Tumor Segmentation in CT Images. *IEEE Trans. Med. Imaging* **2019**, *39*, 1316–1325. [CrossRef] [PubMed]

18.   Peng, C.; Zhang, Y.; Zheng, J.; Li, B.; Shen, J.; Li, M.; Liu, L.; Qiu, B.; Chen, D.Z. IMIIN: An Inter-modality Information Interaction Network for 3D Multi-modal Breast Tumor Segmentation. *Comput. Med. Imaging Graph.* **2021**, *95*, 102021. [CrossRef]

19.   Tian, G.; Liu, J.; Zhao, H.; Yang, W. Small object detection via dual inspection mechanism for UAV visual images. *Appl. Intell.* **2021**, *52*, 4244–4257. [CrossRef]

20.   Chen, C.; Zhong, J.; Tan, Y. Multiple-Oriented and Small Object Detection with Convolutional Neural Networks for Aerial Image. *Remote Sens.* **2019**, *11*, 2176. [CrossRef]

21.   Amudhan, A.N.; Sudheer, A.P. Lightweight and computationally faster Hypermetropic Convolutional Neural Network for small size object detection. *Image Vis. Comput.* **2022**, *119*, 104396. [CrossRef]

22.   Zhang, Q.; Yang, G.; Zhang, G. Collaborative Network for Super-Resolution and Semantic Segmentation of Remote Sensing Images. *IEEE Trans. Geosci. Remote Sens.* **2021**, *60*, 21546971. [CrossRef]

23.   Da Wang, Y.; Blunt, M.J.; Armstrong, R.T.; Mostaghimi, P. Deep learning in pore scale imaging and modeling. *Earth-Sci. Rev.* **2021**, *215*, 103555. [CrossRef]

24.   Guo, Z.; Wu, G.; Song, X.; Yuan, W.; Chen, Q.; Zhang, H.; Shi, X.; Xu, M.; Xu, Y.; Shibasaki, R.; et al. Super-Resolution Integrated Building Semantic Segmentation for Multi-Source Remote Sensing Imagery. *IEEE Access* **2019**, *7*, 99381–99397. [CrossRef]

25.   Aboobacker, S.; Verma, A.; Vijayasenan, D.; Suresh, P.K.; Sreeram, S. Semantic Segmentation on Low Resolution Cytology Images of Pleural and Peritoneal Effusion. In Proceedings of the 2022 National Conference on Communications (NCC 2022), Mumbai, India, 24–27 May 2022. [CrossRef]

26.   Fromm, M.; Berrendorf, M.; Faerman, E.; Chen, Y.; Schüss, B.; Schubert, M. XD-STOD: Cross-Domain Super resolution for Tiny Object Detection. In Proceedings of the 2019 International Conference on Data Mining Workshops (ICDMW 2019), Beijing, China, 8–11 November 2019.

27.   Wei, S.; Zeng, X.; Zhang, H.; Zhou, Z.; Shi, J.; Zhang, X. LFG-Net: Low-Level Feature Guided Network for Precise Ship Instance Segmentation in SAR Images. *IEEE Trans. Geosci. Remote Sens.* **2022**, *60*, 21865424. [CrossRef]

28.   Wei, J.; Xia, Y.; Zhang, Y. M3Net: A multi-model, multi-size, and multi-view deep neural network for brain magnetic resonance image segmentation. *Pattern Recognit.* **2019**, *91*, 366–378. [CrossRef]

29.   Zhou, X.; Takayama, R.; Wang, S.; Hara, T.; Fujita, H. Deep learning of the sectional appearances of 3D CT images for anatomical structure segmentation based on an FCN voting method. *Med. Phys.* **2017**, *44*, 5221–5233. [CrossRef]

30.   Liu, Y.; Kwak, H.-S.; Oh, I.-S. Cerebrovascular Segmentation Model Based on Spatial Attention-Guided 3D Inception U-Net with Multi-Directional MIPs. *Appl. Sci.* **2022**, *12*, 2288. [CrossRef]

31. Sander, J.; de Vos, B.D.; Išgum, I. Automatic segmentation with detection of local segmentation failures in cardiac MRI. *Sci. Rep.* **2020**, *10*, 21769. [CrossRef] [PubMed]

32. Scherr, T.; Löffler, K.; Böhland, M.; Mikut, R. Cell segmentation and tracking using CNN-based distance predictions and a graph-based matching strategy. *PLoS ONE* **2020**, *15*, e0243219. [CrossRef]

33. Larrazabal, A.J.; Martinez, C.; Ferrante, E. Anatomical Priors for Image Segmentation via Post-processing with Denoising Autoencoders. In Proceedings of the International Conference on Medical Image Computing and Computer-Assisted Intervention (MICCAI 2019), Shenzhen, China, 13–17 October 2019. [CrossRef]

34. Chan, R.; Rottmann, M.; Gottschalk, H. Entropy Maximization and Meta Classification for Out-of-Distribution Detection in Semantic Segmentation. In Proceedings of the IEEE/CVF International Conference on Computer Vision (ICCV), Montreal, QC, Canada, 1–17 October 2021; pp. 5128–5137. [CrossRef]

35. Shuvo, B.; Ahommed, R.; Reza, S.; Hashem, M. CNL-UNet: A novel lightweight deep learning architecture for multimodal biomedical image segmentation with false output suppression. *Biomed. Signal Process. Control.* **2021**, *70*, 102959. [CrossRef]

36. Khan, S.A.; Gulzar, Y.; Turaev, S.; Peng, Y.S. A Modified HSIFT Descriptor for Medical Image Classification of Anatomy Objects. *Symmetry* **2021**, *13*, 1987. [CrossRef]

37. Hamid, Y.; Elyassami, S.; Gulzar, Y.; Balasaraswathi, V.R.; Habuza, T.; Wani, S. An improvised CNN model for fake image detection. *Int. J. Inf. Technol.* **2022**, *2022*. [CrossRef]

38. Zhang, W.; Itoh, K.; Tanida, J.; Ichioka, Y. Parallel distributed processing model with local space-invariant interconnections and its optical architecture. *Appl. Opt.* **1990**, *29*, 4790–4797. [CrossRef]

39. Nguyen, N.-D.; Do, T.; Ngo, T.D.; Le, D.-D. An Evaluation of Deep Learning Methods for Small Object Detection. *J. Electr. Comput. Eng.* **2020**, *2020*, 3189691. [CrossRef]

40. Liu, Y.; Sun, P.; Wergeles, N.; Shang, Y. A survey and performance evaluation of deep learning methods for small object detection. *Expert Syst. Appl.* **2021**, *172*, 114602. [CrossRef]

41. Sun, C.; Ai, Y.; Wang, S.; Zhang, W. Mask-guided SSD for small-object detection. *Appl. Intell.* **2020**, *51*, 3311–3322. [CrossRef]

42. Flament, F.; Francois, G.; Qiu, H.; Ye, C.; Hanaya, T.; Batisse, D.; Cointereau-Chardon, S.; Seixas, M.D.G.; Belo, S.E.D.; Bazin, R. Facial skin pores: A multiethnic study. *Clin. Cosmet. Investig. Dermatol.* **2015**, *8*, 85–93. [CrossRef] [PubMed]

43. National Cancer Institute. Common Moles, Dysplastic Nevi, and Risk of Melanoma. Available online: https://www.cancer.gov/types/skin/moles-fact-sheet (accessed on 4 January 2023).

44. He, K.; Gkioxari, G.; Dollár, P.; Girshick, R. Mask R-CNN. In Proceedings of the 2017 IEEE International Conference on Computer Vision (ICCV), Venice, Italy, 22–29 October 2017; pp. 2961–2969. [CrossRef]

45. Ledig, C.; Theis, L.; Huszár, F.; Caballero, J.; Cunningham, A.; Acosta, A.; Aitken, A.P.; Tejani, A.; Totz, J.; Wang, Z.; et al. Photo-Realistic Single Image Super-Resolution Using a Generative Adversarial Network. In Proceedings of the 2017 IEEE Conference on Computer Vision and Pattern Recognition (CVPR), Honolulu, HI, USA, 21–26 July 2017.

46. Alzubaidi, L.; Zhang, J.; Humaidi, A.J.; Al-Dujaili, A.; Duan, Y.; Al-Shamma, O.; Santamaría, J.; Fadhel, M.A.; Al-Amidie, M.; Farhan, L. Review of deep learning: Concepts, CNN architectures, challenges, applications, future directions. *J. Big Data* **2021**, *8*, 53. [CrossRef]

47. Indolia, S.; Goswami, A.K.; Mishra, S.; Asopa, P. Conceptual Understanding of Convolutional Neural Network- A Deep Learning Approach. *Procedia Comput. Sci.* **2018**, *132*, 679–688. [CrossRef]

48. Luo, C.; Li, X.; Wang, L.; He, J.; Li, D.; Zhou, J. How Does the Data set Affect CNN-based Image Classification Performance? In Proceedings of the 2018 5th International Conference on Systems and Informatics (ICSAI 2018), Nanjing, China, 10–12 November 2018. [CrossRef]

49. Zheng, X.; Qi, L.; Ren, Y.; Lu, X. Fine-Grained Visual Categorization by Localizing Object Parts With Single Image. *IEEE Trans. Multimedia* **2020**, *23*, 1187–1199. [CrossRef]

50. Avianto, D.; Harjoko, A. Afiahayati CNN-Based Classification for Highly Similar Vehicle Model Using Multi-Task Learning. *J. Imaging* **2022**, *8*, 293. [CrossRef]

51. Ju, M.; Moon, S.; Yoo, C.D. Object Detection for Similar Appearance Objects Based on Entropy. In Proceedings of the 2019 7th International Conference on Robot Intelligence Technology and Applications (RiTA 2019), Daejeon, Republic of Korea, 1–3 November 2019. [CrossRef]

52. Jang, W.; Lee, E.C. Multi-Class Parrot Image Classification Including Subspecies with Similar Appearance. *Biology* **2021**, *10*, 1140. [CrossRef] [PubMed]

53. Facial Landmarks Shape Predictor. Available online: https://github.com/codeniko/shape_predictor_81_face_landmarks (accessed on 4 January 2023).

54. Wu, X.; Wen, N.; Liang, J.; Lai, Y.K.; She, D.; Cheng, M.; Yang, J. Joint Acne Image Grading and Counting via Label Distribution Learning. In Proceedings of the 2019 IEEE/CVF International Conference on Computer Vision (ICCV 2019), Seoul, Republic of Korea, 27 October–2 November 2019.

55. Karras, T.; Laine, S.; Aila, T. A Style-Based Generator Architecture for Generative Adversarial Networks. *IEEE Trans. Pattern Anal. Mach. Intell.* **2020**, *43*, 4217–4228. [CrossRef] [PubMed]

56. Wu, X. Pytorch Implementation of Joint Acne Image Grading and Counting via Label Distribution Learning. Available online: https://github.com/xpwu95/LDL (accessed on 4 January 2023).

57. NVDIA Research Lab. FFHQ Datase. Available online: https://github.com/NVlabs/ffhq-dataset (accessed on 4 January 2023).

58. Thomaz, C.E. FEI Face Database. Available online: https://fei.edu.br/~{}cet/facedatabase.html (accessed on 4 January 2023).
59. COCO Annotator. Available online: https://github.com/jsbroks/coco-annotator (accessed on 4 January 2023).

*Article*

# Model Predictive Control of Quadruped Robot Based on Reinforcement Learning

**Zhitong Zhang [†], Xu Chang [†], Hongxu Ma, Honglei An and Lin Lang ***

College of Intelligence Science and Technology, National University of Defense Technology, Changsha 410073, China
* Correspondence: langlin_8502@nudt.edu.cn
† These authors contributed equally to this work.

**Abstract:** For the locomotion control of a legged robot, both model predictive control (MPC) and reinforcement learning (RL) demonstrate powerful capabilities. MPC transfers the high-level task to the lower-level joint control based on the understanding of the robot and environment, model-free RL learns how to work through trial and error, and has the ability to evolve based on historical data. In this work, we proposed a novel framework to integrate the advantages of MPC and RL, we learned a policy for automatically choosing parameters for MPC. Unlike the end-to-end RL applications for control, our method does not need massive sampling data for training. Compared with the fixed parameters MPC, the learned MPC exhibits better locomotion performance and stability. The presented framework provides a new choice for improving the performance of traditional control.

**Keywords:** model predictive control; reinforcement learning; parameter adaptive; quadruped robot

## 1. Introduction

The quadrupeds are outstanding in high-speed running, weight-bearing, and terrain passing. For example, cheetahs' sprints have been measured at a maximum of 114 km/h, and they routinely reach velocities of 80–100 km/h while pursuing prey. Goliath frogs can jump up to 5 m high while their body length is only from 17 to 32 cm. Blue sheep are usually found near cliffs, in preparation to run toward rugged slopes to avoid danger. Camels are known as the "ships of the desert", they are able to carry hundreds of kilos of cargo for long trips in harsh desert environments. These striking features have inspired researchers' enthusiasm for the bionic quadruped robots. It is hoped that one day such quadruped robots can surpass animals in movement skills and perform tasks in challenging environments. These visions put forward higher requirements for the performance of robot controllers.

Half a century has witnessed the development of legged robots, and many excellent quadruped robots have emerged. The issue of controller design has shifted from static position planning [1] to highly dynamic optimization [2,3]. Among numerous optimization algorithms, Model Predictive Control (MPC) has emerged as the most widely used control algorithm in the robot field.

There are three main elements of MPC [4]: the predictive model, the reference trajectory, and the control algorithm. This is now more clearly stated as model-based prediction, receding horizontal optimization, and feedback correction. The vast literature invariably says that the greatest attraction of predictive control is its ability to deal explicitly with control and state quantity constraints, this ability arises from the predictions of future dynamic behavior based on the analytical model of the control object, by adding the constraints to the future inputs, outputs, and state. The receding horizontal optimization ensures that the system can quickly respond to the uncertainty from the internal system or external environment.

**Citation:** Zhang, Z.; Chang, X.; Ma, H.; An, H.; Lang, L. Model Predictive Control of Quadruped Robot Based on Reinforcement Learning. *Appl. Sci.* **2023**, *13*, 154. https://doi.org/10.3390/app13010154

Academic Editors: Yue Wu, Xinglong Zhang and Pengfei Jia

Received: 8 November 2022
Revised: 16 December 2022
Accepted: 20 December 2022
Published: 22 December 2022

Since MIT Biomimetic Robotics Lab has open-sourced Cheetah-Software [5], MPC has become a new baseline method for the locomotion control of quadruped robots. Bledt proposed the policy-regularized MPC [6] to stabilize different gaits with several heuristic reference policies. Ding presented the representation-free MPC [7] framework that directly represents orientation using the rotation matrix and can stabilize dynamic motions that involve the singularity. Liu proposed a design approach of gait parameters with minimum energy consumption [8]. Chang presented the Proportional Differential MPC (PDMPC) controller [9] that has the ability to compensate for the unmodeled leg mass or payload. Although MPC has made significant progress in the field of legged-robot control, it still faces many challenges. The precise predictive model is difficult to establish; the simplified model will introduce model mismatch, and the fixed parameters controller does not have strong generalization ability.

In recent years, with the revival of artificial intelligence, achieving autonomous learning control for robots has become one of the research highlights. Kolter [10] learned parameters through policy search based on fixed strategies and realized the action of jumping from the ground to obstacles on Boston Dynamics' little dag. Tuomas Haarnoja [11] developed a variant of the soft actor-critic algorithm to realize the level walking of a quadruped robot with only 8 Degree-of-Freedom (DOF). Yao proposed a video imitation adaptation network [12] that can imitate the action of animals and adapt it to the robot from a few seconds of video. Hwangbo applied model-free reinforcement learning on ANYmal, realized omnidirectional motion on flat ground, and fall recovery. The trained controller can follow the command speed well in any direction. In order to enable the simulation results to be transferred to the real robots, domain randomization, actuator modeling, and random noise during training were used [13]. Additionally, the two-stage training process makes the student policy reconstruct the latent information which not directly observable, such as contact states, terrain shape, and external disturbances. Without vision and environmental information, the robot successfully passed through various complex terrains only with proprioceptive information [14]. Training of reinforcement learning requires a large number of samples, the end-to-end controller policy lack of interpretability, the curse of dimensionality, and the curse of goal specification challenge the usage of reinforcement learning [15]. More importantly, conventional model-based control methods concentrate the intelligence of human researchers, and cannot be discarded roughly.

This work is an extension of our previous work [9]. In this work, we proposed a novel approach to combine the advantages of model predictive control and reinforcement learning. PDMPC with a dozen parameters is considered a parametric controller that provides stable control to generate samples, reinforcement learning training the policy networks to modify parameters online. Compared with the fixed parameters controller, the learned controller has better performance in command tracking and equilibrium stability.

The rest of this paper is organized as follows. Section 2 briefly presents the MPC with Proportional Differential (PD) compensator. Section 3 presents the details of the reinforcement learning framework for PDMPC. The simulation setup and results are illustrated in Section 4. Finally, Section 5 concludes this paper.

## 2. PDMPC Formulation

Our quadruped robot is Unitree A1, as shown in Figure 1. It has four elbow-like legs, and each leg has three degrees of freedom, called hip side, hip front and knee respectively according to the order of attachment. The first two joints are directly driven, and the knee joint is driven by a connecting rod. This design concentrates the majority of quality on the body.

**Figure 1.** Unitree A1 robot.

In order to reduce the computational consumption and the difficulty of optimization solution, the dynamic model of the quadruped robot used as a predictive model is simplified to a single rigid body model with four massless variable-length rods:

$$\ddot{p} = \frac{\sum_{i=1}^{k} f_i}{m} + g \tag{1}$$

$$\frac{d}{dt}(I\omega) = I\dot{\omega} + \omega \times I\omega = \sum_{i=1}^{k}(r_i \times f_i) \tag{2}$$

where $m$ and $I$ are the mass and body inertia matrix, $p$ and $\omega$ are the positions and angular velocity of the body, $g$ is the gravitational acceleration vector, $f_i$ and $r_i$ are the foot reaction force and the position vector from the foot end to the center of mass (CoM) in the world coordinate system. $\omega \times I\omega$ is neglected under the assumption that small angular velocity during the robot locomotion.

The control framework of MPC is shown in Figure 2. The estimated states of the quadruped robot are 6-DOF pose and corresponding 6-DOF velocity, the inputs are gait pattern, desired speed, and attitude angle, the reference trajectory generator plans a desired path within the prediction horizon based on the user inputs, and the current state, the swing planner schedules the legs' phase and the foot trajectory, the MPC controller outputs the desired ground reaction force $f_i^0$, at last, the low-level joint torque/position controller executes the control commands to drive the robot.

$$\tau_i = J_i^T R^T f_i^0 + k_p\left(q^{des} - q\right) + k_d\left(\dot{q}^{des} - \dot{q}\right) \tag{3}$$

where $\tau$ is the joint torque vector, $R$ is the rotation matrix of body, the leg jacobian matrix $J$ maps the force in operation space to the torque in joint space. $q^{des}$ and $\dot{q}^{des}$ are the desired trajectory of joint, $k_p$ and $k_d$ are the gain of the PD controller.

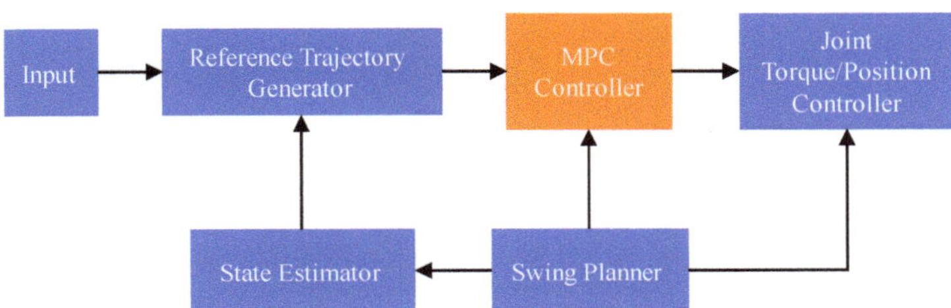


**Figure 2.** Control framework.

Due to the above simplification, model mismatch will inevitably occur, especially when the leg mass is large or the dynamic parameter estimation is inaccurate, and the conventional MPC is no longer effective to the uncertainty. Therefore, PDMPC is designed to solve this problem. The structure of the PDMPC controller is shown in Figure 3.

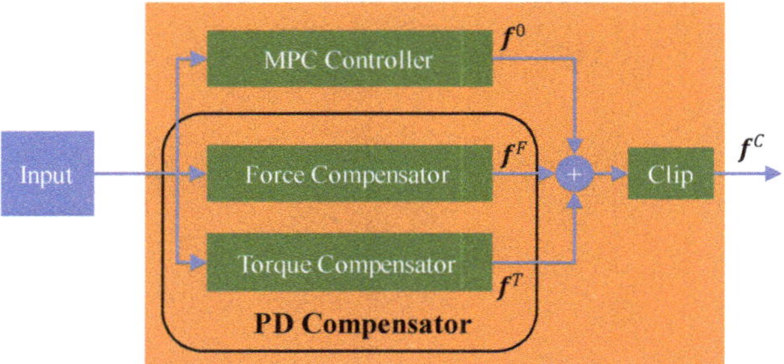


**Figure 3.** PDMPC architecture.

Based on the above-simplified model in Equations (1) and (2), the MPC controller obtains the expected force $f^0$ of the stance leg through linear optimization. As well, our PD compensator is proposed to compensate for the model uncertainty. We divide the compensator into two parts based on force and torque.

$$F_e = K_f(p_{des} - p) + D_f(\dot{p}_{des} - \dot{p}) \tag{4}$$

$$\tau_e = K_\tau(\Theta_{des} - \Theta) + D_\tau(\omega_{des} - \omega) \tag{5}$$

where $\Theta$ is the Euler angle, $F_v$ and $\tau_v$ are the additional virtual force and torque acting on the body. $K$ and $D$ are the diagonal gain matrix with corresponding dimensions.

$$K_f = \begin{bmatrix} k_{fx} & & \\ & k_{fy} & \\ & & k_{fz} \end{bmatrix} \qquad D_f = \begin{bmatrix} d_{fx} & & \\ & d_{fy} & \\ & & d_{fz} \end{bmatrix} \tag{6}$$

$$K_\tau = \begin{bmatrix} k_{\tau\varphi} & & \\ & k_{\tau\theta} & \\ & & k_{\tau\psi} \end{bmatrix} \qquad D_\tau = \begin{bmatrix} d_{\tau\varphi} & & \\ & d_{\tau\varphi} & \\ & & d_{\tau\varphi} \end{bmatrix} \tag{7}$$

The force compensator is used to strengthen the tracking of linear motion. In order to reasonably distribute the virtual force to each stance leg, we describe this problem as the following optimal control problem.

$$\min_{u^F} J_F = f^{F^T} L^T L f^F + \gamma f^{F^T} f^F \tag{8}$$

$$s.t. \ F_e = Q f^F \tag{9}$$

$$\underline{c} \le f^F \le \bar{c} \tag{10}$$

where $L = [[r_1], [r_2], [r_3], [r_4]]$ and $[r]$ is the cross-product matrix. $\gamma$ is the regulatory factor to adjust the uniform distribution of foot force. The first item of objective function aiming resultant the whole moment as zero as possible, and the second item for reducing effort. $Q = [1_3, 1_3, 1_3, 1_3]$. $\underline{c}$ and $\bar{c}$ are the lower and upper bounds of force. By solving a quadratic convex programming problem, we obtain additional foot force $f^F$. So as the torque compensator generate $f^T$ for rotational motion.

At last, based on the consideration of the limited joint torque and friction constrain, a clipper makes the desired ground reaction force $f^C$ to meet the physical feasibility.

## 3. Reinforcement Learning for PDMPC

Controllers with fixed parameters make it difficult to adapt the robot to different states of motion. For example, the gait cycle of the robot will be different when it moves at diverse speeds, so the parameters need to change adaptively. Manual adjustment for the control framework with large-scale parameters is laborious and time-consuming, and the results of parameter adjustment are sometimes tricky to achieve the intended goals.

In this section, the reinforcement learning method is used to establish the relationship between the robot states and the controller parameters, so that the multiple parameters can be automatically adjusted. Figure 4 shows the framework of reinforcement learning based on PDMPC.

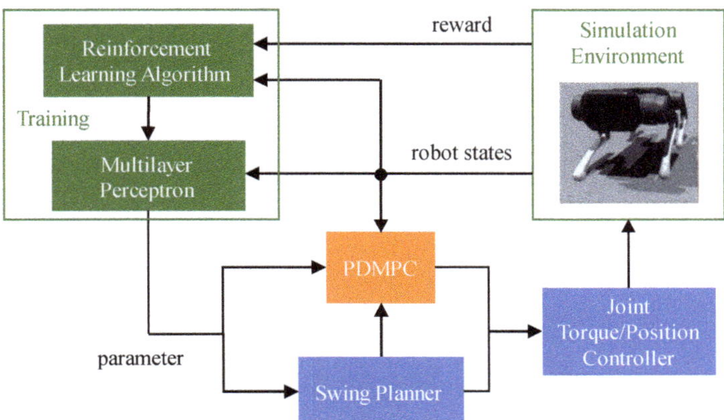


**Figure 4.** Framework of reinforcement learning based on PDMPC.

We utilize the open-source PPO algorithm [16] to train the policy according to the states and the reward provided by the simulation environment. The action policy network is an MLP neural network that receives the current robot states and outputs the parameters for the PDMPC controller and the swing planner. According to the current parameters, the swing planner determines the gait frequency and the target position of the swing leg, the PDMPC controller determines the desired ground reaction force for the stance leg. Finally, the joint controller performs joint control.

### 3.1. Parameters to Be Larned

3.1.1. Swing Planner

The swing planner is used to choose the gait pattern, and determines the phase relation and lift-off schedule. The duty cycle is 0.5, and the result schedule information will be transmitted to the PDMPC controller. The foot point position vector $^wh$ in the world coordinate system is as follows.

$$\begin{cases} ^wh_{xy} = {}^wh_{0xy} + (\alpha_x V_x + \alpha_y V_y)t_s \\ ^wh_z = {}^wh_{0z} + \mathbf{1}(\varphi)l_z \end{cases} \tag{11}$$

where $h_0$ is the initial foot point position vector at the lift-off event. The foot trajectory is determined by the heuristic of single inverted pendulum model, $t_s$ is the duration of the support period, $\alpha_x$ and $\alpha_y$ is the heuristic coefficient in different motion directions (normally set to 0.5). $l_z$ is the maximum height of foot in vertical direction during swing phase. The gait phase variable $\varphi \in [0,\ 2\pi]$, $\mathbf{1}(\varphi)$ is an indicator function, 1 stands for swing period, 0 stands for support period.

$$\mathbf{1}(\varphi) = \begin{cases} 0 & \varphi \in (0,\ \pi] \\ 1 & \varphi \in (\pi,\ 2\pi] \end{cases} \tag{12}$$

So, we have 7 parameters to be learned for swing planner, including the four lift heights of each leg $l_{z1}$, $l_{z2}$, $l_{z3}$ and $l_{z4}$, the two heuristic coefficients $\alpha_x$ and $\alpha_y$, the half support duration $t_s/2$.

### 3.1.2. PDMPC Controller

The PDMPC controller solves the required ground reaction force according to the robot states and the schedule information provided by the swing planner. In our previous work, the manual parameter adjustment takes a long time and has no adaptability. Therefore, it is helpful to improve control performance by incorporating these parameters into the learning process.

There are nine parameters to be learned for PDMPC, including the vertical force coefficients $k_{fz}$ and $d_{fz}$, the horizontal velocity coeffificients $d_{fx}$ and $d_{fy}$, the roll and picth torque coefficients $k_{\tau\varphi}$, $k_{\tau\theta}$, $d_{\tau\varphi}$ and $d_{\tau\varphi}$, the yaw angluar velocity coefficient $d_{\tau\psi}$.

For the low-level joint torque/position controller, the PD gain for position tracking is eazy to turn, therefore, it is unnecessary to put it into our learning process.

### 3.2. Policy Network

The action policy obeys a multidimensional normal distribution $\pi \sim N(\boldsymbol{\mu},\ \boldsymbol{\Sigma})$, The mean vector is $\boldsymbol{\mu}$, and the covariance matrix is $\boldsymbol{\Sigma}$. The covariance matrix is used for exploration, and its elements can be gradually reduced with time to make the training converge. $\boldsymbol{\mu} = f(\boldsymbol{\theta})$, $f$ consists of a full connect neural network, $\theta$ are the network parameters.

Figure 5 shows the schematic diagram of the neural network structure of the action policy, including three hidden layers, the orange balls, the number of units is 256, 128 and 64, respectively. Batch regularization processing is used for inputs [17], and exponential linear units (ELU) are used as activation function.

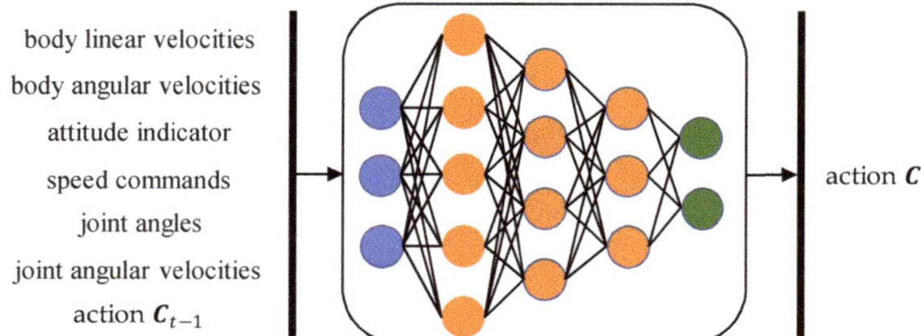


**Figure 5.** The structure of action policy neural network.

The input is a 52-dimensional vector, including a 3-dimensional robot body linear velocity vector, a 3-dimensional body angular velocity vector, a 3-dimensional attitude indicator vector, a 3-dimensional speed command vector, a 12-dimensional joint position vector, a 12-dimensional joint angular velocity vector and a 16-dimensional network output action vector at the previous step $C_{t-1}$. The attitude indicator vector refers to the projection of the unit vector in the gravity direction under the body coordinate system. The speed command includes two linear speeds in the horizontal direction and the yaw angular velocity. $C_t$ is not used directly, its elements are converted to the required parameters for swing planner and PDMPC controller through appropriate mapping, as shown in Table 1.

**Table 1.** The transform between the network action and the parameters for PDMPC.

| Parameter | Transform | Parameter | Transform |
|---|---|---|---|
| $l_{z1}$ | $0.03 + c_1/100$ | $d_{fy}$ | $50 + c_9 \cdot 10$ |
| $l_{z2}$ | $0.03 + c_2/100$ | $k_{fz}$ | $200 + c_{10} \cdot 50$ |
| $l_{z3}$ | $0.03 + c_3/100$ | $d_{fz}$ | $20 + c_{11} \cdot 5$ |
| $l_{z4}$ | $0.03 + c_4/100$ | $k_{\tau\varphi}$ | $50 + c_{12} \cdot 10$ |
| $t_s/2$ | $200 + c_5 \cdot 10$ | $d_{\tau\varphi}$ | $8 + c_{13} \cdot 2$ |
| $\alpha_x$ | $0.5 + c_6/10$ | $k_{\tau\theta}$ | $50 + c_{14} \cdot 10$ |
| $\alpha_y$ | $0.5 + c_7/10$ | $d_{\tau\theta}$ | $8 + c_{15} \cdot 2$ |
| $d_{fx}$ | $50 + c_8 \cdot 10$ | $d_{\tau\psi}$ | $(1 + c_{16}) \cdot 5$ |

The pseudo code of reinforcement learning based on PDMPC is as follow (Algorithm 1).

---

**Algorithm 1 Reinforcement learning based on PDMPC**

---

**Input** $\theta_0$ initial parameter of action network
$x_0$ initial parameter of state value function $V$
**for** $k = 0, 1, 2, \ldots$ **do**
sample parameter vector $C$ from $\pi_{\theta_k}$
$C_{gait} \rightarrow$ swing planner, assign leg states $1(\varphi)$ and target positions $^wh$
$C_{mpc} \rightarrow$ PDMPC, desired ground reaction force $f^F$
joint Controll $\rightarrow \tau$
sample trajectory $\Phi_k = \{\tau_i\}_{i=1,2,\ldots,n}$
**if** reset_flag **then**
reset robot
**end if**
**if** data sufficient **then**
compute reward $\hat{R}_t$
estimate advantage function $\hat{A}_t$ based on state value function $V_{x_k}$
compute Clipped Surrogate Objective (PPO) $M$
update policy by gradient ascent algorithm (G):
$$\theta_{k+1} = arg \max_{\theta} \frac{1}{|\Phi_k|T} \sum_{T \in \Phi_k} \sum_{t=0}^{T} M$$
fitting $V_{x_k}$ by quadratic mean square regression, update parameter:
$$x_{k+1} = arg \min_{x} \frac{1}{|\Phi_k|T} \sum_{T \in \Phi_k} \sum_{t=0}^{T} \left(V_{x_k}(s_t) - \hat{R}_t\right)^2$$
**end if**
**end for**

---

## 4. Simulation and Result

### 4.1. Simulation Platform

We constructed a quadruped robot control algorithm software platform, and its architecture is shown in Figure 6. The platform can be divided into three layers according to functions, namely control architecture layer, conversion layer and training layer.

The control layer runs the traditional manually designed controllers, the code is written in C++ and integrates many open-source mathematical libraries (GSL, Eigen, and qpOASES), which improve the efficiency of algorithm development. The algorithm code has good transplant characteristics and can be rapidly deployed to different dynamic environments, such as Gazebo and other robot simulation environments, as well as real quadruped robots.

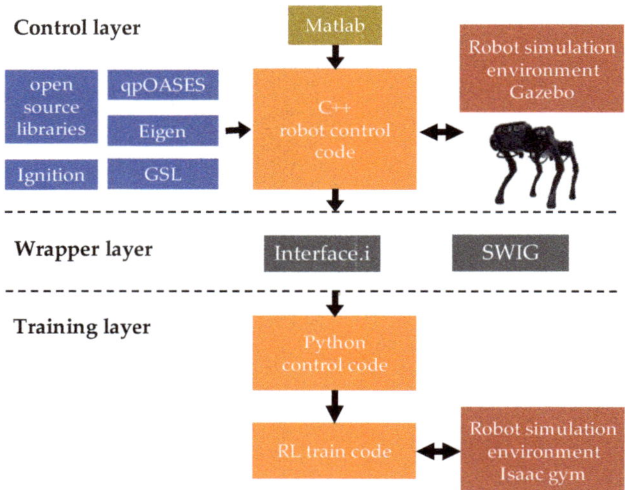


**Figure 6.** The software architecture of quadruped robot control.

The wrapper layer converts the C++ code of the control layer to other types of programming languages to meet the needs of different environments. SWIG-4.0.2 (Simplified Wrapper and Interface Generator) is a software development tool used to build C and C++ program script language interfaces. It can quickly package C/C++ code into Python, Perl, Ruby, Java and other languages. The reinforcement learning we used requires Python language. Declare the C++ control algorithm function in the interface file Interface.i as required, and then convert it by SWIG.

The training layer consists of the converted Python control code library, the learning and training algorithm and Isaac Gym, where Isaac Gym is a physical simulation environment specially developed for reinforcement learning research [18].

*4.2. Training and Rewards*

The simulation was conducted on a desktop laptop with eight CPU cores (Intel Core I7-7700HQ) and single GPU (NVIDIA GeForce GTX 1070). In Isaac Gym environment, we train 20 quadruped robots in parallel. The robot uses a diagonal trotting gait, the simulation time step is 0.005 s, the control frequency of PDMPC is 100 Hz, the maximum alive duration is 15 s, the linear speed command range is [0, 1] m/s, and the angular speed command range is [−1, 1]rad/s.

Algorithm 1 and Adam optimizer [19] are used to train the policy network, and the corresponding hyper-parameters are listed in Table 2.

**Table 2.** Hyper-parameters in Algorithm 1.

| Hyper-Parameters | Value |
| --- | --- |
| decay factor | 0.99 |
| clip factor | 0.2 |
| KL divergence threshold | 0.0008 |
| learning rate | $3e-4$ |
| batch size | 960 |

The reward function we designed is as follows:

- Task target reward

For the task of locomotion with the command speed, we encourage the robot with smaller speed error respect to expectation. The nonlinear exponential function makes the robot obtain much more score when the speed tracking performance improved a little, especially when the robot have medium tracking ability.

$$r_{rew} = \beta_{lvxy} \exp\left(-3\left\| {}^B V_{xy} - V_{cxy}\right\|_2\right) + \beta_{a\omega} \exp\left(-3\left({}^B\omega - \omega_c\right)^2\right) \tag{13}$$

- Balance punishment

On the other hand, we punish the robot with Equation (14).

$$r_{pun} = -\beta_{lvz} v_z^2 - \beta_{a\varphi\theta}\left(\varphi^2 + \theta^2\right) - \beta_g\left\|\left({}^B T e_g\right)_{xy}\right\|_2 \tag{14}$$

where ${}^B T$ is the rotation matrix from the world coordinate system to the body coordinate system, and $e_g$ is the unit vector in gravity direction. The first item punishing the robot is unable to maintain the body height stable; the remaining item punishes the robot with unnecessary roll and pitch movement.

All $\beta$ coefficients adjust each item to form the whole reward to evaluate the current policy. Figure 7 shows that the training reward gradually increases with the increase of the number of samples, and the number of samples required is $10^6$.

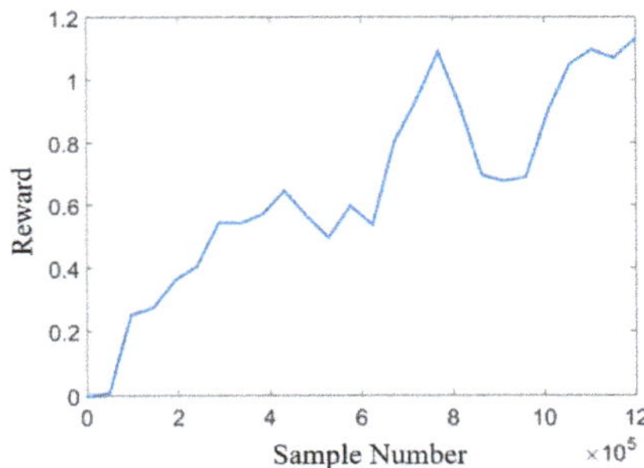

**Figure 7.** Training reward.

*4.3. Result and Discussion*

We first compare the locomotion performance of the fixed parameter PDMPC and the learned PDMPC. Figure 8 shows the velocity tracking performance of each controller. Obviously, the robot under the learned controller has smaller tracking error, and the motion is more smooth.

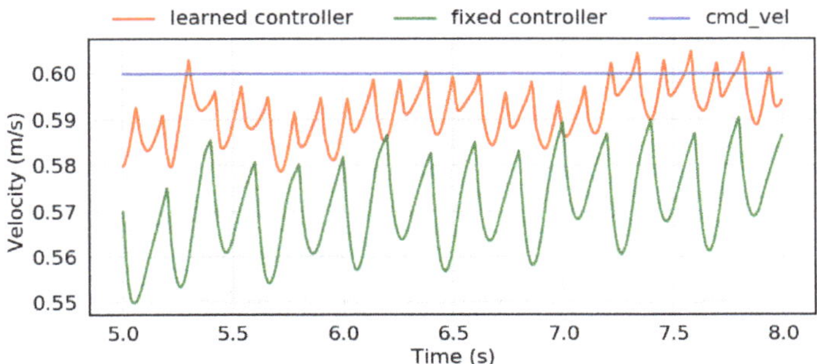


**Figure 8.** Forward velocity tracking performance.

Next, we check the gait and stability at different speeds. The forward velocity command gradually increases from 0 to 1 m/s. As shown in Figures 9 and 10, when the command speed increases, the fixed parameter controller holds the same gait, but the learned controller reduces its phase time. This change is in line with the biological norm that animals have a higher frequency of gait at high speeds. In addition, the attitude angle of the robot under the learned controller is closer to 0, and its oscillation amplitude is smaller. This phenomenon indicates that the learned controller is more capable of absorbing the impact of the swing leg when it touches down.

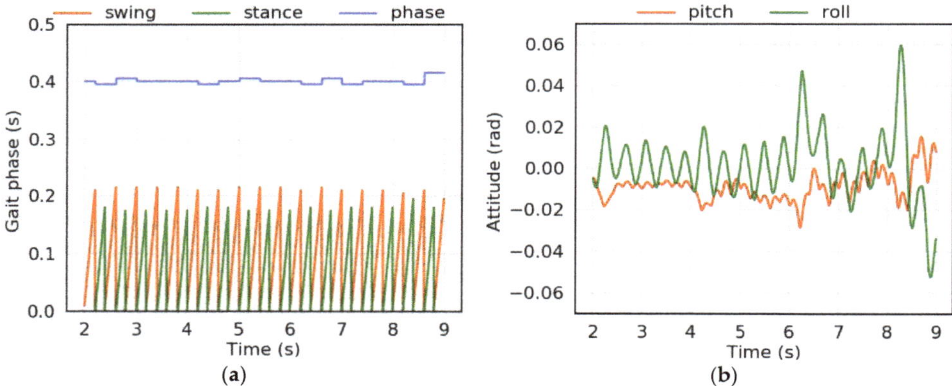

**Figure 9.** Motion controlled by fixed parameter PDMPC. (**a**) Gait; (**b**) Attitude angle.

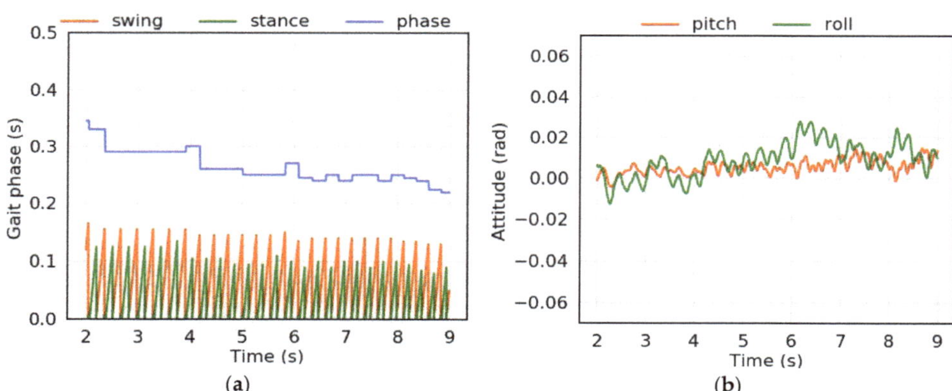

**Figure 10.** Motion controlled by learned parameter PDMPC. (**a**) Gait; (**b**) Attitude angle.

Figure 11 shows the adaptive changes of parameters of PDMPC during the acceleration of the quadruped robot. Figure 11a shows the adaptive heuristic coefficients obtained through reinforcement learning. The forward and lateral coefficients increase with the increase of speed. Manual-designed swing trajectory usually sets the heuristic coefficient of the inverted pendulum heuristic to a fixed value of 0.5. Because the model cannot be strictly regarded as a single-stage inverted pendulum during the motion, it is not reasonable to use fixed coefficients.

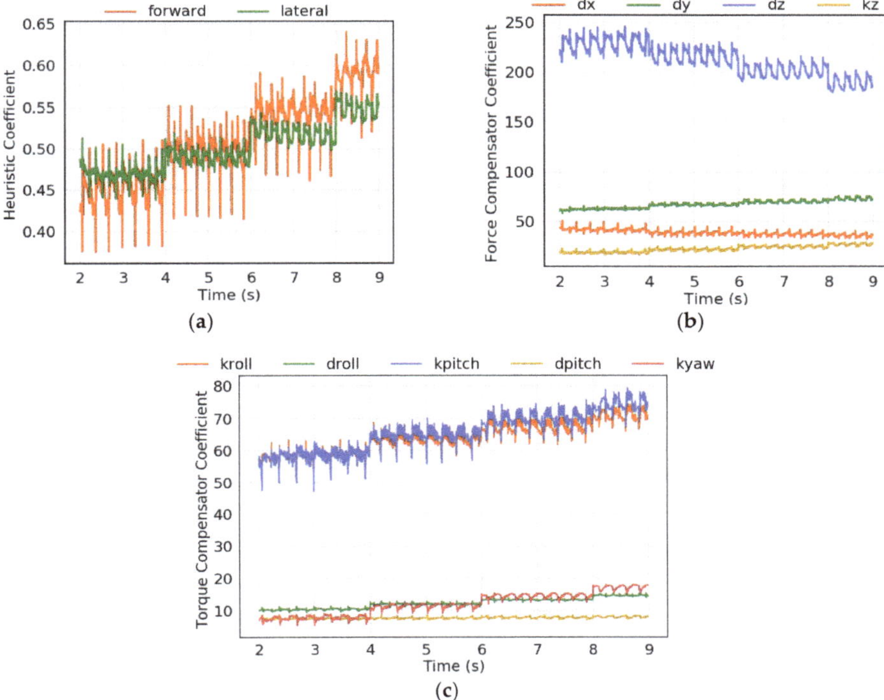

**Figure 11.** The adaptive variation of parameters during the acceleration of the quadruped robot. (**a**) Parameters for swing planner; (**b**) Parameters for force compensator; (**c**) Parameters for torque compensator.

It can be seen in Figure 11b that when the robot speed increases, the decreasing parameter $d_x$ makes the forward speed tracking of the robot softer and the increasing $dy$ strengthens the lateral control at the same time; the decreasing parameter $k_z$ and $d_z$ reduce the position compensation in the height direction of the robot and increase the damping, making the position of the center of mass more flexible. At the same time, the parameters of the torque compensator are all increased, which is conducive to making the robot's attitude more stable during acceleration.

## 5. Conclusions and Future Work

In this work, we proposed a novel locomotion control algorithm for a quadruped robot which combines the advantages of model-based MPC and model-free reinforcement learning. PDMPC controller performs a fundamental locomotion capability that provides a safe exploration region, and reinforcement learning endows robots with evolutionary ability. The trained policy chooses the parameters for the PDMPC controller adaptively according to the current state. The simulation results show the effectiveness of our algorithm, compared with the fixed parameters controller, the adaptive parameters make the learned controller have better performance in command tracking and equilibrium stability, and the gait of robot changes with speed just as quadrupeds do.

Model-free RL learns an end-to-end control policy based on the reward function to maximize performance. This learning from scratch requires massive samples, which means a large number of robots and time consumption. In this work, with the model-based controller to generate good samples, our training process has been greatly shortened. It requires fewer robots and has a faster learning speed. On the other hand, the manually designed controller will guide the policy into a local optimal region, limiting the powerful exploration, discovery and learning ability of reinforcement learning. As two major frameworks for solving optimal control problems, conventional control and learning-based control have their own advantages and can promote and develop together. We will explore more ways to merge them and make online quick learning possible on the physical platform.

**Author Contributions:** Z.Z. and X.C. designed this research methods and wrote the manuscript; H.M., H.A. and L.L. edited and revised the manuscript. All authors have read and agreed to the published version of the manuscript.

**Funding:** This research was funded by National Science Foundation of China (No. 61903131) and the China Postdoctoral Science Foundation (No. 2020M683715).

**Institutional Review Board Statement:** Not applicable.

**Informed Consent Statement:** Not applicable.

**Data Availability Statement:** Not applicable.

**Conflicts of Interest:** The authors declare no conflict of interest.

# References


1. Kimura, H.; Fukuoka, Y.; Hada, Y.; Takase, K. Three-dimensional Adaptive Dynamic Walking of a Quadruped rolling motion feedback to CPGs controlling pitching motion. In Proceedings of the 2002 IEEE International Conference on Robotics and Automation, Washington, DC, USA, 11–15 May 2002.
2. Farshidian, F.; Neunert, M.; Winkler, A.W.; Rey, G.; Buchli, J. An efficient optimal planning and control framework for quadrupedal locomotion. In Proceedings of the IEEE International Conference on Robotics and Automation (ICRA), Marina Bay Sands, Singapore, 29 May–3 June 2017; pp. 93–100.
3. Neunert, M.; Farshidian, F.; Winkler, A.W.; Buchli, J. Trajectory Optimization Through Contacts and Automatic Gait Discovery for Quadrupeds. *IEEE Robot. Autom. Lett.* **2017**, *2*, 1502–1509. [CrossRef]
4. Richalet, J.; Rault, A.; Testud, J.; Papon, J. Model predictive heuristic control: Applications to an industrial process. *Automatica* **1978**, *14*, 413–428. [CrossRef]
5. Carlo, J.D.; Wensing, P.M.; Katz, B.; Bledt, G.; Kim, S. Dynamic Locomotion in the MIT Cheetah 3 Through Convex Model-Predictive Control. In Proceedings of the 2018 IEEE/RSJ International Conference on Intelligent Robots and Systems (IROS), Madrid, Spain, 1–8 October 2018; pp. 1–9.
6. Bledt, G.; Wensing, P.M.; Kim, S. Policy-regularized model predictive control to stabilize diverse quadrupedal gaits for the MIT cheetah. In Proceedings of the 2017 IEEE/RSJ International Conference on Intelligent Robots and Systems (IROS), Vancouver, BC, Canada, 24–28 September 2017; pp. 4102–4109.
7. Ding, Y.; Pandala, A.; Li, C.; Shin, Y.-H.; Park, H.-W. Representation-Free Model Predictive Control for Dynamic Motions in Quadrupeds. *IEEE Trans. Robot.* **2021**, *37*, 1154–1171. [CrossRef]
8. Liu, X.; Ma, H.; An, H.; Wei, Q.; Lang, L. Gait Parameters Design Method of Trotting Gait Based on Energy Consumption. In Proceedings of the 2021 6th IEEE International Conference on Advanced Robotics and Mechatronics (ICARM), Chongqing, China, 3–5 July 2021; pp. 174–179.
9. Chang, X.; Ma, H.; An, H. Quadruped Robot Control through Model Predictive Control with PD Compensator. *Int. J. Control. Autom. Syst.* **2021**, *19*, 3776–3784. [CrossRef]
10. Kolter, J.Z.; Ng, A.Y. Policy search via the signed derivative. In *Robotics: Science and Systems*; The MIT Press: Cambridge, MA, USA, 2009.
11. Haarnoja, T.; Zhou, A.; Ha, S.; Tan, J.; Tucker, G.; Levine, S. Learning to Walk via Deep Reinforcement Learning. *arXiv* **2018**, arXiv:1812.11103.
12. Yao, Q.; Wang, J.; Yang, S.; Wang, C.; Zhang, H.Y.; Zhang, Q.F.; Wang, D.L. Imitation and Adaptation Based on Consistency: A Quadruped Robot Imitates Animals from Videos Using Deep Reinforcement Learning. *arXiv* **2022**, arXiv:2203.05973.
13. Hwangbo, J.; Lee, J.; Dosovitskiy, A.; Bellicoso, D.; Tsounis, V.; Koltun, V.; Hutter, M. Learning agile and dynamic motor skills for legged robots. *Sci. Robot.* **2019**, *4*, eaau5872. [CrossRef] [PubMed]
14. Lee, J.; Hwangbo, J.; Wellhausen, L.; Koltun, V.; Hutter, M. Learning quadrupedal locomotion over challenging terrain. *Sci. Robot.* **2020**, *5*, eabc5986. [CrossRef] [PubMed]
15. Kober, J.; Bagnell, J.A.; Peters, J. Reinforcement learning in robotics: A survey. *Int. J. Robot. Res.* **2013**, *32*, 1238–1274. [CrossRef]
16. Schulman, J.; Wolski, F.; Dhariwal, P. Proximal Policy Optimization Algorithms. *arXiv* **2017**, arXiv:1707.06347.
17. Ioffe, S.; Szegedy, C. Batch Normalization: Accelerating Deep Network Training by Reducing Internal Covariate Shift. In Proceedings of the International Conference on Machine Learning (ICML), Lille, France, 6–11 July 2015.

18. Rudin, N.; Hoeller, D.; Reist, P.; Hutter, M. Learning to Walk in Minutes Using Massively Parallel Deep Reinforcement Learning. In Proceedings of the 5th Annual Conference on Robot Learning, London, UK, 8 November 2021.
19. Kingma, D.P.; Ba, J. Adam: A Method for Stochastic Optimization. In Proceedings of the 3rd International Conference for Learning Representations, San Diego, CA, USA, 7–9 May 2015.

*applied sciences*

MDPI

*Article*

# A Deep Learning Approach for Credit Scoring Using Feature Embedded Transformer

Chongren Wang [1,2,*] and Zhuoyi Xiao [1]

1   School of Management Science and Engineering, Shandong University of Finance and Economics, Jinan 250014, China
2   Digital Economy Research Institute, Shandong University of Finance and Economics, Jinan 250014, China
*   Correspondence: wangchongren@sdufe.edu.cn

**Abstract:** In this paper, we introduce a transformer into the field of credit scoring based on user online behavioral data and develop an end-to-end feature embedded transformer (FE-Transformer) credit scoring approach. The FE-Transformer neural network is composed of two parts: a wide part and a deep part. The deep part uses the transformer deep neural network. The output of the deep neural network and the feature data of the wide part are concentrated in a fusion layer. The experimental results show that the FE-Transformer deep learning model proposed in this paper outperforms the LR, XGBoost, LSTM, and AM-LSTM comparison methods in terms of area under the receiver operating characteristic curve (AUC) and the Kolmogorov–Smirnov (KS). This shows that the FE-Transformer deep learning model proposed in this paper can accurately predict user default risk.

**Keywords:** credit scoring; machine learning; deep learning; transformer

## 1. Introduction

**Citation:** Wang, C.; Xiao, Z. A Deep Learning Approach for Credit Scoring Using Feature Embedded Transformer. *Appl. Sci.* **2022**, *12*, 10995. https://doi.org/10.3390/app122110995

Academic Editor: Habib Hamam

Received: 29 September 2022
Accepted: 28 October 2022
Published: 30 October 2022

**Publisher's Note:** MDPI stays neutral with regard to jurisdictional claims in published maps and institutional affiliations.

With the development of financial technology, big data and artificial intelligence technology have been paid increasingly more attention by financial enterprises. For financial enterprises, such as banks and P2P lending platforms, the most important risk is credit risk, that is, user default risk. Therefore, an increasing number of enterprises are trying to apply artificial intelligence technology, i.e., deep learning, to user credit risk assessment so as to reduce the loan default rate and to improve the ability of enterprises to resist risks [1,2]; this problem has attracted increasing attention.

Credit scoring is essentially a classification problem in machine learning. With the help of a credit risk assessment model, applicants can be divided into "good" customers and "bad" customers. Financial institutions can make loan approval decisions and risk pricing based on the credit scoring results.

With the development of financial technology, some loan businesses are carried out on online platforms, from basic websites to the current mobile application (APP), which has accumulated massive amounts of user online behavioral data, such as data on user registration behavior, user login behavior, user click behavior, and user authentication behavior. These online behavioral data have important mining value. In recent years, with the maturity of deep learning technology, it has become feasible to mine these data.

Based on the online behavioral data of users and the credit data of financial enterprises, this study proposed an end-to-end transformer credit scoring system, which can accurately predict users' default risk.

The main contributions of this study are as follows:

1.   This paper introduces transformer into the field of credit scoring based on user online behavioral data, and the experimental results show that the transformer used in this study outperforms LSTM and traditional machine learning models.

2. We make use of credit feature data and user behavioral data and develop a novel end-to-end deep learning credit scoring framework. The framework is composed of two parts, a wide part and a deep part, and it can automatically learn from user behavioral data and feature data.

The structure of this study is as follows: Section 2 summarizes the literature relevant to this study, Section 3 introduces the relevant theories and the transformer method proposed in this study, Section 4 analyzes the experimental results, and Section 5 summarizes this study.

## 2. Related Work

At the early stage of the development of credit scoring methods, due to the lack of comprehensive historical data in financial institutions, credit scoring mainly depended on the personal experience of experts. Later, with an increase in credit data, many statistical models and credit scoring methods gradually emerged. Altman [3] built a Z-score credit scoring model based on multivariate discriminant analysis technology, and Parnes [4] verified the superiority of the Z-score credit scoring method through detailed comparative analysis experiments. Logistic regression models are the most representative of statistical models. They are widely used because of their high prediction accuracy, simple calculation, and strong interpretation ability [5].

At present, a large number of scholars are introducing machine learning methods into the field of credit scoring research [6,7]. The traditional machine learning methods in this research field can be divided into individual classifier methods and ensemble learning methods. Individual classifiers that have been studied and applied in credit scoring include decision trees (DTs) [8] and SVM [9]. In addition, some recent studies have also proposed some improved individual classifiers [10]. Munkhdalai et al. [11] proposed a credit scoring approach that combines linear (softmax regression) and non-linear (neural network) methods.

Ensemble learning improves model performance by building and combining base learners, which can be further divided into homogeneous ensemble learning and heterogeneous ensemble learning. Homogeneous ensemble learning methods only use one kind of base learner for ensemble, such as random forest (RF) [12] and extreme gradient boosting (XGBOOST) [13]. Heterogeneous ensemble learning combines several kinds of base learners to improve model performance. Wang et al. [2] proposed a two-stage credit scoring model. The first stage is credit scoring, and the second stage is profit scoring. They used stacked generalization (stacking) to build the model, and the base learner includes LR, DT, and SVM. However, the features of the experimental data in these studies were usually low-dimensional and designed by experts [14].

In recent years, deep learning has shown remarkable results in many application fields, such as text sentiment classification [15], image classification [16], and recommendation systems [17]. Similarly, many studies have applied deep learning to the field of credit scoring, and research has proven the abilities of deep learning algorithms, which can automatically learn features from data. Tomczak and Zi ę Ba [18] proposed a new RBM-like credit risk prediction approach and proved the advantages of this credit scoring method through experiments. Yu et al. [19] proposed a new multi-level deep belief network (DBN) credit risk prediction method based on limit learning machine (ELM), which improved the credit risk prediction performance of this approach. Zhang et al. [20] proposed a hybrid model that combines transformer networks with CatBoost decision trees, and their experimental data came from a bank, but they were low-dimensional feature data.

With the development of the Internet industry, people's lives are becoming increasingly Internet-based, resulting in a large amount of user online behavioral data. Considering the large volume, high dimension, and sequential characteristics of user online behavioral data, for these kinds of data, the learning ability of traditional machine learning algorithms is limited; therefore, researchers have begun to use deep learning methods to deeply mine user online behavioral data. Some researchers have attempted to apply deep learning methods based on user online behavioral data to recommendation systems. Hidasi et al. [21] built a recommendation system using a recurrent neural network (RNN) based on users' online operation behavioral data. The experimental results show that this recommendation method is superior to existing methods. Lang and Rettenmeier [22] introduced a long short-term memory network (LSTM) to predict consumer behavior on e-commerce websites using user behavioral data, and the experimental results show that this approach has good prediction effects.

Similarly, some studies have attempted to apply deep learning methods based on user behavioral data to the field of credit scoring. Wang et al. [1] made use of borrowers' online operation behavioral data and proposed a consumer credit scoring method based on an attention mechanism LSTM. This method only uses user behavioral data, and the research results show that this approach has advantages over existing methods.

To sum up, credit scoring methods based on machine learning and deep learning are increasingly becoming a research hotspot. The research on deep learning methods based on user behavioral data is still relatively scarce, and there are still some research gaps in the research field of deep learning credit scoring models based on user online behavioral data. On the one hand, the LSTM model has long-term dependence and cannot be parallelized, and further research on deep learning algorithms is required. On the other hand, existing studies have only built deep learning credit scoring models based on user behavioral data, and they have not used feature data to build an end-to-end neural network model. Therefore, further research combining user behavioral data and feature data to build deep learning credit scoring models needs to be carried out.

## 3. Theory and Method

### 3.1. LSTM

LSTM, which was proposed by Hochreiter and Schmidhuber [23], is widely used to process sequence information, such as text classification [24] and machine translation [25], because it can alleviate long-term dependencies. LSTM can realize the remembering and forgetting of long-term historical states through different gate structures.

As shown in Figure 1, suppose $x_t$ is the parameter information of the new incoming training process, and $h_{t-1}$ is the staged result of the last iteration process. The input $x_t$, the memory state $C_{t-1}$, and the intermediate output $h_{t-1}$ in the forget gate determine the forgetting part of the memory state. $x_t$ in the input gate is changed by sigmoid and tan h functions, and then, it determines the reserved vector in the memory state. Finally, the effective information is output by the output gate control, and a performance model with better prediction can be obtained by iterating the error correction many times. However, LSTM can only calculate in sequence, which leads to two problems. On the one hand, the calculation of each time period depends on the calculation results of the previous time period, so the model cannot calculate in parallel. On the other hand, although the gate structure of LSTM alleviates the problem of long-term dependence, LSTM still cannot solve this problem.

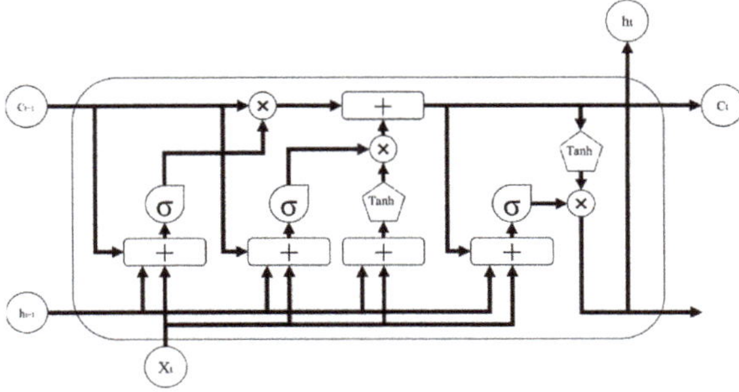


**Figure 1.** Structure of the LSTM model.

### 3.2. Transformer

The transformer model proposed by Google was first applied to the task of machine translation [26]. In this research, a transformer is an encoder–decoder structure. The transformer consists of an encoder and a decoder, which are stacked with 6 layers in total. This model does not use a recurrent structure. After passing through the 6-layer encoder in the model, the input data are output to the decoder of each layer in order to calculate the attention. The architecture of a transformer consists of four modules: an input module, an encoding module, a decoding module, and an output module.

A transformer is a deep neural network based on the self-attention mechanism and parallel data processing. It outperforms RNNs and convolutional neural networks (CNNs) in machine translation tasks, and it has become the current mainstream feature extractor. At the same time, the transformer solves two problems of LSTM. On the one hand, it uses an attention mechanism to reduce the distance between any two positions in a sequence to a constant. On the other hand, the transformer can be computed in parallel unlike the sequential structure of LSTM. The transformer is obviously superior to LSTM in terms of comprehensive feature extraction ability. Therefore, in the task of machine translation, the traditional attention-mechanism-based LSTM has migrated to the network structure based on the transformer model.

### 3.3. Feature Embedded Transformer

In this study, we introduce a transformer into the field of credit scoring and develop an end-to-end deep learning credit scoring framework; we named this framework the feature embedded transformer (FE-Transformer). The architecture of this method is shown in Figure 2. The FE-Transformer neural network is composed of two parts: a wide part and a deep part. The deep part uses the transformer neural network; the output of the transformer neural network and the feature data of the wide part are concentrated in the fusion layer; and finally, the prediction results are output. The FE-Transformer can automatically learn from user behavioral data and feature data.

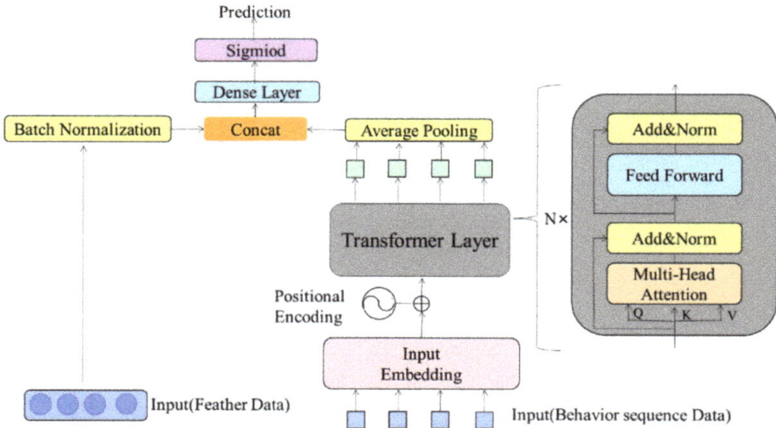


**Figure 2.** Network architecture of the FE-Transformer.

### 3.3.1. Input Data and Data Coding

There are two kinds of input data in this model: one is feature data, and the other is behavioral data. Feature data include users' gender, age, credit record, and other credit data. The users' behavioral data mainly include the users' online operation behavioral data, such as click behavior and input behavior. After the feature data are processed, they are used as the input of the model.

For the behavioral data, inspired by NLP, each kind of behavior event can be regarded as a word. The behavioral data of each user are composed of a series of events, which constitute a sequence of events and can be regarded as a sentence. We process the raw online operation behavior record data and convert these behaviors into event sequences in chronological order. Then, we encode the input behavioral data via embedding and position encoding.

An event is the basic unit of model processing. First, the input event needs to be converted into a vector through a word embedding algorithm. In order to understand a sequence of events, the model needs to know the position of the event in the sentence in addition to understanding the meaning of the event. Since the calculation of the transformer abandons the recursion and convolution of the cyclic structure, it cannot simulate the positional information of the events in the sequence, so it is necessary to obtain the positional vectors of the events through positional encoding. The position vector is then added to the event vector to obtain the input to the model. We take the sine function to generate the position vector for each event:

$$PE_{(pos,2i)} = \sin\left(pos/10000^{2i/d_{model}}\right) \tag{1}$$

$$PE_{(pos,2i+1)} = \cos\left(pos/10000^{2i/d_{model}}\right) \tag{2}$$

where $pos$ is the position of the event in the behavior sequence of events, $d_{model}$ is the dimension of positional encoding, $2i$ is the even dimension, and $2i + 1$ is the odd dimension ($2i \leq d_{model}, 2i + 1 \leq d_{model}$). After data coding, the data are used as the input of the transformer layer.

### 3.3.2. Transformer Encoding Layer

The transformer encoding layer is composed of one or more layers of stacked encoders. Each layer of the encoder is mainly composed of a multi-head attention layer and a fully connected feed-forward layer. Layer normalization [27] is used in front of each sublayer, and residual connection is used behind each sublayer. The transformer encoding layer structure is shown in Figure 2.

The event vector matrix obtained by the embedding layer is passed into the encoder through the multi-head attention layer and into the fully connected feed-forward layer, and then, the output is passed up to the next encoder. After one or more encoders, the encoding information matrix of all events in the behavior sequence is obtained.

The self-attention mechanism is an improvement in the attention mechanism, and it has the advantages of reducing the network's dependence on external information and being good at capturing internal correlations in data. The transformer architecture introduces a self-attention mechanism, which avoids the use of recursive structures in neural networks and completely relies on the self-attention mechanism to draw the global dependencies between the input and output [28].

The attention layer uses scaled dot-product attention. Compared with general attention, scaled dot-product attention uses the dot product for similarity calculation, which has the advantages of a faster calculation speed and being more space-saving. The basic structure is shown in Figure 3.

The self-attention mechanism is used to calculate the degree of relatedness between events. When calculating, each event in the input is first linearly projected into three different spaces to obtain a query vector (Q), a key vector (K), and a value vector (V). When obtaining self-attention information, the Q vector is used to query all candidate positions. Each candidate position has a pair of K and V vectors. The query process is the processing of dot products between the Q vector and the K vector of all candidate positions [29]. The product result is divided by the scaling factor (the square root of the dimension of the key vector) to improve the convergence speed. The result is normalized using the softmax function and then weighted to the respective V vector, and the summation determines the final self-attention result. The calculation formula is shown in Formula (3):

$$\text{Attention}(Q, K, V) = \text{softmax}\left(\frac{QK^T}{\sqrt{d_k}}\right)V \tag{3}$$

$d_k$ is the number of columns of matrices Q and K, that is, the vector dimension.

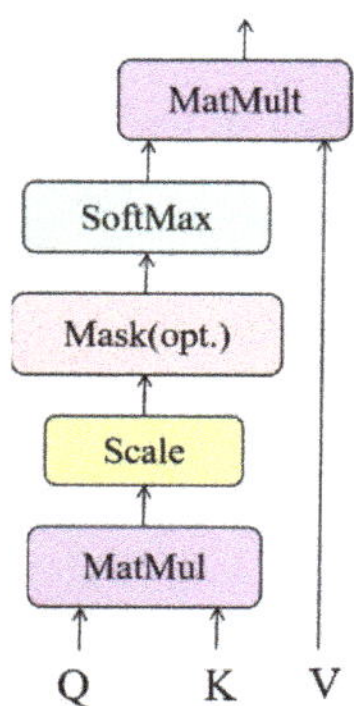


**Figure 3.** Scaled dot-product attention.

Multi-head self-attention enables the model to jointly learn the representation information of different locations from different representation subspaces. It is equivalent to a collection of different self-attention heads. As shown in Figure 4, after Q, K, and V are subjected to different linear projections, the scaled dot-product attention calculation is performed so that different parts of the input can be paid attention to and different semantic information can be learned. After multiple operations in parallel, the attention information in all subspaces is finally merged. The calculation process of each multi-head module shown in Equations (4) and (5) indicates that the results of multiple self-attention heads are spliced and converted into an output vector of a specific dimension.

$$\text{head}_i = \text{Attention}\left(QW_i^Q, KW_i^K, VW_i^V\right) \tag{4}$$

$$\text{MultiHead}(Q, K, V) = \text{Concat}(\text{head}_1, \ldots, \text{head}_h)W^O \tag{5}$$

$W_i^Q$, $W_i^K$, and $W_i^V$ are the weight matrices after the linear transformation of Q, K, and V, respectively; $W^O \in R^{d_{model} \times d_k}$ is the weight matrix for the multi-head self-attention mechanism; and h is the number of self-attention heads.

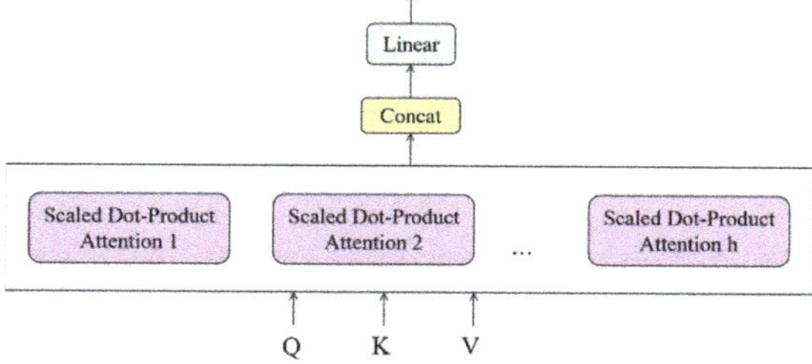


**Figure 4.** Multi-head attention.

The multi-head self-attention mechanism is the key to the transformer model, as it enriches the relationship between events and can even understand the semantic and syntactic structure information of sequences of events.

### 3.3.3. Concatenate Layer and Output Layer

The output of the transformer encoding layer is connected to an average pooling layer and output as a vector. In the concatenate layer, the vector output by the transformer encoding layer and the feature data are concatenate. In order to make the dimension of the data consistent, a batch normalization layer is added behind the feature data.

On this basis, following the full connection layer is the output layer. The output layer uses the sigmoid activation function to obtain the output, and the output result is the user's possibility of default. The formula of the output layer is as follows:

$$y = sigmoid(Wx + b) \tag{6}$$

In the process of model training, we choose cross-entropy as the loss function: cross-entropy represents the gap between the actual category of the model and the probability of the category predicted by the model. The smaller the value of the cross-entropy loss, the closer the model prediction probability and the real value. The loss function is calculated as follows:

$$L = -\frac{1}{N} \sum_{i=1}^{N} [y_i \log(p_1) + (1 - y_i) \log(1 - p_i)] \tag{7}$$

where $y_i$ represents the real label of the sample, $p_i$ represents the prediction probability of the model, and $N$ represents the number of samples.

Finally, we select the back propagation (BP) algorithm to update the model parameters.

### 3.4. Evaluation Metrics

In order to test the validity of the model, we choose two commonly used indicators of credit scoring to evaluate the performance of the model: area under the receiver operating characteristic curve (AUC) and Kolmogorov–Smirnov (KS).

Let TP be the real status of the customer classified as non-default and who is judged to be non-default. FN is the real status of the customer classified as non-default and who is judged to be default. TN is the real status of the customer who is judged to be default. FP is the actual status of the customer classified as default and who is judged to be non-default. Define the True Positive Rate (TPR) as the number of TPs divided by the total number of positive customers, and define the False Positive Rate (FPR) as the number of FPs divided by the total number of negative customers; the formulae of TPR and FPR is as follows:

$$TPR = \frac{TP}{TP + FN} \times 100\% \tag{8}$$

$$FPR = \frac{FP}{FP + TN} \times 100\% \tag{9}$$

Taking TPR as the abscissa and FPR as the ordinate, we draw the receiver operating characteristic (ROC) curve of the model. The closer the ROC curve is to the upper left corner, the better the performance of the classifier. However, since "close to the upper left corner" is only an intuitive description of the graph, it is generally only chosen to calculate the AUC value to better quantify the degree of proximity. AUC also considers the model's ability to discriminate between defaulting customers and non-defaulting customers, avoiding the problem of model evaluation criteria failure caused by sample imbalance. The larger the AUC value, the stronger the ability of the model to Identify defaults.

The Kolmogorov–Smirnov (KS) is a commonly used credit score evaluation index, and it is mainly used to measure the model's ability to distinguish default users. After the model predicts the default probability of all samples, we sort the samples according to the predicted default probability, calculate the cumulative TPR value and the cumulative FPR value under each default rate, and then calculate the sum of the two values under each default rate. Then, after obtaining the absolute value of the difference, we take the maximum value of these absolute values as the KS value. The larger the value of KS, the better the ability of the model to distinguish between defaulting borrowers and on-time borrowers.

## 4. Experimental Results

### 4.1. Experimental Set

Our experimental environment is a server with a Ubuntu 16 operating system, and the programming language is Python. The Python libraries used in this experiment mainly include Numpy, Pandas, Scikit-learn, Matplotlib, and Keras. Numpy is a scientific computing library of Python, and it provides the function of matrix operation; Pandas is a library mainly used for data processing and data analyses; Scikit-learn is a machine learning library; and Matplotlib is a drawing library. The deep learning framework used in this experiment is Keras (Tensorflow as the back end).

The dataset used in this research comes from an anonymous P2P lending company in China. The dataset includes feature data and behavioral data, with a total of 100,000 borrowers. The label of the data is whether the borrower defaults. If the borrower defaults, the label is 1; otherwise, the label is 0. The dataset includes five months of user loan data. To verify the stability of the model in predicting future loans, we first sort the loans according to the loan date, and then, we divide the data with the loan date in the last month into test sets. The test set data account for about 20% of the dataset, and the remaining data are divided into training sets. The training set is mainly used for training the model, and the test set is mainly used for testing model performance.

Then, data preprocessing and data coding are carried out. For user behavioral data, considering that the length of each user's behavior sequence is different, this paper converts all sequences into fixed length sequences. After a series of experiments, the length of the time series is fixed to 100, the sequences whose length exceeds 100 intercept the first 100 events, and the sequences whose length is less than 100 are filled with 0. For the FE-Transformer credit scoring model proposed in this paper, the number of transformer coding layers is set to 2, and the number of headers of the multi-head attention mechanism is set to 4. In order to alleviate the overfitting problem, Dropout [30], which is a method of dropping neural units with a certain probability from the network while training the neural network, is added to the transformer coding layer, and the dropout ratio is set to 0.3. The model training adopts mini-batch random gradient descent, the learning rate is set to 0.001, the parameter update adopts adaptive motion estimation (Adam) rules, and the early stopping strategy is adopted in the process of deep learning model training to alleviate overfitting problem.

In order to evaluate the FE-Transformer credit scoring method proposed in this paper and to prove the superiority of this approach, we conducted a detailed comparative analysis, and the comparison methods are as follows:

Logistic regression (LR): Logistic regression is the most representative of statistical models, and the input of this model is feature data.

XGBoost: XGBoost [31] is an ensemble learning algorithm, and the input data processing method of the XGBoost model is the same as that of the LR model.

LSTM: In the deep part of the model, LSTM is adopted. Firstly, user behavioral data are converted into event sequences as the input of LSTM; then, the output of LSTM is fused with the feature data; and finally, the sigmoid function is used for classification.

AM-LSTM: Using the method proposed by Wang et al. [2], the attention mechanism is added to the LSTM approach.

FE-Transformer: The approach proposed in this study.

In order to demonstrate the performance advantages of the FE-Transformer approach proposed in this study, we conducted three types of experiments, which used different datasets: one dataset only comprises feature data, one dataset only comprises behavioral data, and one dataset comprises all data.

The first type of experiment only used feature data to examine the effect of the machine learning models on the traditional credit data. Considering that the feature data only contain feature data and have low dimensions, they are not suitable for training deep learning models, so we chose two traditional models, namely, logical regression and XGBoost. The second type of experiment used the dataset with only behavioral data. For the deep learning models of LSTM, AM-LSTM, and the transformer, user event sequences can be directly used as model input, but for the traditional machine learning models of LR and XGBoost, sequence data cannot be used as input, so we manually extracted features and selected the frequency of each event as the feature. The third type of experiment used the dataset with all the data, and the five models LR, XGBoost, LSTM, AM-LSTM, and FE-Transformer were selected for the experiment.

*4.2. Performance Analysis*

The results of the models only using the feature data are shown in Table 1. As can be seen in the experimental results, the AUC and KS values of the XGBoost model are higher than those of LR, indicating that the performance of the ensemble learning algorithm is superior to that of the single linear model.

**Table 1.** Results of models only using credit data.

| Models | Training Set | | Test Set | |
|:---:|:---:|:---:|:---:|:---:|
| | KS | AUC | KS | AUC |
| LR | 0.23 | 0.622 | 0.23 | 0.621 |
| XGBoost | 0.248 | 0.634 | 0.241 | 0.63 |

The results of the models only using the behavioral data are shown in Table 2 and Figure 5. For the models only using the behavioral data, the performance of the deep learning models (LSTM and transformer) exceed that of the traditional machine learning algorithms (LR and XGBoost). This is because the deep learning algorithm can extract higher-level feature information. At the same time, consistent with the results of existing research, the effect of the AM-LSTM model is better than that of the basic LSTM model. The transformer model used in this study performs better than LSTM, AM-LSTM, and traditional machine learning models, and it achieves the highest AUC and KS values.

**Table 2.** Results of models only using behavioral data.

| Models | Training Set | | Test Set | |
|:---:|:---:|:---:|:---:|:---:|
| | KS | AUC | KS | AUC |
| LR | 0.092 | 0.57 | 0.09 | 0.54 |
| XGBoost | 0.1 | 0.58 | 0.095 | 0.553 |
| LSTM | 0.203 | 0.631 | 0.198 | 0.62 |
| AM-LSTM | 0.243 | 0.66 | 0.238 | 0.661 |
| Transformer | 0.26 | 0.679 | 0.25 | 0.672 |

The input data of the models using all data include the behavioral data and feature data. The results of the models using all data are shown in Table 3 and Figure 6. From the experimental results, it can be seen that the performance of the LR and XGBoost models is better when using all data than when only using feature data. This indicates that user behavioral data can improve the prediction effect of the credit scoring model. The performance of the deep learning models (LSTM and the transformer) exceeds that of the traditional machine learning models (LR and XGBoost). The performance of the FE-Transformer model proposed in this study is better than that of the other machine learning models, and it also achieved the highest AUC (0.72) and the highest KS values (0.32) on the test dataset.

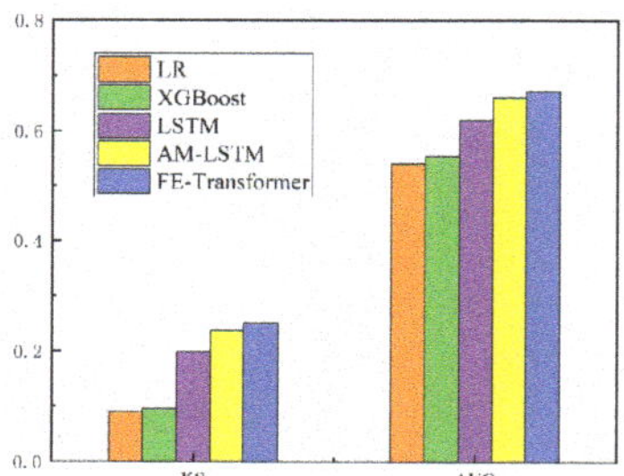

**Figure 5.** Performance comparison of models only using behavioral data.

**Table 3.** Results of models using all data.

| Models | Train Set | | Test Set | |
|---|---|---|---|---|
| | KS | AUC | KS | AUC |
| LR | 0.25 | 0.670 | 0.251 | 0.658 |
| XGBoost | 0.262 | 0.679 | 0.26 | 0.665 |
| LSTM | 0.273 | 0.7 | 0.26 | 0.682 |
| AM-LSTM | 0.31 | 0.707 | 0.313 | 0.71 |
| FE-Transformer | 0.33 | 0.731 | 0.32 | 0.72 |

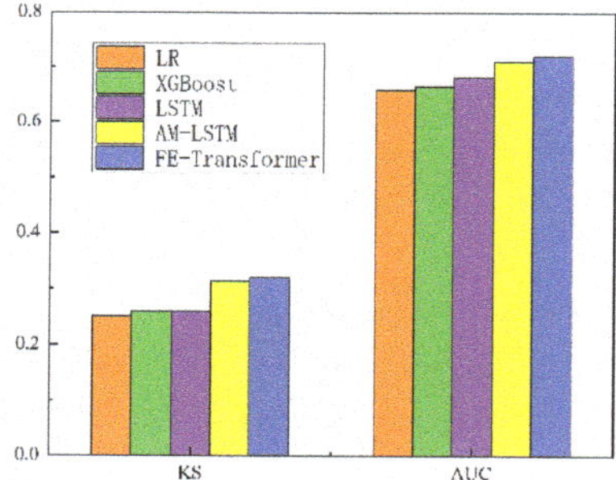

**Figure 6.** Performance comparison of models with all data.

## 4.3. Parameter Analysis

In this section, we analyzed the influence of different hyper-parameters on the performance of the FE-Transformer model. We selected two important parameters for analysis, the number of heads and the number of transformer layers. As can be seen in Figure 7, the experimental results show that, with an increase in parameters, the performance of the model increases first and then decreases. When the number of heads is set to 4, the KS and AUC of the FE-Transformer achieve the highest values, and when the number of transformer layers is set to 2, the KS and AUC of the FE-Transformer achieve the highest values. The reason for this may be that, when the hyper-parameters value is very small, the model training is not enough, so the performance of the model is general, and when the hyper-parameter values are too large, overfitting problems occur, which affect the performance of the FE-Transformer.

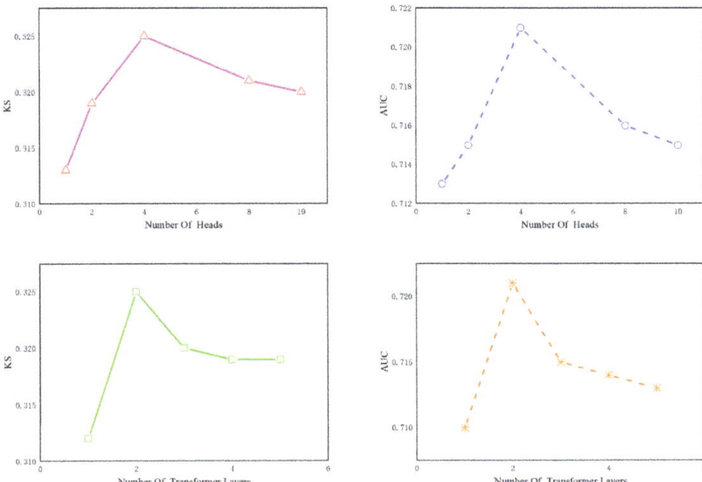

**Figure 7.** Influence of different hyper-parameters on the performance of FE-Transformer model.

For the deep learning model containing all data, when we fuse the feature data with the data output from the deep learning model, we added a batch normalization layer. Batch normalization can normalize the data and improve the generalization ability of neural networks [32]. In order to verify the impact of batch normalization on the performance of the model, we conducted a comparative experiment on whether to conduct batch normalization. The results are represented in Table 4 and Figure 8. For the three deep learning models LSTM, AM-LSTM, and FE-Transformer, the performance of the models with batch normalization significantly exceeds that of the models without batch normalization. The reason for this may be that batch normalization can make the output of the deep learning model be consistent with the dimension of the feature data, which is conducive to the use of the gradient descent algorithm to optimize the model.

**Table 4.** Performance comparison of deep learning models with normalization and without normalization.

| Models | With Normalization | | Without Normalization | |
|---|---|---|---|---|
| | **KS** | **AUC** | **KS** | **AUC** |
| LSTM | 0.26 | 0.682 | 0.175 | 0.61 |
| AM-LSTM | 0.313 | 0.71 | 0.21 | 0.6 |
| FE-Transformer | 0.32 | 0.72 | 0.24 | 0.63 |

Finally, to analyze the impact of behavioral data on credit scores, we chose the XGBoost model using only the behavioral data for analysis. For this model, the input of the model is the frequency of each event. After building the XGBoost model, we extracted the Top 15 important features. The feature importance score represents the usefulness of the input feature to the user's credit default prediction; the results are shown in Figure 9. In consideration of commercial confidentiality requirements, we desensitized the event names. The results show that the feature importance of different events varies greatly, and some events have a significant prediction effect on user default risk.

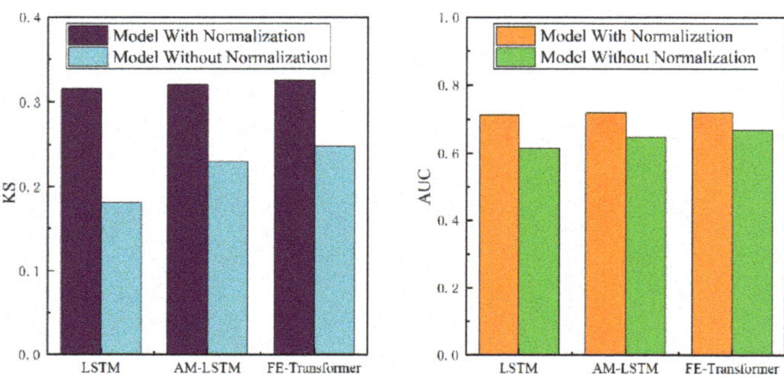

**Figure 8.** Performance comparison of models with normalization and models without normalization.

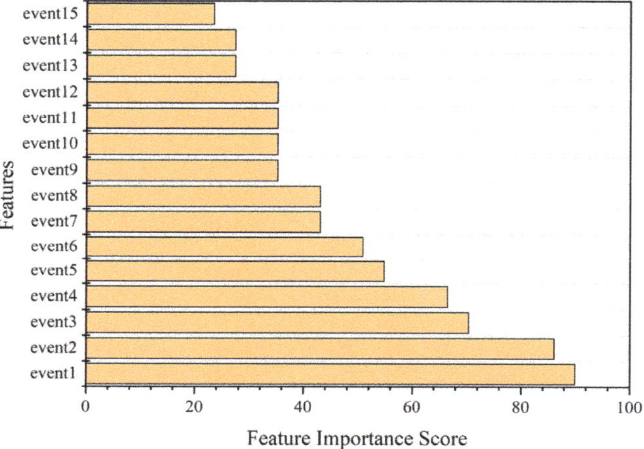

**Figure 9.** Feature importance of XGBoost model.

The FE-Transformer model proposed in this research outputs the predicted user default probability. The probability value is between 0 and 1. Based on this probability value, the user's credit score can be calculated. The credit score is used as the basis for loan approval and pricing. If the APP of financial institutions is upgraded, the events of user behavior will change. Therefore, after the APP is upgraded, the deep learning model needs to be updated.

To sum up, the experimental results show that the FE-Transformer model proposed in this study outperforms the LR, XGBoost, LSTM, and AM-LSTM comparison methods in terms of AUC and KS. This shows that the FE-Transformer deep learning model proposed in this research can accurately predict user default risk, which is conducive to reducing the loan default rate of financial enterprises, reducing the credit risk of financial enterprises, and maintaining the healthy and sustainable development of financial enterprises.

## 5. Conclusions

With the development of big data and artificial intelligence technology, deep learning models have become the research focus of credit scoring. We study the credit scoring methods of financial enterprises and propose a FE-Transformer neural network model.

The main conclusions of this study are as follows:

On the one hand, user online behavioral data provide a novel credit scoring data source. The research results show that user online behavioral data can help improve the effect of user default prediction models. On the other hand, the performance of the

FE-Transformer model proposed in this paper is better than that of the other comparison methods, and this proves the effectiveness and feasibility of this method in the field of credit scoring. The user default probability output by the model can provide the basis for loan approval decisions and the risk pricing of financial institutions, and it can help financial institutions improve their credit risk management levels and abilities.

For future research, several issues can be considered. On the one hand, due to the difficulty of data acquisition, this experiment only uses the datasets of one enterprise, and we will continue to look for other enterprise datasets for research. On the other hand, the credit scoring model in this study is a static model, and the dynamic update of credit scoring models is a research hotspot. On the basis of this study, the dynamic update of the model proposed in this research can be further studied.

**Author Contributions:** Conceptualization, C.W. and Z.X.; methodology, C.W. and Z.X.; experiment, C.W.; writing—original draft preparation, C.W. and Z.X.; writing—review and editing, C.W.; project administration, C.W. All authors have read and agreed to the published version of the manuscript.

**Funding:** This research was supported by Key R&D Plan funded by the Science and Technology Department of Shandong Province, China (No. 2019GSF108222).

**Institutional Review Board Statement:** Not applicable.

**Informed Consent Statement:** Not applicable.

**Data Availability Statement:** Not applicable.

**Conflicts of Interest:** The authors declare no conflict of interest.

## References

1. Wang, C.; Han, D.; Liu, Q.; Luo, S. A Deep Learning Approach for Credit Scoring of Peer-to-Peer Lending Using Attention Mechanism LSTM. *IEEE Access* **2018**, *7*, 2161–2168. [CrossRef]
2. Wang, C.; Liu, Q.; Li, S. A two-stage credit risk scoring method with stacked-generalisation ensemble learning in peer-to-peer lending. *Int. J. Embed. Syst.* **2022**, *15*, 158–166. [CrossRef]
3. Altman, E.I. Financial ratios, discriminant analysis and the prediction of corporate bankruptcy. *J. Financ.* **1968**, *23*, 589–609. [CrossRef]
4. Parnes, D. Applying Credit Score Models to Multiple States of Nature. *J. Fixed Income* **2007**, *17*, 57–71. [CrossRef]
5. Bolton, C. *Logistic Regression and Its Application in Credit Scoring*; University of Pretoria: Pretoria, South Africa, 2010.
6. Lessmann, S.; Baesens, B.; Seow, H.-V.; Thomas, L.C. Benchmarking state-of-the-art classification algorithms for credit scoring: An update of research. *Eur. J. Oper. Res.* **2015**, *247*, 124–136. [CrossRef]
7. Bhatia, S.; Sharma, P.; Burman, R.; Hazari, S.; Hande, R. Credit scoring using machine learning techniques. *Int. J. Comput. Appl.* **2017**, *161*, 1–4. [CrossRef]
8. Mandala, I.G.N.N.; Nawangpalupi, C.B.; Praktikto, F.R. Assessing Credit Risk: An Application of Data Mining in a Rural Bank. *Procedia Econ. Financ.* **2012**, *4*, 406–412. [CrossRef]
9. Harris, T. Credit scoring using the clustered support vector machine. *Expert Syst. Appl.* **2015**, *42*, 741–750. [CrossRef]
10. Abellán, J.; Castellano, J.G. A comparative study on base classifiers in ensemble methods for credit scoring. *Expert Syst. Appl.* **2017**, *73*, 1–10. [CrossRef]
11. Munkhdalai, L.; Ryu, K.; Namsrai, O.-E.; Theera-Umpon, N. A Partially Interpretable Adaptive Softmax Regression for Credit Scoring. *Appl. Sci.* **2021**, *11*, 3227. [CrossRef]
12. Malekipirbazari, M.; Aksakalli, V. Risk assessment in social lending via random forests. *Expert Syst. Appl.* **2015**, *42*, 4621–4631. [CrossRef]
13. Xia, Y.; Liu, C.; Li, Y.; Liu, N. A boosted decision tree approach using Bayesian hyper-parameter optimization for credit scoring. *Expert Syst. Appl.* **2017**, *78*, 225–241. [CrossRef]
14. Kang, Y.; Chen, L.; Jia, N.; Wei, W.; Deng, J.; Qian, H. A CWGAN-GP-based multi-task learning model for consumer credit scoring. *Expert Syst. Appl.* **2022**, *206*, 117650. [CrossRef]
15. Ghorbanali, A.; Sohrabi, M.K.; Yaghmaee, F. Ensemble transfer learning-based multimodal sentiment analysis using weighted convolutional neural networks. *Inf. Process. Manag.* **2022**, *59*, 102929. [CrossRef]
16. Bae, J.-H.; Yu, G.-H.; Lee, J.-H.; Vu, D.T.; Anh, L.H.; Kim, H.-G.; Kim, J.-Y. Superpixel Image Classification with Graph Convolutional Neural Networks Based on Learnable Positional Embedding. *Appl. Sci.* **2022**, *12*, 9176. [CrossRef]
17. Liu, Q.; Mu, L.; Sugumaran, V.; Wang, C.; Han, D. Pair-wise ranking based preference learning for points-of-interest recommendation. *Knowl.-Based Syst.* **2021**, *225*, 107069. [CrossRef]

18. Tomczak, J.M.; Zięba, M. Classification restricted Boltzmann machine for comprehensible credit scoring model. *Expert Syst. Appl.* **2015**, *42*, 1789–1796. [CrossRef]
19. Yu, L.; Yang, Z.; Tang, L. A novel multistage deep belief network based extreme learning machine ensemble learning paradigm for credit risk assessment. *Flex. Serv. Manuf. J.* **2015**, *28*, 576–592. [CrossRef]
20. Zhang, Z.; Wang, Z. Research on Credit Scoring Based on Transformer-CatBoost Network Structure. In Proceedings of the 2022 IEEE 12th International Conference on Electronics Information and Emergency Communication (ICEIEC), Beijing, China, 15–17 July 2022; pp. 75–79.
21. Hidasi, B.; Quadrana, M.; Karatzoglou, A.; Tikk, D. Parallel Recurrent Neural Network Architectures for Feature-rich Session-based Recommendations. In Proceedings of the 10th ACM Conference on Recommender Systems, Boston, MA, USA, 15–19 September 2016; pp. 241–248. [CrossRef]
22. Lang, T.; Rettenmeier, M. Understanding consumer behavior with recurrent neural networks. In Proceedings of the Workshop on Machine Learning Methods for Recommender Systems, Houston, TX, USA, 27–29 April 2017.
23. Hochreiter, S.; Schmidhuber, J.U.R. Long short-term memory. *Neural Comput.* **1997**, *9*, 1735–1780. [CrossRef]
24. Liu, G.; Guo, J. Bidirectional LSTM with attention mechanism and convolutional layer for text classification. *Neurocomputing* **2019**, *337*, 325–338. [CrossRef]
25. Guo, D.; Zhou, W.; Li, H.; Wang, M. Hierarchical lstm for sign language translation. *Proc. AAAI Conf. Artif. Intell.* **2018**, *32*, 6845–6852. [CrossRef]
26. Vaswani, A.; Shazeer, N.; Parmar, N.; Uszkoreit, J.; Jones, L.; Gomez, A.N.; Kaiser, L.U.; Polosukhin, I. Attention is all you need. In *Advances in Neural Information Processing Systems*; The MIT Press: Cambridge, MA, USA, 2017; p. 30.
27. Ba, J.L.; Kiros, J.R.; Hinton, G.E. Layer normalization. *arXiv* **2016**, arXiv:1607.06450.
28. Zhang, X.; Gao, T. Multi-head attention model for aspect level sentiment analysis. *J. Intell. Fuzzy Syst.* **2020**, *38*, 89–96. [CrossRef]
29. Jing, H.; Yang, C. Chinese text sentiment analysis based on transformer model. In Proceedings of the 2022 3rd International Conference on Electronic Communication and Artificial Intelligence (IWECAI), Zhuhai, China, 14–16 January 2022; pp. 185–189.
30. Srivastava, N.; Hinton, G.; Krizhevsky, A.; Sutskever, I.; Salakhutdinov, R. Dropout: A simple way to prevent neural networks from overfitting. *J. Mach. Learn. Res.* **2014**, *15*, 1929–1958.
31. Chen, T.; Guestrin, C. Xgboost: A scalable tree boosting system. In Proceedings of the 22nd ACM SIGKDD International Conference on Knowledge Discovery and Data Mining, San Francisco, CA, USA, 13–17 August 2016; pp. 785–794.
32. Ioffe, S.; Szegedy, C. Batch normalization: Accelerating deep network training by reducing internal covariate shift. In Proceedings of the International Conference on Machine Learning, Lille, France, 6 July 2015; pp. 448–456.

*applied sciences*



*Article*

# Vehicle-Following Control Based on Deep Reinforcement Learning

Yong Huang [1], Xin Xu [2], Yong Li [3], Xinglong Zhang [2], Yao Liu [1] and Xiaochuan Zhang [1,*]

[1] School of Artificial Intelligence, Chongqing University of Technology, Chongqing 401135, China
[2] College of Computer Science, National University of Defense Technology, Changsha 400015, China
[3] China Coal Technology Engineering Group Chongqing Research Institute, Chongqing 400039, China
* Correspondence: zxc@cqut.edu.cn

**Abstract:** Intelligent vehicle-following control presents a great challenge in autonomous driving. In vehicle-intensive roads of city environments, frequent starting and stopping of vehicles is one of the important cause of front-end collision accidents. Therefore, this paper proposes a subsection proximal policy optimization method (Subsection-PPO), which divides the vehicle-following process into the start–stop and steady stages and provides control at different stages with two different actor networks. It improves security in the vehicle-following control using the proximal policy optimization algorithm. To improve the training efficiency and reduce the variance of advantage function, the weighted importance sampling method is employed instead of the importance sampling method to estimate the data distribution. Finally, based on the TORCS simulation engine, the advantages and robustness of the method in vehicle-following control is verified. The results show that compared with other deep learning learning, the Subsection-PPO algorithm has better algorithm efficiency and higher safety than PPO and DDPG in vehicle-following control.

**Keywords:** subsection proximal policy optimization; weighted importance sampling; TORCS; vehicle-following; autonomous driving

**Citation:** Huang, Y.; Xu, X.; Li, Y.; Zhang, X.; Liu, Y.; Zhang, X. Vehicle-Following Control Based on Deep Reinforcement Learning. *Appl. Sci.* **2022**, *12*, 10648. https://doi.org/10.3390/app122010648

Academic Editor: Giancarlo Mauri

Received: 7 August 2022
Accepted: 22 September 2022
Published: 21 October 2022

**Publisher's Note:** MDPI stays neutral with regard to jurisdictional claims in published maps and institutional affiliations.

## 1. Introduction

Nowadays, with the rapid development of autonomous driving technology, an increasing number of enterprises and universities are investing in the research and development of autonomous driving technology, and the future mode of travel will bear great changes. However, the current autonomous driving technology is not mature enough, and there are many areas that need to be developed. Especially in vehicle-intensive roads of city environments, traffic congestion is a frequently encountered situation, and the frequent start and stop of vehicles and instabilities in vehicle speed lead to a large number of front-end collision accidents. Therefore, a safe vehicle-following control method is of great significance for driving safety and alleviating traffic congestion.

Vehicle following is the most basic microscopic driving behavior in vehicle driving. It mainly deals with the interaction between the front and rear vehicles when the vehicles are platooning in a single lane [1]. It includes longitudinal control and lateral control. There are various methods of vehicle-following control, such as Model-Predictive-Control (MPC) [2], Proportional Integral Derivative (PID) control [3], fuzzy control method [4] and methods based on deep neural networks [5,6]. The deep network-based learning control method has been studied in dealing with complex road scenes. The methods based on deep networks can be roughly divided into two categories, supervised learning methods that use expert data to train deep networks [7], and deep reinforcement learning methods that explore and find high-reward strategies continuously during interacting with the environment [8]. The former trains the controller by collecting a large amount of expert driving data, while the latter trains the control policy by continuous exploration and trial-and-error in the environment.

The data for training using expert data need to be prepared with much human efforts. It is difficult to manually assess and screen unsafe driving data, which can lead to extreme security risks. Deep reinforcement learning can obtain optimized driving policies in a self-learning way. It learns through continuous exploration in the environment. Exploratory ability determines the ability to learn from the environment.

Most accidents that occur when vehicle-following on densely packed roads in a city environment are mainly in the stop–start phase, while the accident rate is lower in the slow-moving phase. Therefore, this paper proposes to divide the entire vehicle-following process into two stages: start–stop, steady driving. And we modify the training of the single policy network of the PPO algorithm to train two policy networks corresponding to different vehicle-following stages, respectively. Furthermore, the PPO algorithm uses the importance sampling method to estimate the distribution of the advantage function. We found that using the importance sampling method to estimate the distribution can lead to inconsistency, resulting in a large variance, which is detrimental to the optimization efficiency of the policy. Hence this paper proposes to use weighted importance sampling instead of importance sampling, which can effectively reduce the variance between the resampled data distribution and the real distribution, and improve the training efficiency of the policy network. Our work contributions can be summarized as follows:

- According to the characteristics of vehicle-following in dense urban roads, the vehicle-following process is divided into two stages: start–stop, steady driving. In order to improve the PPO algorithm, a subsection policy optimization algorithm (Subsection-PPO) is proposed to use two different policy networks for training in different following stages.
- The weighted importance sampling method is used instead of the importance sampling method when estimating the objective function.
- In order to evaluate the effectiveness of the Section-PPO method in vehicle-following control, this paper uses the TORCS (The Open Racing Car Simulator) simulation environment to simulate urban traffic flow for verification. The experimental results show that the method has a very good effect in vehicle-following scenarios.

## 2. Related Work

Vehicle-following research has been an important research direction in traffic flow analysis and autonomous driving research. The related research on vehicle-following model can be traced back to the middle of the last century, and there have been many research advances since then. In terms of vehicle-following model research, the first vehicle-following model was proposed by Pips [9] and was widely used in the description of vehicle flow. Later, different types of vehicle-following models based on different directions and fields were also proposed. Gazis et al. [10] jointly proposed the GM model, which is based on the driver's stimulus response while taking into account the safe distance. The subsequent stimulus-response vehicle-following models are mostly based on this model. The Gipps [11] model as the earliest safe distance model was proposed by Kometani and Sasaki. Based on the Gipps model, AyresTJ et al. [12] proposed a safety distance model based on the headway. Jamson et al. [13] proposed a driver state vehicle distance model based on kinematics analysis, and a safe distance model based on the difference of driver response delay in emergency state. Treiber, Helbing et al. proposed the classic IDM model [14]. The IDM model can fully describe the change of vehicle-following behavior from free flow to congested state with a unified structure. With the development of neural networks, the use of fuzzy logic to establish a neural network vehicle-following model has been realized. The results of the neural network vehicle-following model proposed by Mathew et al. [15] after comparing with Gipps show that its prediction accuracy is higher than the latter. The vehicle-following model based on sequence-to-sequence proposed by Sharma et al. [16] has memory effect and response delay capabilities, which further expands the spatial expectation and improves the accuracy of platoon simulation and the stability of traffic flow. Li et al. [17] proposed a novel platoon formation and optimization model combining graph theory and safety potential field (G-SPF) theory which can form a collision-free platoon in a short time. Zhu and Zhang [18] proposed an improved forward-considered vehicle-following model that uses an average expected velocity field to describe the flow of autonomous vehicles. The new model has three key parameters: adjustable sensitivity, intensity factor, and average expected velocity field size, which in general bear a large impact on the stability and congestion state of autonomous vehicle flow.

The purpose of vehicle-following decision-making is to form a vehicle-following decision that can ensure the safety and rationality of vehicle-following according to the state information of the vehicle-following vehicle or the vehicle-following queue and certain decision-making methods. In terms of vehicle-following decision research, Li et al. [19] proposed an adaptive hierarchical control structure, in which the upper control layer is used to obtain the sliding mode control law of the required acceleration according to the inter-vehicle state information. In the lower control layer, switching logic with hysteresis boundaries is developed to ensure ride comfort, and the desired torque is calculated in real-time based on an inverse dynamics model to track the desired acceleration planned by the upper control layer. Zhang et al. [20] proposed a behavior estimation method based on contextual traffic information to identify and predict lane change intentions, and optimize the acceleration sequence by combining the lane change intentions of other vehicles. The above methods are based on traditional control methods and do not have the robustness to adapt to most scenarios, so many teams have turned their attention to deep reinforcement learning. The intelligent vehicle-following process can be abstracted into a state transition process that conforms to the Markovian property [21], so it is possible to use the deep reinforcement learning realize the vehicle-following. Guerrieri et al. [22] proposed a new automatic traffic data acquisition method mom-dl based on the deep learning method and yolov3 algorithm. This method can automatically detect vehicles in the traffic flow and estimate the traffic variable flow, spatial average speed and vehicle density of the expressway under static and uniform traffic conditions. Masmoudi et al. [23] used the vision algorithm YOLO to identify the current state, the reinforcement learning algorithm Q-learning and the DQN algorithm to control the following vehicles. After conducting a simulation experiment, they concluded that the following vehicle can make reasonable decisions. However, the Q-learning algorithm has the problem of dimension explosion in continuous problems such as vehicle-following, so Zhu et al. [24] and others chose the DDPG [25] algorithm that can output continuous actions to improve and verify it in the vehicle-following scene, while showing good generalization ability. Reinforcement learning shows great potential in sequential decision optimization problems, but there are certain difficulties in the design of reward functions. Gao et al. [26] used an inverse reinforcement learning algorithm to establish a reward function for each driver's data, and analyzed the driving characteristics and following policy, and subsequent simulations in a highway environment demonstrated the effectiveness of the method.

### 3. Problem Formulation

Vehicle-following on city roads is different from cruise control since the vehicle will start, stop, and shift frequently. Therefore, it is necessary to adjust the accelerator or brake according to the state of the vehicle in front. The ability to maintain a safe driving distance is the most important indicator of vehicle-following control. Following is a random and interactive process, so vehicle-following control can be modeled as a Markov decision process (MDP). During the MDP process, the following vehicle needs to continuously observe the current state and make decisions. MDP can be represented by a tuple $\{S, A, P, R\}$, where $S$ is a set of states. We divide state $S$ into two states as the input of the policy network, namely the start–stop state $S_{start-stop}$ and steady state $S_{\text{steady}}$. $A$ is the action set and we divide $A$ into start–stop phase action $A_{start-stop}$ and stable phase action $A_{\text{steady}}$. $P$ is the state transition probability, and $R$ is the immediate reward obtained after performing action $A$. Figure 1 shows a schematic diagram of car following control. Below, we will introduce the state space and action space in the process of vehicle-following.

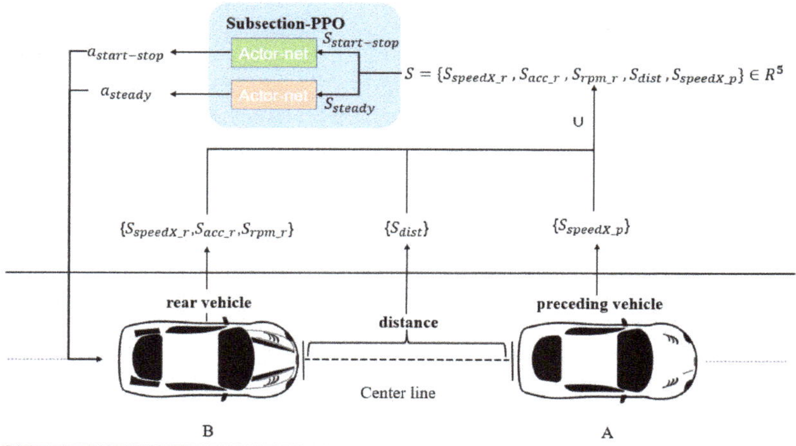



**Figure 1.** Vehicle-following control.

### 3.1. State Space

Vehicle-following constantly explores the environment for learning, so it needs to continuously obtain the current state as input. The simulation environment in this paper is TORCS, and Table 1 shows the state space.

Because the state space data obtained by the sensor bear different dimensions, we adopt normalization to [0,1] in the experimental stage to eliminate the adverse influence caused by the singular sample data.

**Table 1.** State space.

| Sensor | Range | Description |
|---|---|---|
| $speedX\_r$ | $(-\infty, +\infty)\,(\mathrm{km/h})$ | Speed along the longitudinal axis of the vehicle (driving direction of the rear vehicle) |
| $speedX\_p$ | $(-\infty, +\infty)\,(\mathrm{km/h})$ | Speed along the longitudinal axis of the vehicle (direction of travel of the vehicle in front) |
| $acc\_r$ | $(-\infty, +\infty)\,(\mathrm{km/h})$ | Acceleration of rear vehicle |
| $dist$ | $(-\infty, +\infty)\,(\mathrm{m/s^2})$ | Spacing between vehicles |
| $rpm\_r$ | $(0, +\infty)\,(\mathrm{rpm})$ | Speed along the Z-axis of the vehicle |

### 3.2. Action Space

Intelligent vehicle-following as longitudinal control requires reasonable control of the accelerator opening and braking force to maintain a safe and stable vehicle spacing. The accelerator opening and braking force constitute the action space vector, as shown in Table 2.

**Table 2.** Action Space.

| Action | Range | Description |
|---|---|---|
| *acceleration* | $[0,1]$ | Throttle force 0 means not to step on the throttle, 1 means fully depressed. |
| *brake* | $[0,1]$ | Braking force, 0 means no braking, 1 means fully depressed. |

*3.3. Reward Function*

The reward function $R : S \times A \times S' \rightarrow \mathbb{R}$ in reinforcement learning is an incentive mechanism that enables the agent to learn a behavioral strategy to meet the ultimate goal. Two policy networks are used in this paper but need to maintain a consistent estimate of the advantage function, so a reward function is used uniformly: $R = \gamma_1 \times \alpha - \gamma_2 \times \beta - \eta \times acc\_r$

- where $\alpha = \begin{cases} dist, \text{ if } speedX\_r \times T_{mth} \leq dist \leq speedX\_r \times T_{mth} + 2 \\ 0, \text{ others} \end{cases}$, $\gamma_1 = 1$. A positive reward is given when the vehicle maintains a safe spacing $[speedX\_r \times T_{mth}, speedX\_r \times T_{mth} + 2]$, and $T_{mth} = 2$ s is the Minimum Time Headway.
- $\beta = | speedX\_r - speedX\_p |$, $\gamma_2 = 0.2$. On vehicle-intensive roads, the vehicle speed is generally less than 30 km/h. In this paper, the maximum speed of the preceding vehicle is set as $speedX\_p = 25$ km/h $\approx 7$ m/s. Since the vehicle uses the sensor to obtain the current state information, the accuracy and speed are much higher than that of a human driver. Maintaining a similar speed can stably follow the distance to ensure safety. Therefore, the speed of the vehicle is kept as consistent as possible in the experiment.
- $acc\_r$ is the vehicle acceleration, $\eta = 0.05$.

## 4. Methodology

Intelligent vehicle-following control can be described as a Markov decision process. This paper proposes the Subsection-PPO algorithm, which divides a set of trajectories into a start–stop part and a steady part and uses a weighted importance sampling method to calculate the objective function. In order to provide the training vehicle with initial power and exploration capability during the training phase, this paper uses Ornstein–Uhlenbeck (OU) [27] noise with random process.

*4.1. Noise*

Since the use of time-series-related noise can increase the exploration efficiency, this paper adopts the time-series-related Ornstein–Uhlenbeck (OU) noise. OU noise is a stochastic process and its differential equation is as follows:

$$\mathcal{N}_t = -\theta(x_t - \mu)dt + \sigma dW_t \tag{1}$$

where $x_t$ is usually one dimension of agent action, $\mu \in \mathbb{R}$ represents the mean value of action, $\theta > 0$ and $\sigma > 0$, $dW_t = W_t - W_s \sim (0, (t - s))$ is Wiener process. The magnitude of $\theta$ is directly proportional to the degree that $a_t$ tends to $\mu$, and $\sigma$ is the magnification of the disturbance in the Wiener process.

*4.2. Proximal Policy Optimization Algorithm*

The proximal policy optimization algorithm is a reinforcement learning algorithm based on policy gradient, which is evolved from the trust region policy optimization (TRPO) [28]. If the agent's reward in the environment is higher, it means that they have a stronger ability to complete tasks, and the ultimate goal of all policy gradient methods is to maximize the cumulative reward, that is, to maximize $\eta(\pi) = \mathbb{E}_{s_0,a_0,s_1,a_1\ldots}\left[\sum_{t=0}^{\infty} \gamma^t r(s_t)\right]$, among them, $\gamma$ is the discount factor, indicating that the farther away from the current state, the smaller the impact on the current state, $\eta(\pi)$ refers to the cumulative reward obtained when performing actions according to the policy $\pi$. $S_0, S_1, S_2 \ldots$ represents the state transition of the agent in the environment, treat state transitions as a given distribution $s \sim \rho(s)$. The TRPO algorithm proposes to use the advantage function $A_\pi(s,a) = Q_\pi(s,a) - V_\pi(s)$ to evaluate the quality of executing an action, where $Q_\pi(s,a)$ is the value-action pair, and $V_\pi(s)$ is the state value. It has been proved that the cumulative reward of a new policy $\pi_{\text{new}}$ can be expressed as:

$$\eta(\pi_{\text{new}}) = \eta(\pi) + \mathbb{E}_{s_0,a_0,\cdots \sim \pi_{\text{nev}}} \left[ \sum_{t=0}^{\infty} \gamma^t A_{\pi_{\text{old}}}(s_t, a_t) \right] \tag{2}$$

That is, the cumulative reward of the old policy plus the cumulative advantage function of the new policy.

Therefore, if $\mathbb{E}_{s_0,a_0,\cdots \sim \pi_{\text{nev}}} \left[ \sum_{t=0}^{\infty} \gamma^t A_{\pi_{\text{old}}}(s_t, a_t) \right]$ can be guaranteed to be greater than or equal to 0, the monotonic increase of the cumulative reward can be guaranteed, that is, the optimization of the strategy. Since the cumulative advantage function cannot be calculated directly, TRPO uses the importance sampling method to estimate the advantage function and uses the KL divergence to limit the update range of the policy to ensure the monotony of the cumulative reward, that is, to ensure the continuous optimization of policy. The core problem of TRPO algorithm is defined as:

$$\text{maximize}_\theta \ \mathbb{E}_{s \sim \pi_{\theta_{old}}, a \sim \pi_{\theta_{old}}} \left[ \frac{\pi_\theta(a|s)}{\pi_{\theta_{old}}(a|s)} A_{\theta_{old}}(s,a) \right]$$
$$\text{subject to } \mathbb{E}_{s \sim \pi_{\theta_{old}}} \left[ D_{KL} \left( \pi_{\theta_{old}}(\cdot \mid s) \| \pi_\theta(\cdot \mid s) \right) \right] \leq \delta \tag{3}$$

The TRPO algorithm uses the KL divergence method to calculate the confidence region, and the update range of the control policy network is within a certain range to ensure that the value of the cumulative advantage function is greater than or equal to 0. However, the method of updating the policy in the trust region is complex and inefficient, so the PPO algorithm proposes to use the method of clipping the importance weight to limit the updating range of the policy. The maximizing objective "replacement" function of the PPO algorithm is:

$$L^{CPI}(\theta) = \mathbb{E}_t \left[ \frac{\pi_\theta(a_t \mid s_t)}{\pi_{\theta_{old}}(a_t \mid s_t)} A_t \right] = \mathbb{E}_t[r_t(\theta) A_t] \tag{4}$$

where, the superscript $CPI$ indicates conservative strategy iteration [29] , $r_t(\theta) = \frac{\pi_\theta(a_t|s_t)}{\pi_{\theta_{old}}(a_t|s_t)}$ is the importance weight obtained by importance sampling. The objective function of the PPO algorithm is finally:

$$L^{CLIP}(\theta) = \mathbb{E}_t[\min(r_t(\theta) A_t, \text{clip}(r_t(\theta), 1 - \epsilon, 1 + \epsilon) A_t)] \tag{5}$$

The value of $r_t(\theta)$ is limited to $[1 - \epsilon, 1 + \epsilon]$, where $\epsilon$ is a hyperparameter.

*4.3. Subsection-PPO*

In urban congested roads, due to the high density of vehicles and the slow speed, frequent starts and stops are often caused, and accidents also occur. Therefore, this paper proposes to divide vehicle-following into two stages: start–stop and steady driving. Therefore, the collected trajectories $\tau = (s_1, a_1, r_1, s_2, a_2, r_2, \cdots, s_n, a_n, r_n)$ data need to be divided into two categories and processed separately. The original data are divided into two categories, the algorithm also needs to make corresponding changes. Hence this paper proposes the Subsection-PPO algorithm on the basis of the proximal policy optimization algorithm (PPO), using two Actor networks to learn different stages policy, as shown in Figure 2. It is worth noting that in order to ensure consistent state value estimation, we only use one Critic network to estimate the state value of the two stages. Afterwards, simulation experiments show that this method is effective. The following will introduce the division method of different stages:

- start-stop stage $\begin{cases} \text{start, when } acc\_p > 0 \ \&\& \\ dist \notin [speedX\_r * T_{mth}, speedX\_r * T_{mth} + 2] \\ \text{stop, when } acc\_p \leq 0 \ \&\& \\ dist \notin [speedX\_r * T_{mth}, speedX\_r * T_{mth} + 2] \end{cases}$

  The trajectories generated in this stage are start–stop data.
- steady stage: when $dist \in [speedX\_r * T_{mth}, speedX\_r * T_{mth} + 2]$

  The trajectories generated in this stage are steady driving data.

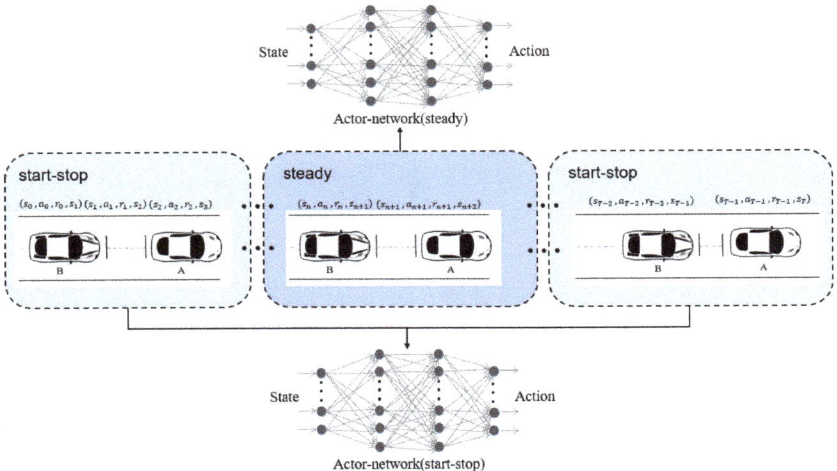


**Figure 2.** Actor-network at different stages.

In the PPO algorithm, the importance sampling method changes the algorithm from on-policy to off-policy, which improves the utilization of data, and controls the update region of the policy network by clipping the importance weight to ensure that the update process is monotonous unabated. Although the importance sampling method is unbiased in estimating the distribution of the data, using a new distribution to estimate the old distribution will lead to large variance, so the model training efficiency is not as good as the sampling method that is both unbiased and consistent. To solve this problem, this paper proposes to use the weighted importance sampling method instead of the importance sampling method to estimate the actual objective function. Then, the final objective function is transformed from Equation (5) to:

$$L^{CPI}(\theta) = \mathbb{E}_t\left[\sum_{n=1}^{N} \frac{\omega^n}{\sum_{m=1}^{N} \omega^n} A_t\right] = \mathbb{E}_t[r_t(\theta)A_t] \tag{6}$$

where $\omega^n = \frac{\pi_{\theta_{old}}(a_t|s_t)}{\pi^n\theta(a_t|s_t)}$. The weighted importance sampling is both unbiased and consistent. Hence as the sampling volume increases, the method can render the estimated value increasingly close to the true distribution of the objective function. Figure 3 shows the overall framework of the subsection-PPO algorithm.

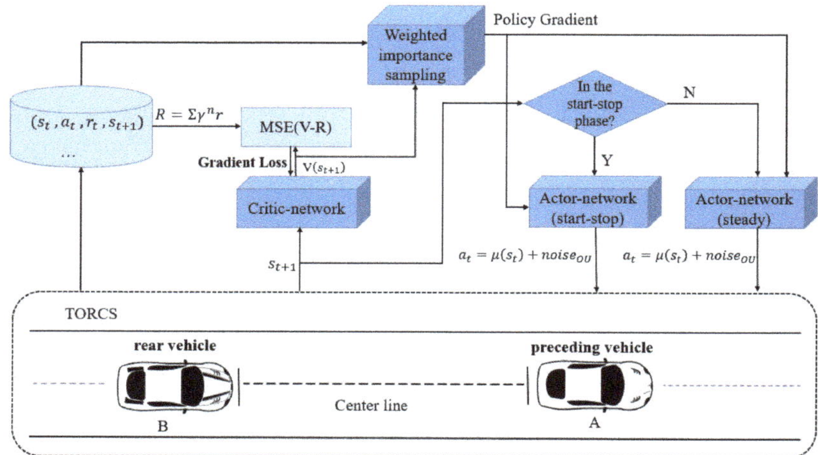



**Figure 3.** Subsection-PPO.

## 5. Experimental Simulation

This experiment is based on the TORCS (The Open Racing Car Simulator) simulation platform, which provides rich road data and comprehensive vehicle radars. We use the python language to implement the reinforcement learning code, use UDP protocol to achieve data interaction with the simulation platform and control the simulated vehicle. To verify the effectiveness of the algorithm proposed in this paper, we simulate the actual urban traffic flow in the TORCS simulation platform, and only consider the longitudinal control of the vehicle in the experiment. Our experiments include:

- The experiment compares the cumulative reward of the Subsection-PPO algorithm proposed in this paper with PPO and DDPG.
- The total distance of vehicle-following by different algorithms while maintaining a safe spacing is compared.
- The effect of the proposed algorithm in vehicle-following control is described using the relationship between speed and distance.

### 5.1. Hardware Configuration

The experimental platform employs an Ubuntu20.04 operating system with 16GB DDR4, the processor is Intel Core i5-10200h CPU @ 2.40 GHz sixteen core, and the graphics card is NVIDIA Quadro RTX 5000. The learning rates of actor_network (start–stop) and actor_network (steady) are both $lr_a = 3 \times 10^{-4}$. The learning rate of critic_network $lr\_c = 1 \times 10^{-3}$. Training timesteps are 10,000.

### 5.2. Experiment and Comparison

We choose to compare with two reinforcement learning algorithms, namely: PPO and DDPG. Both of their algorithms are based on the basic Actor–Critic architecture. In simple terms, the Actor network is used for action output, and the Critic network is used for state or action value evaluation.

Firstly, the comparison of cumulative rewards is essential. The change of cumulative rewards shows the exploration ability and learning ability of the reinforcement learning algorithm. The cumulative reward is opposite to the loss value. The higher the cumulative reward obtained after convergence, the better the learning of the agent. The perception of the environment is also richer.

In this paper, the method of random acceleration and braking is used to longitudinally control the preceding vehicle to simulate the state of the preceding vehicle, and the reinforcement learning method is used to control the following vehicle. The reward function is introduced in Section 3.3. The variation of cumulative reward during training phase is shown in Figure 4.

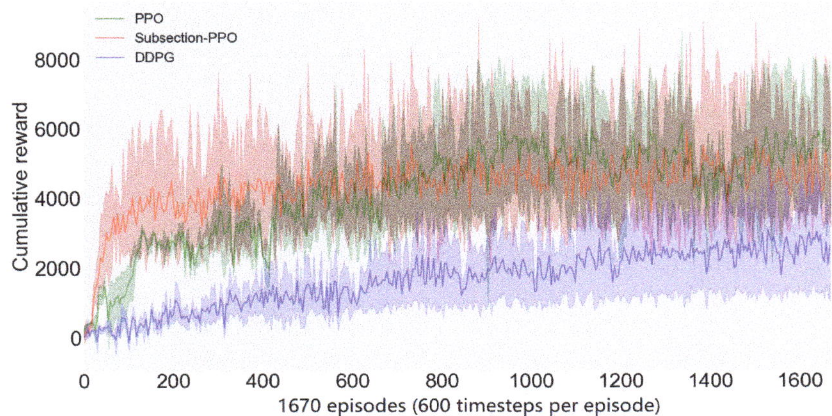


**Figure 4.** Cumulative rewards.

The experiment simulates vehicle-following in a crowded road, so the Minimum Time Headway (MTH) in the strong vehicle-following state is selected to calculate the safe vehicle-following spacing. It is considered safe when the vehicle-following state spacing is within the range $[speedX_{-}r \times T_{mth}, speedX_{-}r \times T_{mth} + 2]$. Figure 5 shows the driving distance of different control methods under the condition of maintaining safe spacing. The whole journey is 1600 m. Figure 6 shows the relationship between vehicle velocity and spacing.

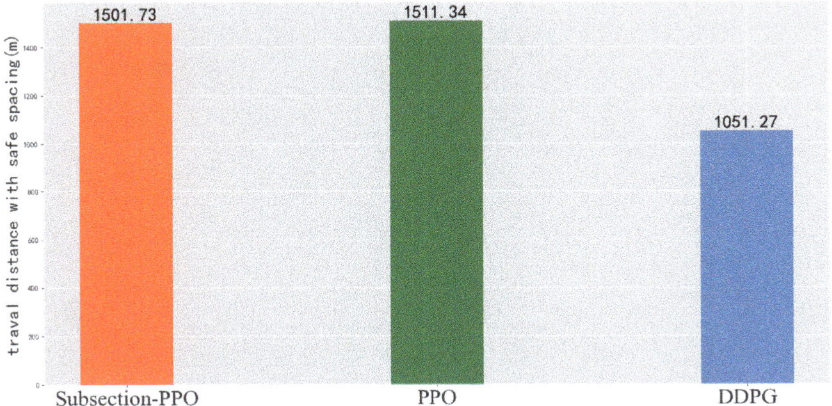


**Figure 5.** Driving distance with safe spacing.

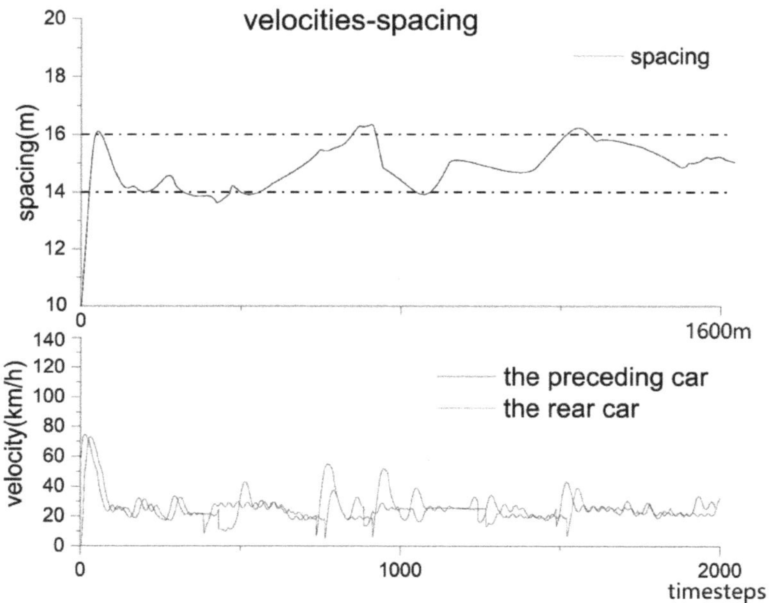


**Figure 6.** The dotted line indicates the spacing range in the ideal state. Initial vehicle spacing is 10 m. The distance traveled using the Subsection-PPO algorithm while maintaining a safe spacing accounted for 93.8% of the total mileage.

## 6. Conclusions

Based on the PPO algorithm, according to the characteristics of different stages of vehicle-following control we divide the trajectories into two parts: stop–start, steady driving. And we use the weighted importance sampling method instead of the importance sampling method. To sum up, we propose the Subsection-PPO algorithm for vehicle-following control. Subsection-PPO algorithm uses a dual actor network, but in order to avoid the training non convergence caused by inconsistent value estimates, we choose to employ a critic network for value estimation. The action vectors of different vehicle-following stages are calculated by the corresponding actor network, which makes our method well applicable to vehicle-following problems. Furthermore, the weighted importance sampling method improves the training efficiency. We simulate the vehicle-following situation of urban roads in the TORCS simulation environment and subsequently compare and verify the methods we propose. These results prove the feasibility and advantages of our proposed vehicle-following safety of the method. However, there are still shortcomings in our work. At this stage, the technology of autonomous driving is constantly developing. In the case of ensuring safety, it is necessary to consider the acceleration changes of the vehicle, which affects the energy consumption and ride comfort of the vehicle. This will be the direction of our future work.

**Author Contributions:** Y.H. wrote the manuscript and designed research methods; X.Z. (Xinglong Zhang), X.X. and X.Z. (Xiaochuan Zhang) edited and revised the manuscript; Y.L. (Yao Liu) and Y.L. (Yong Li) analyzed the data. All authors have read and agreed to the published version of the manuscript.

**Funding:** This work is supported by research on key technologies of internet of things platform for smart city (Grant No. 2020ZDXM12) of the key program and research on the basic support system of urban management comprehensive law enforcement (Grant No. 2021ZDXM17) of China Coal Technology Engineering Group Chongqing Research Institute.

*Appl. Sci.* **2022**, *12*, 10648

**Institutional Review Board Statement:** Not applicable.

**Informed Consent Statement:** Not applicable.

**Data Availability Statement:** The data used to support the findings of this study are available from the corresponding author upon request.

**Conflicts of Interest:** The authors declare no conflict of interest.

# References


1. Paschalidis, E.; Choudhury, C.F.; Hess, S. Combining driving simulator and physiological sensor data in a latent variable model to incorporate the effect of stress in car-following behaviour. *Anal. Methods Accid. Res.* **2019**, *22*, 100089. [CrossRef]
2. Liu, Z.; Yuan, Q.; Nie, G.; Tian, Y. A multi-objective model predictive control for vehicle adaptive cruise control system based on a new safe distance model. *Int. J. Automot. Technol.* **2021**, *22*, 475–487. [CrossRef]
3. Farag, W. Complex Trajectory Tracking Using PID Control for Autonomous Driving. *Int. J. Intell. Transp. Syst. Res.* **2019**, *18*, 356–366. [CrossRef]
4. Choomuang, R.; Afzulpurkar, N. Hybrid Kalman filter/fuzzy logic based position control of autonomous mobile robot. *Int. J. Adv. Robot. Syst.* **2005**, *2*, 20. [CrossRef]
5. Fayjie, A.R.; Hossain, S.; Oualid, D.; Lee, D.J. Driverless car: Autonomous driving using deep reinforcement learning in urban environment. In Proceedings of the IEEE 2018 15th International Conference on Ubiquitous Robots (UR), Honolulu, HI, USA, 26–30 June 2018; pp. 896–901. [CrossRef]
6. Colombaroni, C.; Fusco, G.; Isaenko, N. Modeling car following with feed-forward and long-short term memory neural networks. *Transp. Res. Procedia* **2021**, *52*, 195–202. [CrossRef]
7. Bhattacharyya, R.; Wulfe, B.; Phillips, D.; Kuefler, A.; Morton, J.; Senanayake, R.; Kochenderfer, M. Modeling human driving behavior through generative adversarial imitation learning. *arXiv* **2020**, arXiv:2006.06412. [CrossRef]
8. Lin, Y.; McPhee, J.; Azad, N.L. Longitudinal dynamic versus kinematic models for car-following control using deep reinforcement learning. In Proceedings of the 2019 IEEE Intelligent Transportation Systems Conference (ITSC), Auckland, New Zealand, 27–30 October 2019; pp. 1504–1510. [CrossRef]
9. Pipes, L.A. An operational analysis of traffic dynamics. *J. Appl. Phys.* **1953**, *24*, 274–281. [CrossRef]
10. Gazis, D.C.; Herman, R.; Potts, R.B. Car-following theory of steady-state traffic flow. *Oper. Res.* **1959**, *7*, 499–505. [CrossRef]
11. Cattin, J.; Leclercq, L.; Pereyron, F.; El Faouzi, N.E. Calibration of Gipps' car-following model for trucks and the impacts on fuel consumption estimation. *IET Intell. Transp. Syst.* **2019**, *13*, 367–375. [CrossRef]
12. Ayres, T.; Li, L.; Schleuning, D.; Young, D. Preferred time-headway of highway drivers. In Proceedings of the ITSC 2001, Oakland, CA, USA, 25–29 August 2001; 2001 IEEE Intelligent Transportation Systems. Proceedings (Cat. No. 01TH8585); pp. 826–829. [CrossRef]
13. Jamson, A.H.; Merat, N. Surrogate in-vehicle information systems and driver behaviour: Effects of visual and cognitive load in simulated rural driving. *Transp. Res. Part F Traffic Psychol. Behav.* **2005**, *8*, 79–96. [CrossRef]
14. Treiber, M.; Kesting, A. Traffic flow dynamics: data, models and simulation. *Phys. Today* **2014**, *67*, 54.
15. Mathew, T.V.; Ravishankar, K. Neural Network Based Vehicle-Following Model for Mixed Traffic Conditions. *Eur. Transp.-Trasp. Eur.* **2012**, *52*, 1–15.
16. Sharma, O.; Sahoo, N.; Puhan, N. Highway Discretionary Lane Changing Behavior Recognition Using Continuous and Discrete Hidden Markov Model. In Proceedings of the 2021 IEEE International Intelligent Transportation Systems Conference (ITSC), Indianapolis, IN, USA, 19–22 September 2021; pp. 1476–1481. [CrossRef]
17. Li, L.; Gan, J.; Qu, X.; Mao, P.; Yi, Z.; Ran, B. A novel graph and safety potential field theory-based vehicle platoon formation and optimization method. *Appl. Sci.* **2021**, *11*, 958. [CrossRef]
18. Zhu, W.X.; Zhang, L.D. A new car-following model for autonomous vehicles flow with mean expected velocity field. *Phys. A: Stat. Mech. Its Appl.* **2018**, *492*, 2154–2165. [CrossRef]
19. Li, W.; Chen, T.; Guo, J.; Wang, J. Adaptive car-following control of intelligent electric vehicles. In Proceedings of the 2018 IEEE 4th International Conference on Control Science and Systems Engineering (ICCSSE), Wuhan, China, 21–23 August 2018; pp. 86–89. [CrossRef]
20. Zhang, Y.; Lin, Q.; Wang, J.; Verwer, S.; Dolan, J.M. Lane-change intention estimation for car-following control in autonomous driving. *IEEE Trans. Intell. Veh.* **2018**, *3*, 276–286. [CrossRef]
21. Kamrani, M.; Srinivasan, A.R.; Chakraborty, S.; Khattak, A.J. Applying Markov decision process to understand driving decisions using basic safety messages data. *Transp. Res. Part Emerg. Technol.* **2020**, *115*, 102642. [CrossRef]
22. Guerrieri, M.; Parla, G. Deep learning and yolov3 systems for automatic traffic data measurement by moving car observer technique. *Infrastructures* **2021**, *6*, 134. [CrossRef]
23. Masmoudi, M.; Friji, H.; Ghazzai, H.; Massoud, Y. A Reinforcement Learning Framework for Video Frame-based Autonomous Car-following. *IEEE Open J. Intell. Transp. Syst.* **2021**, *2*, 111–127. [CrossRef]
24. Zhu, M.; Wang, X.; Wang, Y. Human-like autonomous car-following model with deep reinforcement learning. *Transp. Res. Part C Emerg. Technol.* **2018**, *97*, 348–368. [CrossRef]

25. Silver, D.; Lever, G.; Heess, N.; Degris, T.; Wierstra, D.; Riedmiller, M. Deterministic policy gradient algorithms. In Proceedings of the International Conference on Machine Learning (PMLR), Beijing, China, 21–26 June 2014; pp. 387–395.
26. Gao, H.; Shi, G.; Xie, G.; Cheng, B. Car-following method based on inverse reinforcement learning for autonomous vehicle decision-making. *Int. J. Adv. Robot. Syst.* **2018**, *15*, 1729881418817162. [CrossRef]
27. Ngoduy, D.; Lee, S.; Treiber, M.; Keyvan-Ekbatani, M.; Vu, H. Langevin method for a continuous stochastic car-following model and its stability conditions. *Transp. Res. Part C Emerg. Technol.* **2019**, *105*, 599–610. [CrossRef]
28. Schulman, J.; Levine, S.; Abbeel, P.; Jordan, M.; Moritz, P. Trust region policy optimization. In Proceedings of the International Conference on Machine Learning, PMLR, Lille, France, 6–11 July 2015; pp. 1889–1897. [CrossRef]
29. Kakade, S.; Langford, J. Approximately optimal approximate reinforcement learning. In Proceedings of the 19th International Conference on Machine Learning, Citeseer, Sydney, Australia, 8–12 July 2002.

*applied*
*sciences*

*Article*

# A Novel Mixed-Attribute Fusion-Based Naive Bayesian Classifier

Guiliang Ou[1], Yulin He [1,2,*], Philippe Fournier-Viger [1,2] and Joshua Zhexue Huang [1,2]

[1] Big Data Institute, College of Computer Science and Software Engineering, Shenzhen University, Shenzhen 518060, China
[2] Guangdong Laboratory of Artificial Intelligence and Digital Economy (SZ), Shenzhen 518107, China
* Correspondence: yulinhe@gml.ac.cn; Tel.: +86-185-3131-5747

**Abstract:** The Naive Bayesian classifier (NBC) is a well-known classification model that has a simple structure, low training complexity, excellent scalability, and good classification performances. However, the NBC has two key limitations: (1) it is built upon the strong assumption that condition attributes are independent, which often does not hold in real-life, and (2) the NBC does not handle continuous attributes well. To overcome these limitations, this paper presents a novel approach for NBC construction, called mixed-attribute fusion-based NBC (MAF-NBC). It alleviates the two aforementioned limitations by relying on a mixed-attribute fusion mechanism with an improved autoencoder neural network for NBC construction. MAF-NBC transforms the original mixed attributes of a data set into a series of encoded attributes with maximum independence as a pre-processing step. To guarantee the generation of useful encoded attributes, an efficient objective function is designed to optimize the weights of the autoencoder neural network by considering both the encoding error and the attribute's dependence. A series of persuasive experiments was conducted to validate the feasibility, rationality, and effectiveness of the designed MAF-NBC approach. Results demonstrate that MAF-NBC has superior classification performance than eight state-of-the-art Bayesian algorithms, namely the discretization-based NBC (Dis-NBC), flexible naive Bayes (FNB), tree-augmented naive (TAN) Bayes, averaged one-dependent estimator (AODE), hidden naive Bayes (HNB), deep feature weighting for NBC (DFW-NBC), correlation-based feature weighting filter for NBC (CFW-NBC), and independent component analysis-based NBC (ICA-NBC).

**Keywords:** naive Bayesian classifier; attribute independence assumption; mixed-attribute classification; conditional probability; Bayesian network; attribute transformation

**Citation:** Ou, G.; He, Y.; Fournier-Viger, P.; Huang, J.Z. A Novel Mixed-Attribute Fusion-Based Naive Bayesian Classifier. *Appl. Sci.* **2022**, *12*, 10443. https://doi.org/10.3390/app122010443

Academic Editor: Xinglong Zhang, Pengfei Jia and Yue Wu

Received: 26 September 2022
Accepted: 10 October 2022
Published: 17 October 2022

**Publisher's Note:** MDPI stays neutral with regard to jurisdictional claims in published maps and institutional affiliations.

## 1. Introduction

As one of the top 10 algorithms in the fields of data mining and machine learning [1], the naive Bayesian classifier (NBC) has been used in numerous domains. The main advantage of the NBC is its simple model structure that makes it easy to implement and its good theoretical interpretability. In recent years, the NBC also received much attention from the industry and academia since it can be easily deployed in distributed environments to process big data. Despite possessing several desirable properties, the NBC is built upon a strong assumption, called the attribute independence assumption, which states that condition attributes must be mutually independent with respect to the decision attribute. This assumption simplifies the calculation of posterior probabilities (the probabilities that attribute values of a sample are all observed for a given class). Rather than computing a posterior probability as a joint probability of condition attributes, the NBC calculates it as the product of multiple marginal probabilities. This makes the computation of NBC very efficient and allows the estimation of probabilities even with small data sets. However, the attribute independence assumption does not hold in many real-life data sets, which substantially limits the prediction performance of the NBC. Recent studies on the NBC mainly focused on finding ways to relax the independence assumption so as to further improve its generalization performance. There are two main approaches for improving the testing ability of the NBC [2]: (1) improving the structure of models and (2) performing data transformations. The former approach introduces complex Bayesian network structures to model attribute dependencies, while the latter applies feature selection or feature extraction methods to select independent attributes. Some representative studies of these two approaches are described next.

Model structure-oriented improvement methods use different estimation strategies (e.g., density estimation, Bayesian network, and attribute weighting) to estimate the conditional probability that a sample's attribute values are observed given its class. The flexible naive Bayes (FNB) [3] relies on kernel density estimations to determine a class's conditional probabilities. However, FNB can only handle continuous attributes and does not offer a solution to relax the attribute independence assumption. The tree-augmented naive (TAN) Bayes [4] algorithm is a semi-naïve Bayesian learning method that relaxes the attribute independence assumption by employing a tree structure, where each condition attribute only depends on the decision attribute and at most one condition attribute. The averaged one-dependent estimator (AODE) [5] is an ensemble classifier that applies an attribute selection algorithm to construct a series of one-dependent classifiers. Each classifier is a simple Bayesian network that is obtained by averaging all one-dependence classifiers. The hidden naive Bayes (HNB) classifier [6] uses the weighted sum of two-attribute dependencies to represent multiple-attribute dependence, where the weights are determined based on the mutual information between two condition attributes. The deep feature weighting for NBC (DFW-NBC) [7] technique and correlation-based feature weighting filter for NBC (CFW-NBC) [8] are two attribute weighting-based NBCs. DFW-NBC estimates each conditional probability of the NBC by deeply computing attribute weighted frequencies from training data while CFW-NBC weights the condition attributes by considering both the mutual relevance and the average mutual redundancy.

Data transformation-oriented improvement methods focus on the deep exploration of the training data to avoid complex and time-consuming structure learning. The NBCs obtained by this approach still rely on the independence assumption; thus, the performance may be degraded if the training data do not satisfy that assumption. Feature extraction techniques are the most commonly used to transform data with attribute dependencies into data that have no dependencies. Bressan and Vitria [9] applied class-conditional independent component analysis (CC-ICA) to improve the classification performance of the NBC. Qin et al. [10] proposed an ICA-based NBC (ICA-NBC) to improve the classification performance of the NBC. Fan and Poh [11] evaluated the classification performance of the NBC built using three feature extraction methods, namely principal component analysis (PCA), ICA, and CC-ICA. Experimental results have shown that PCA, ICA, and CC-ICA can slightly increase the testing accuracies of the NBC on the selected data sets. Jayanthi and Sasikala [12] trained an NBC based on website attributes extracted by PCA to perform web link spam detection. Zhang et al. [13] applied PCA to extract key attributes in network data and then trained the NBC to conduct network intrusion detection.

Although the aforementioned methods can improve the classification performance of the NBC for specific application scenarios, these methods do not provide a good solution for handling mixed attributes (continuous and categorical attributes). Some studies [14,15] revealed that the discretization of continuous attributes can lead to losing precious information about the original data set, while the determination of an optimal Bayesian network structure is an NP-hard problem [16]. To address the above limitations, this paper proposes a new strategy to enhance the generalization capability of the NBC for the mixed-attribute classification problem. The proposed approach not only retains as much information as possible about the original data but also keeps the model structure as simple as possible. The contributions of this paper are summarized as follows. A new mixed attribute fusion-based NBC (MAF-NBC) is proposed for mixed-attribute classification problems. To relax the attribute independence assumption, an autoencoder neural network (ANN) based on the minimization of both encoding error and attribute dependence is iteratively trained to transform the original mixed attributes into a series of independent and encoded attributes. Extensive experiments conducted on multiple data sets demonstrate the feasibility, rationality, and effectiveness of the MAF-NBC.

The remainder of this paper is organized as follows. Section 2 reviews the principles of the NBC. Section 3 presents the proposed MAF-NBC. Section 4 reports results from experiments to assess the performance of MAF-NBC. Finally, Section 5 concludes this paper and describes future studies.

## 2. Naive Bayesian Classifier for Mixed-Attribute Classification

Let there be a classification data set $\mathbb{D} = \bigcup_{m=1}^{\mathcal{M}} \mathbb{D}^{(m)}$ containing $\mathcal{N}$ samples. Each sample has $\mathcal{D}$ condition attributes including $\mathcal{D}_1$ discrete attributes and $\mathcal{D}_2$ continuous attributes. Moreover, all samples are divided into $\mathcal{M}$ different classes, where $\mathcal{N}_m$ denotes the number of samples that belong to the $m$-th ($m = 1, 2, \cdots, \mathcal{M}$) class.

$$\mathbb{D}^{(m)} = \left\{ \begin{array}{c} \left( x_n^{(m)}, y_n^{(m)} \right) \Big| x_n^{(m)} = \left( a_{n1}^{(m)}, \cdots, a_{n\mathcal{D}_1}^{(m)}, b_{n1}^{(m)}, \cdots, b_{n\mathcal{D}_2}^{(m)} \right), \\ y_n^{(m)} = w_m, n = 1, 2, \cdots, \mathcal{N}_m \end{array} \right\}, \tag{1}$$

$\sum_{m=1}^{\mathcal{M}} \mathcal{N}_m = \mathcal{N}, \mathcal{D}_1 + \mathcal{D}_2 = \mathcal{D}, a_{ni}^{(m)}, i \in \{1, 2, \cdots, \mathcal{D}_1\}$ is the $i$-th discrete attribute value, $b_{nj}^{(m)}, j \in \{1, 2, \cdots, \mathcal{D}_2\}$ is the $j$-th continuous attribute value, and $\{w_1, w_2, \cdots, w_{\mathcal{M}}\}$ is the class label set. The next paragraphs explain the principles of the naive Bayesian classifier (NBC) for the mixed-attribute classification problem on data set $\mathbb{D}$.

Assume that there is a new sample $x = \left( a_1, \cdots, a_{\mathcal{D}_1}, b_1, \cdots, b_{\mathcal{D}_2} \right)$. The NBC determines its class label by using this equation:

$$\begin{aligned} y &= \underset{m=1,2,\cdots,\mathcal{M}}{\arg\max} \ \mathrm{P}(w_m|x) \\ &= \underset{m=1,2,\cdots,\mathcal{M}}{\arg\max} \ \frac{\mathrm{P}(x|w_m)\mathrm{P}(w_m)}{\mathrm{P}(x)}, \\ &\propto \underset{m=1,2,\cdots,\mathcal{M}}{\arg\max} \ \mathrm{P}(x|w_m)\mathrm{P}(w_m) \end{aligned} \tag{2}$$

where $\mathrm{P}(w_m|x)$ is the posterior probability, $\mathrm{P}(w_m)$ is the prior probability, and $\mathrm{P}(x|w_m)$ is the conditional probability. Generally, the prior probability in Equation (1) can be calculated as follows.

$$\mathrm{P}(w_m) = \frac{\mathcal{N}_m}{\mathcal{N}}. \tag{3}$$

The key of training an NBC for the mixed-attribute classification problem is to calculate the conditional probability based on the independent attribute assumption as follows.

$$\begin{aligned} \mathrm{P}(x|w_m) &= \mathrm{P}\left( a_1, \cdots, a_{\mathcal{D}_1}, b_1, \cdots, b_{\mathcal{D}_2}|w_m \right) \\ &= \left[ \prod_{i=1}^{\mathcal{D}_1} \mathrm{P}(a_i|w_m) \right] \left[ \prod_{j=1}^{\mathcal{D}_2} \mathrm{P}(b_j|w_m) \right]. \end{aligned} \tag{4}$$

Conditional probability $\mathrm{P}(a_i|w_m)$ corresponding to a discrete attribute value $a_i$ is calculated as follows:

$$\mathrm{P}(a_i|w_m) = \frac{\sum_{n=1}^{\mathcal{N}_m} \mathrm{I}\left( a_i, a_{ni}^{(m)} \right)}{\mathcal{N}_m}, \tag{5}$$

where

$$\mathrm{I}(u,v) = \left\{ \begin{array}{ll} 1, & \text{if } u = v \\ 0, & \text{if } u \neq v \end{array} \right. \tag{6}$$

is an indicator function that is used to count the frequency of $a_i$ in the $i$-th continuous attribute values $a_{1i}^{(m)}, a_{2i}^{(m)}, \cdots, a_{\mathcal{N}_{m},j}^{(m)}$ corresponding to samples belonging to the $m$-th class. John and Langley [3] constructed a flexible NBC (FNBC), which used the kernel density estimation technique [17] to calculate the term of $P(b_j|w_m)$ for the continuous attribute value in Equation (3) as follows:

$$P(b_j|w_m) \propto p(b_j|w_m) = \frac{1}{\mathcal{N}_m} \sum_{n=1}^{\mathcal{N}_m} \frac{1}{\sqrt{2\pi}h_j} \exp\left[-\frac{1}{2}\left(\frac{b_j - b_{nj}^{(m)}}{h_j}\right)^2\right] \tag{7}$$

where $p(b_j|w_m)$ is the estimated probability density function (*p.d.f.*) value of $b_j$ based on the $j$-th continuous attribute values $b_{1j}^{(m)}, b_{2j}^{(m)}, \cdots, b_{\mathcal{N}_{m},j}^{(m)}$ corresponding to samples from the $m$-th class, and $h_j > 0$ ($h_j = \frac{1}{\sqrt{\mathcal{N}_m}}$ in [3]) is the bandwidth parameter. In addition, continuous attribute discretization can also be used to determine $p(b_j|w_m)$ as follows:

$$P(b_j|w_m) = P(c_j|w_m) = \frac{\sum\limits_{n=1}^{\mathcal{N}_m} I\left(c_j, c_{nj}^{(m)}\right)}{\mathcal{N}_m} \tag{8}$$

by transforming the continuous attribute values $b_j, b_{1j}^{(m)}, b_{2j}^{(m)}, \cdots, b_{\mathcal{N}_{m},j}^{(m)}$ into the discrete attribute values $c_j, c_{1j}^{(m)}, c_{2j}^{(m)}, \cdots, c_{\mathcal{N}_{m},j}^{(m)}$. This form of discretization-based NBC is called dis-NBC in this study.

### 3. Mixed-Attribute Fusion-Based Naive Bayesian Classifier

As stated in the introduction, continuous attribute discretization and the attribute independence assumption limit the generalization performance of the NBC for the continuous attribute classification problem. To cope with these issues, recent studies either introduced complex structures to represent attribute dependencies or applied discretization techniques to transform mixed-attribute values into discrete attribute values [18–20]. This section presents a novel solution to the NBC-based mixed-attribute classification problem by considering attribute dependence and mixed-attribute transformation simultaneously. A mixed-attribute fusion strategy is designed to construct an NBC that can be trained based on the transformed continuous attributes with the minimum attribute dependence.

For discrete attributes of an original mixed-attribute (OMA) data set $\mathbb{D}$, the one-hot encoding technique [21] is applied to transform them into 0-1 numerical attributes; i.e., $a_{ni}^{(m)}, i = 1, 2, \cdots, \mathcal{D}_1$ is encoded as $\left(e_{n1}^{(mi)}, \cdots, e_{n\mathcal{K}_i}^{(mi)}\right)$ for the discrete attribute value of the $n$-th sample $x_n^{(mi)}, n = 1, 2, \cdots, \mathcal{N}_m$, where the following is the case:

$$e_{nk}^{(m)} = \begin{cases} 1, & \text{if } a_{ni}^{(m)} = A_k^{(i)} \\ 0, & \text{if } a_{ni}^{(m)} \neq A_k^{(i)} \end{cases} \tag{9}$$

and $\left\{A_1^{(i)}, A_2^{(i)}, \cdots, A_{\mathcal{K}_i}^{(i)}\right\}$ is $\mathcal{K}_i$ categorical values corresponding to the $i$-th discrete attribute. Then, the one-hot encoded form of the original sample $x_n^{(m)}$ can be expressed as follows.

$$\overline{x}_n^{(m)} = \left(e_{n1}^{(m1)}, \cdots, e_{n\mathcal{K}_1}^{(m1)}, \cdots, e_{n1}^{(m\mathcal{D}_1)}, \cdots, e_{n\mathcal{K}_{\mathcal{D}_1}}^{(m\mathcal{D}_1)}, b_{n1}^{(m)}, \cdots, b_{n\mathcal{D}_2}^{(m)}\right). \tag{10}$$

In fact, the one-hot encoding technique causes attribute redundancy when extending a discrete attribute into multiple 0-1 numerical attributes. For example, the essence of the following transformation

$$\begin{bmatrix} A_1 \\ A_1 \\ A_2 \end{bmatrix} \rightarrow \begin{bmatrix} 1 & 0 \\ 1 & 0 \\ 0 & 1 \end{bmatrix} \tag{11}$$

is to represent a discrete attribute using two "repetitive" numerical attributes. The main role of one-hot encoding is to transform the discrete attribute values into numbers. The attribute independence assumption is not alleviated when constructing an NBC based on a one-hot encoded attribute (OHEA) data set.

$$\overline{X} = \begin{bmatrix} \overline{x}_1^{(1)} & \cdots & \overline{x}_{\mathcal{N}_1}^{(1)} & \overline{x}_1^{(2)} & \cdots & \overline{x}_{\mathcal{N}_2}^{(2)} & \cdots & \overline{x}_1^{(\mathcal{M})} & \cdots & \overline{x}_{\mathcal{N}_\mathcal{M}}^{(\mathcal{M})} \end{bmatrix}^{\mathrm{T}} . \tag{12}$$

Thus, an autoencoder neural network (ANN) is meticulously designed to solve the problems of attribute redundancy and attribute dependence mentioned above.

An ANN is a special single hidden-layer feed-forward neural network that has the same input matrix and output matrix. The main purpose of training an ANN is to find the optimal input layer weight matrix $\alpha = (\alpha_{rl})_{\mathcal{R} \times \mathcal{L}} = \begin{bmatrix} \alpha_1 & \alpha_2 & \cdots & \alpha_\mathcal{L} \end{bmatrix}$ and output layer matrix $\beta = (\beta_{lr})_{\mathcal{L} \times \mathcal{R}}$ so that the practical output matrix $\overline{X}'$ approximates the true output matrix $\overline{X}$ as closely as possible, where $\mathcal{R} = \sum_{i=1}^{\mathcal{D}_1} \mathcal{K}_i + \mathcal{D}_2$ is the number of input layer nodes of the ANN and $\mathcal{L}$ is the number of hidden layer nodes of the ANN.

The following objective function for ANN is designed to transform one-hot encoded attributes into independent encoded attributes:

$$L(\overline{X}, \alpha, \beta) = \lambda L_1(\overline{X}, \alpha, \beta) + (1 - \lambda) L_2(\overline{X}, \alpha, \beta), \tag{13}$$

where $\lambda \in (0, 1)$ is a balance factor. Optimal weight matrices $\overline{\alpha}$ and $\overline{\beta}$ are determined by minimizing the objective function as follows:

$$\overline{\alpha}, \overline{\beta} = \operatorname*{argmax}_{\substack{\alpha_{rl}, \beta_{lr} \in \Re \\ r=1,2,\cdots,\mathcal{R}; l=1,2,\cdots,\mathcal{L}}} \{ L(\overline{X}, \alpha, \beta) \}. \tag{14}$$

Here, it is worthwhile to note that the original intention of using an autoencoder neural network is to fuse the mixed attributes and to relax the attribute independence assumption rather than exploring the usage of deep learning technology. When a shallow learning can already meet the requirement of good NBC construction, it is unnecessary to resort to complex and time-consuming deep learning. The excessive attention on attribute transformation with deep learning is out of the scope of this study. Interested readers can refer to specialized studies on the combination of deep learning and supervised learning for more details on such approaches, e.g., deep support vector machine [22], deep decision tree [23], and deep nearest neighbor [24].

The first term of the objective function is the encoding error, which is used to measure the error between the practical output matrix $\overline{X}'$ and the true output matrix $\overline{X}$. It is defined as follows:

$$L_1(\overline{X}, \alpha, \beta) = \left\| \overline{X}' - \overline{X} \right\|_2^2 = \left\| [\mathrm{sigmoid}(\overline{X}\alpha)]\beta - \overline{X} \right\|_2^2 \tag{15}$$

where

$$\overline{\overline{X}} = \mathrm{sigmoid}(\overline{X}\alpha) = \begin{bmatrix} \overline{\overline{x}}_1^{(1)} & \cdots & \overline{\overline{x}}_{\mathcal{N}_1}^{(1)} & \cdots & \overline{\overline{x}}_1^{(\mathcal{M})} & \cdots & \overline{\overline{x}}_{\mathcal{N}_\mathcal{M}}^{(\mathcal{M})} \end{bmatrix}^{\mathrm{T}} \tag{16}$$

is a $\mathcal{N} \times \mathcal{L}$ hidden-layer output matrix that is a class-independent encoded-attribute (IEA) data set, such that $\overline{\overline{x}}_n^{(m)} = \left( \overline{\overline{x}}_{n1}^{(m)}, \cdots, \overline{\overline{x}}_{n\mathcal{L}}^{(m)} \right) = \left( \text{sigmoid}\left( \overline{x}_n^{(m)} \alpha_1 \right), \cdots, \text{sigmoid}\left( \overline{x}_n^{(m)} \alpha_{\mathcal{L}} \right) \right)$ is the independent encoded form of the original sample $x_n^{(m)}$ and

$$\text{sigmoid}(s) = \frac{1}{1 + \exp(-s)}, s \in (-\infty, +\infty) \tag{17}$$

is the sigmoid activation function. To minimize attribute dependence, the attribute dependence term in the objective function is defined as follows:

$$L_2\left(\overline{X}, \alpha, \beta\right) = \frac{2}{\mathcal{L}(\mathcal{L}-1)} \sum_{i=1}^{\mathcal{L}} \sum_{\substack{j=1 \\ j \neq i}}^{\mathcal{L}} I(h_i, h_j), \tag{18}$$

where

$$h_i = \left( \overline{\overline{x}}_{1i}^{(1)}, \cdots, \overline{\overline{x}}_{\mathcal{N}_1,i}^{(1)}, \cdots, \overline{\overline{x}}_{1i}^{(\mathcal{M})}, \cdots, \overline{\overline{x}}_{\mathcal{N}_{\mathcal{M}},i}^{(\mathcal{M})} \right)^{\mathrm{T}} \tag{19}$$

is the $i$-th independent encoded attribute, and $I(h_i, h_j)$ is the mutual information between independent encoded attributes $h_i$ and $h_j$ and $i, j \in \{1, 2, \cdots, \mathcal{L}\}$ and $i \neq j$.

The updating rules of $\alpha_{rl}$ and $\beta_{lr}$ are derived as follows. The second term of the objective function is unrelated to the output-layer weights. Thus, the updating rule mainly depends on the encoding error. The gradient descent method is used to determine the updating rule of $\beta_{lr}$. The partial derivative of $L\left(\overline{X}, \alpha, \beta\right)$ with respect to $\beta_{lr}$ is calculated as follows:

$$\Delta\beta_{lr} = \frac{\partial L\left(\overline{X}, \alpha, \beta\right)}{\partial \beta_{lr}} = \lambda \frac{\partial L_1\left(\overline{X}, \alpha, \beta\right)}{\partial \beta_{lr}} \tag{20}$$

and then the updating rule of $\beta_{lr}$ is given as follows

$$\beta_{lr} \leftarrow \beta_{lr} - \xi \times \Delta\beta_{lr}, \tag{21}$$

where $\xi > 0$ is the learning rate. For the input-layer weight $\alpha_{rl}$, the updating rule cannot be derived by using the gradient descent method because of the existence of mutual information terms. Here, a new updating strategy based on the Monte Carlo method [25] is designed as follows:

$$\alpha_{rl} \leftarrow \alpha_{rl} - \zeta \times \Delta\alpha_{rl}, \tag{22}$$

where $\zeta > 0$ is the learning rate and

$$\Delta\alpha_{rl} = \frac{1}{\mathcal{N}} \sum_{m=1}^{\mathcal{M}} \sum_{n=1}^{\mathcal{N}_m} L\left( \overline{x}_n^{(m)}, \alpha, \beta \right) \tag{23}$$

is the approximation of the gradient $\frac{\partial L\left(\overline{X}, \alpha, \beta\right)}{\partial \alpha_{rl}}$.

Based on the IEA data set, mixed-attribute fusion-based NBC (MAF-NBC) determines the class label for a given new sample $x = \left( a_1, \cdots, a_{\mathcal{D}_1}, b_1, \cdots, b_{\mathcal{D}_2} \right)$ as follows. First, the one-hot encoded form of $x$ is expressed as follows:

$$\overline{x} = \left( e_1^{(1)}, \cdots, e_{\mathcal{K}_1}^{(1)}, \cdots, e_1^{(\mathcal{D}_1)}, \cdots, e_{\mathcal{K}_{\mathcal{D}_1}}^{(\mathcal{D}_1)}, b_1, \cdots, b_{\mathcal{D}_2} \right), \tag{24}$$

where the following.

$$e_k^{(i)} = \begin{cases} 1, & \text{if } a_i = A_k^{(i)} \\ 0, & \text{if } a_i \neq A_k^{(i)} \end{cases}, k = 1, 2, \cdots, \mathcal{K}_i, i = 1, 2, \cdots, \mathcal{D}_1. \tag{25}$$

Second, $\overline{x}$ is transformed into the following independent encoded expression:

$$\overline{\overline{x}} = \left(\overline{\overline{x}}_1, \overline{\overline{x}}_2, \cdots, \overline{\overline{x}}_{\mathcal{L}}\right) = \left(\text{sigmoid}(\overline{x}\alpha_1), \text{sigmoid}(\overline{x}\alpha_2), \cdots, \text{sigmoid}(\overline{x}\alpha_{\mathcal{L}})\right) \tag{26}$$

based on the trained ANN with input weight matrix $\alpha = (\alpha_{rl})_{\mathcal{R} \times \mathcal{L}}$. Third, the class label of x is determined according to Equation (1), where the conditional probability is calculated as follows:

$$\begin{aligned}
\text{P}(x|w_m) &= \text{P}(\overline{x}|w_m) = \text{P}\left(\overline{\overline{x}}|w_m\right) \\
&= \prod_{l=1}^{\mathcal{L}} \text{P}(\overline{\overline{x}}_l|w_m) = \prod_{l=1}^{\mathcal{L}} \int_{-\infty}^{\overline{\overline{x}}_l} \text{p}(s|w_m)'
\end{aligned} \tag{27}$$

where $\text{p}(s|w_m)$ is approximated with a normal *p.d.f.*

$$f_l^{(m)}(s) = \frac{1}{\sqrt{2\pi}\sigma_l^{(m)}} \exp\left[-\frac{1}{2}\left(\frac{s - \mu_l^{(m)}}{\sigma_l^{(m)}}\right)^2\right], \tag{28}$$

$$s \in (-\infty, +\infty), m = 1, 2, \cdots, \mathcal{M}, l = 1, 2, \cdots, \mathcal{L}$$

with the mean value

$$\mu_l^{(m)} = \frac{1}{\mathcal{N}_m} \sum_{n=1}^{\mathcal{N}_m} \overline{\overline{x}}_{nl}^{(m)} \tag{29}$$

and standard deviation (std).

$$\sigma_l^{(m)} = \sqrt{\frac{1}{\mathcal{N}_m - 1} \sum_{n=1}^{\mathcal{N}_m} \left[\overline{\overline{x}}_{nl}^{(m)} - \mu_l^{(m)}\right]^2}. \tag{30}$$

Here, an in-depth discussion regarding the normal *p.d.f.* $f_l^{(m)}(\bullet)$ is given. The sigmoid activation function is used in the designed ANN; thus, the outputs corresponding to each hidden layer's nodes obey a quasi-normal probability distribution. A visual comparison of the sigmoid activation function and normal probability distribution functions with standard deviations 0.1, 0.5, 1.0, and 2.0, as shown in Figure 1, clearly demonstrates this empirical conclusion.

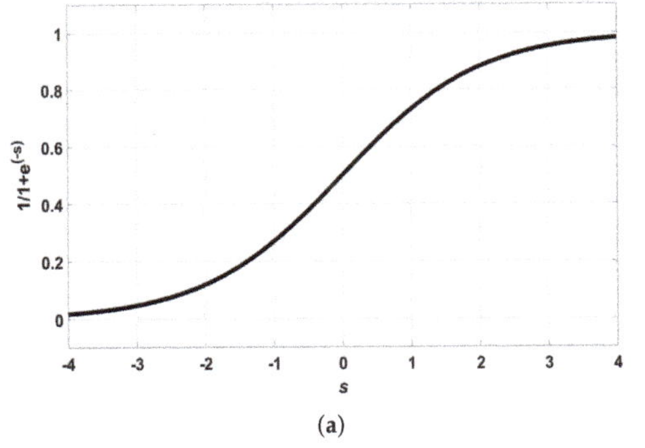

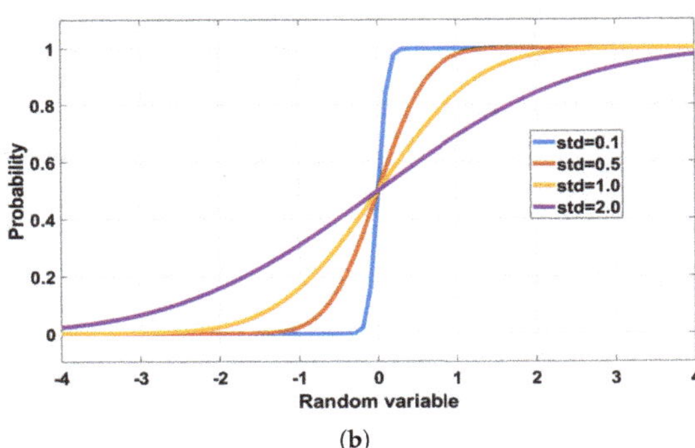


**Figure 1.** Graphical comparison between sigmoid activation and normal distribution. (**a**) Sigmoid activation function. (**b**) Normal probability distribution functions.

For independent attributes, the joint probability distribution is the normal probability distribution if marginal probability distributions are normal probability distributions. Thus, the joint probability $\text{P}\left(\overline{\overline{x}}|w_m\right)$ can be modeled as the product of multiple marginal probabilities $\text{P}(\overline{\overline{x}}_l|w_m)$, $l = 1, 2, \cdots, \mathcal{L}$. For the sake of simplicity, hidden-layer biases are not used in the constructed ANN. The role of hidden-layer biases is to control the bias between the hidden layer's input and the original point of the sigmoid function. When conducting classification or regression tasks, hidden-layer biases are helpful for the generation of high-performance learners. In MAF-NBC, the ANN is designed to transform one-hot encoded attributes into independent encoded attributes that are expected to have minimum dependence. The reasonable objective function shown in Equation (11) is able to guarantee the generation of independent encoded attributes even though hidden-layer biases are not used in ANN. The following experiments will support the aforementioned discussion and conclusion.

## 4. Experimental Settings and Results

A series of experiments are conducted in this section to validate the feasibility, rationality, and effectiveness of the proposed mixed-attribute fusion-based naive Bayesian classifier (MAF-NBC). Experiments were conducted using 20 KEEL [26] mixed-attribute data sets. Their characteristics are listed in Table 1, where the Arabic numerals in parentheses represent the numbers of avaliable categorical values corresponding to the discrete attributes. The data processing strategy proposed by Helal and Otero [27] was used to generate the mixed attributes for the data sets marked without '*' in Table 1. Then, the equal width discretization method was used to transform the continuous attributes into discrete attributes. All data sets as shown in Table 1 can be downloaded from our BaiDuPan or GitHub online storage space. The experiments were run on a PC equipped with an Intel(R) Quadcore 3.00 GHz i5-9400 CPU and 16 GB of RAM.

**Table 1.** Details of 20 mixed-attribute data sets.

| Data sets | Abalone_3_4 | Abalone_9_10 | Adult * | Band * |
|---|---|---|---|---|
| Number of samples | 423 | 1073 | 5000 | 540 |
| Number of continuous attributes | 6 | 6 | 9 | 18 |
| Number of discrete attributes | 2 (2, 4) | 2 (2, 4) | 5 (5, 7, 5, 4, 2) | 5 (3, 5, 2, 3, 3) |
| Number of classes | 2 | 2 | 2 | 2 |
| data sets | Bd | Bp | Heart | Ionosphere |
| Number of samples | 569 | 198 | 270 | 351 |
| Number of continuous attributes | 23 | 25 | 9 | 25 |
| Number of discrete attributes | 7 (4, 6, 5, 3, 2, 4, 3) | 8 (6, 3, 5, 4, 5, 3, 4, 2) | 4 (4, 3, 3, 4) | 7 (4, 6, 5, 3, 4, 3, 5) |
| Number of classes | 2 | 2 | 2 | 2 |
| Data sets | Libras | Page blocks | Parkinsons | Ring |
| Number of samples | 360 | 547 | 195 | 500 |
| Number of continuous attributes | 85 | 7 | 17 | 15 |
| Number of discrete attributes | 15 (4, 5, 7, 5, 8, 4, 5, 6, 4, 5, 6, 7, 3, 4, 5) | 3 (3, 5, 4) | 5 (4, 5, 6, 5, 3) | 5 (3, 4, 5, 3, 4) |
| Number of classes | 15 | 5 | 2 | 2 |
| Data sets | Segment | Sonar | SPECTF | Vehicle |
| Number of samples | 2310 | 208 | 267 | 846 |
| Number of continuous attributes | 14 | 50 | 36 | 6 |
| Number of discrete attributes | 5 (3, 6, 5, 4, 3) | 10 (6, 4, 3, 5, 6, 7, 5, 3, 4, 6) | 8 (4, 6, 5, 5, 3, 6, 7, 4) | 2 (4, 3) |
| Number of classes | 7 | 2 | 2 | 4 |
| Data sets | Vowel | Wine | WineQR | WineQW |
| Number of samples | 528 | 178 | 1599 | 489 |
| Number of continuous attributes | 7 | 10 | 8 | 8 |
| Number of discrete attributes | 3 (3, 3, 4) | 3 (4, 5, 4) | 3 (4, 4, 3) | 3 (3, 3, 3) |
| Number of classes | 11 | 3 | 6 | 6 |

### 4.1. Feasibility Validation of MAF-NBC

A first experiment was performed to validate the feasibility of the MAF-NBC method by checking the convergence of ANN weights by the iterative update process. This experiment was conducted on the representative *Vowel* data set, which has three discrete attributes and seven continuous attributes. The experimental results are the average values corresponding to 10 independent ANN training. Ten ANNs with 50 hidden-layer nodes were constructed with random weights in the $[-1, 1]$ range, a balance factor $\lambda = 0.50$, and learning rates $\xi = \zeta = 0.01$. Figure 2 depicts the variation trends of the 1-norms for the input layer and output layer weights as the iteration number increased. Figure 3 presents the variation trends of encoding error and attribute dependence as the iteration number increased.

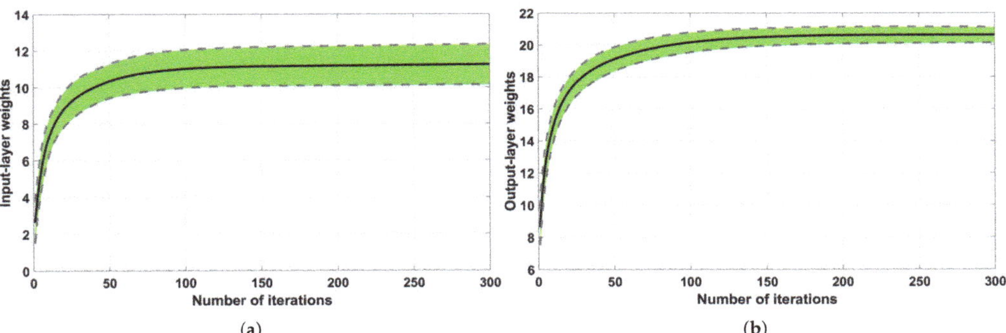

**Figure 2.** Convergence of the ANN's weights. (**a**) Input-layer weights. (**b**) Output-layer weights.

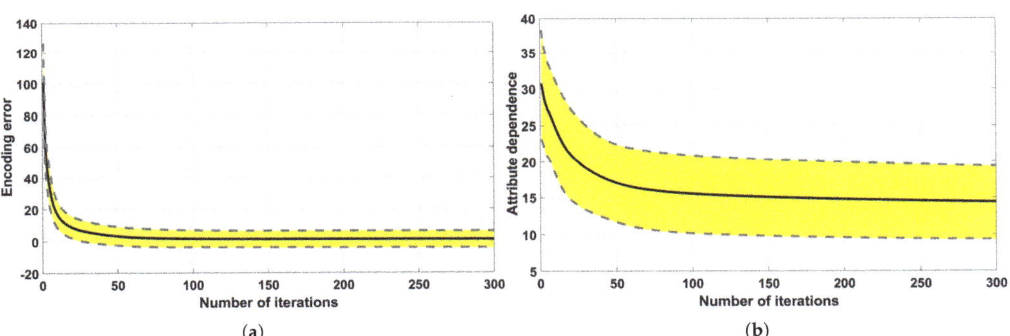

**Figure 3.** Convergence of the ANN's objective function. (**a**) Encoding error. (**b**) Attribute dependence.

Figure 2 shows that the sums of absolute values corresponding to the input layer and output-layer weights become more and more stable with the number of iterations. In addition, Figure 3 shows that the encoding error and attribute dependence decrease gradually and then remain unchanged. These experimental results reveal that the updating rules of Equations (15) and (16) are effective for the determination of optimal ANN weights. In fact, Equation (17) is the approximation of the expected loss $E_{\overline{\mathcal{X}}}\left[\frac{\partial L(\overline{\mathcal{X}},\alpha,\beta)}{\partial \alpha_{rl}}\right]$, where $\overline{\mathcal{X}}$ is the domain of the objective function $L(\bullet, \alpha, \beta)$. Although the gradient of input layer weights cannot be analytically calculated due to the existence of the attribute dependence term, Equation (17) uses the mean of the objective function values corresponding to all samples in the one-hot encoded attribute data set to approximate the gradient. The convergence of output layer weights guarantees the convergence of input layer weights, because the update of $\alpha_{rl}$ depends on the update of $\beta_{lr}$. This experiment demonstrates the feasibility of transforming the one-hot encoded attributes into independent encoded attributes.

### 4.2. Rationality Validation of MAF-NBC

A second experiment was carried out to evaluate the rationality of MAF-NBC in terms of whether the designed ANN can transform the one-hot encoded attributes corresponding to the original mixed attributes into independent encoded attributes. On the representative *Page_small* data set, an ANN with 50 hidden-layer nodes was constructed with random weights in the $[-1,1]$ interval, a balance factor $\lambda = 0.50$, and learning rates $\xi = \zeta = 0.01$. Ten representative encoded attributes were selected from the hidden-layer outputs of ANN. The dependence between the two encoded attributes was measured with the mutual information, which is calculated with *sklearn.metrics.mutual_info_score* (https://scikit-learn.org/stable/modules/generated/sklearn.metrics.mutual_info_score.html, accessed on 14 October 2022) package of the *scikit-learn* machine learning library.

Figure 4 displays a series of heatmaps corresponding to iterations #1, #50, #100, and #150. These heatmaps show the change in attribute dependence as the iteration number increases during the ANN's training. It can be clearly seen that the dependence between encoded attribute gradually decreases as ANN weights are updated. It indicates that independent attributes can be obtained by transforming the original mixed attributes into the encoded attributes. The experimental results shown in Figure 4 are consistent with the experimental results shown in Figure 3b, where the attribute dependence gradually decreases with the increase in iteration number. This experiment demonstrated that the ANN is able to transform the original mixed attributes into independent encoded attributes for NBC construction.

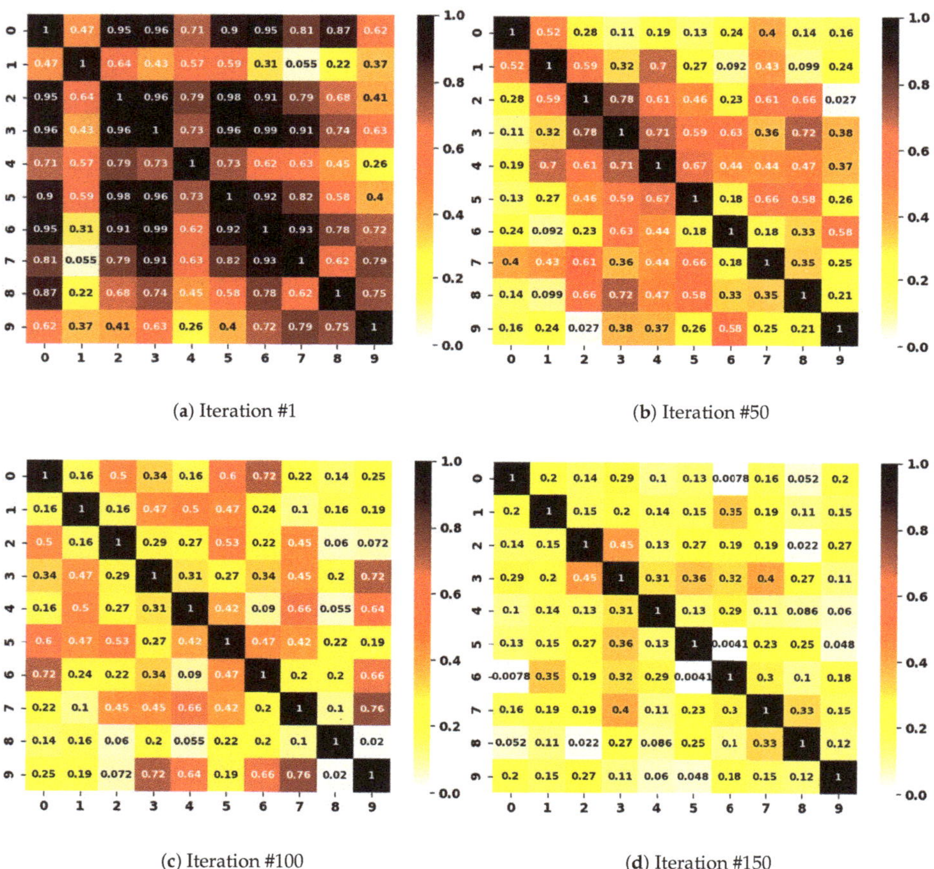

(a) Iteration #1

(b) Iteration #50

(c) Iteration #100

(d) Iteration #150

**Figure 4.** Dependence change of the encoded continuous attributes.

*4.3. Effectiveness Validation of MAF-NBC*

A third experiment was performed to evaluate the effectiveness of MAF-NBC. This was performed by comparing the classification performances of MAF-NBC with Dis-NBC, FNBC [3], TAN [4], AODE [5], HNB [6], DFW-NBC [7], CFW-NBC [8], and ICA-NBC [10]. All Bayesian algorithms were implemented using the Python programming language. Twenty KEEL mixed-attribute data sets (described in Table 1) were selected to test the training and testing accuracies of the compared algorithms. For each data set, independent training and testing were conducted 10 times with random data set partitions, i.e., 70% samples to train the different Bayes algorithms and 30% to test their generalization capabilities. All ANNs were constructed with $2\mathcal{L}$ hidden-layer nodes, initialized with random weights in the $[-1, 1]$ range, and setup with the learning parameters $\lambda = 0.50$ and $\xi = \zeta = 0.01$. The mean and standard derivation of the 10 training and testing accuracies are listed for each algorithm in Tables 2 and 3, respectively.

The training and testing performances of MAF-NBC were statistically validated by comparing with eight other Bayes algorithms on the 20 data sets. For the given significance level of 0.05, the critical difference (CD) value [28] is calculated as follows:

$$\text{CD} = 3.102 \times \sqrt{\frac{9 \times (9 + 1)}{6 \times 20}} \approx 2.686, \tag{31}$$

where the number of compared algorithms is nine, and the number of data sets is 20. In Figure 5, an interval of one CD value can be observed to the left and right of the average rank of MAF-NBC. Any algorithm with a rank outside this area is significantly different from MAF-NBC. It is found that the ranks of MAF-NBC corresponding to the training and testing accuracies are obviously smaller than the other algorithms. In addition, the number of wins for MAF-NBC on 20 data sets is at least $\frac{20}{2} + 1.96 \times \frac{\sqrt{20}}{2} \approx 14$ for the significance level of 0.05. Furthermore, it can be said that MAF-NBC has significantly improved testing accuracies than Dis-NBC (20 wins), FNBC (20 wins), TAN (20 wins), AODE (20 wins), HNB (20 wins), DFW-NBC (19 wins), CFW-NBC (19 wins), and ICA-NBC (18 wins) under a significance level of 0.10. This indicates that MAF-NBC is significantly better for classification than the other algorithms on the selected data sets. The experimental results and statistical analyses demonstrate the effectiveness of MAF-NBC and indicate that MAF-NBC is a viable method to handle mixed-attribute classification tasks. MAF-NBC does not modify the simple model structure of the NBC and preserves the amount of information of the original mixed-attribute data set as much as possible, because the ANN transforms the original mixed attributes into encoded attributes rather than selecting or extracting independent attributes from the original mixed attributes.

**Table 2.** Comparison of the training accuracies of MAF-NBC, Dis-NBC, FNB, TAN, AODE, HNB, DFW-NBC, CFW-NBC, and ICA-NBC.

| Data Set | MAF-NBC | Dis-NBC | FNBC | TAN | AODE | HNB | DFW-NBC | CFW-NBC | ICA-NBC |
|---|---|---|---|---|---|---|---|---|---|
| Abalone_3_4 | **0.803** ± 0.036 | 0.721 ± 0.032 | 0.736 ± 0.026 | 0.767 ± 0.038 | 0.755 ± 0.032 | 0.744 ± 0.029 | 0.747 ± 0.018 | 0.755 ± 0.017 | 0.781 ± 0.045 |
| Abalone_9_10 | 0.607 ± 0.029 | 0.597 ± 0.012 | 0.592 ± 0.021 | 0.595 ± 0.025 | 0.596 ± 0.024 | 0.587 ± 0.041 | 0.617 ± 0.008 | 0.605 ± 0.014 | **0.619** ± 0.017 |
| Adult | **0.870** ± 0.013 | 0.782 ± 0.014 | 0.792 ± 0.015 | 0.807 ± 0.013 | 0.829 ± 0.008 | 0.811 ± 0.017 | 0.805 ± 0.011 | 0.813 ± 0.006 | 0.851 ± 0.015 |
| Band | **0.718** ± 0.027 | 0.669 ± 0.028 | 0.637 ± 0.034 | 0.671 ± 0.035 | 0.663 ± 0.024 | 0.681 ± 0.028 | 0.703 ± 0.015 | **0.718** ± 0.017 | 0.714 ± 0.016 |
| Bd | **0.991** ± 0.002 | 0.919 ± 0.004 | 0.938 ± 0.010 | 0.961 ± 0.009 | 0.959 ± 0.008 | 0.953 ± 0.011 | 0.959 ± 0.007 | 0.970 ± 0.004 | 0.983 ± 0.007 |
| Bp | **0.852** ± 0.022 | 0.788 ± 0.023 | 0.793 ± 0.027 | 0.803 ± 0.029 | 0.801 ± 0.034 | 0.791 ± 0.041 | 0.841 ± 0.036 | 0.824 ± 0.033 | 0.837 ± 0.027 |
| Heart | **0.924** ± 0.014 | 0.699 ± 0.013 | 0.824 ± 0.017 | 0.858 ± 0.015 | 0.849 ± 0.021 | 0.867 ± 0.029 | 0.886 ± 0.013 | 0.905 ± 0.015 | 0.901 ± 0.017 |
| Ionosphere | **0.981** ± 0.005 | 0.901 ± 0.006 | 0.917 ± 0.011 | 0.934 ± 0.009 | 0.932 ± 0.013 | 0.926 ± 0.021 | 0.928 ± 0.012 | 0.941 ± 0.015 | 0.974 ± 0.004 |
| Libras | 0.905 ± 0.007 | **0.938** ± 0.015 | 0.857 ± 0.019 | 0.899 ± 0.015 | 0.847 ± 0.016 | 0.864 ± 0.015 | 0.911 ± 0.016 | 0.910 ± 0.015 | 0.908 ± 0.011 |
| Page_small | **0.952** ± 0.010 | 0.915 ± 0.005 | 0.911 ± 0.007 | 0.909 ± 0.018 | 0.919 ± 0.014 | 0.914 ± 0.017 | 0.931 ± 0.006 | 0.933 ± 0.008 | 0.937 ± 0.007 |
| Parkinsons | **0.872** ± 0.029 | 0.782 ± 0.021 | 0.801 ± 0.022 | 0.808 ± 0.025 | 0.814 ± 0.027 | 0.821 ± 0.021 | 0.827 ± 0.022 | 0.841 ± 0.025 | 0.852 ± 0.024 |
| Ring | 0.937 ± 0.015 | 0.801 ± 0.017 | 0.836 ± 0.015 | 0.855 ± 0.028 | 0.869 ± 0.024 | 0.858 ± 0.027 | 0.916 ± 0.010 | 0.948 ± 0.006 | **0.956** ± 0.008 |
| Segment | 0.940 ± 0.007 | 0.878 ± 0.010 | 0.889 ± 0.019 | 0.897 ± 0.016 | 0.895 ± 0.014 | 0.901 ± 0.016 | 0.928 ± 0.006 | 0.945 ± 0.009 | **0.959** ± 0.005 |
| Sonar | **0.865** ± 0.019 | 0.823 ± 0.029 | 0.834 ± 0.027 | 0.839 ± 0.025 | 0.844 ± 0.018 | 0.840 ± 0.025 | 0.841 ± 0.027 | 0.847 ± 0.029 | 0.851 ± 0.024 |
| Spectf | **0.908** ± 0.011 | 0.768 ± 0.023 | 0.771 ± 0.017 | 0.783 ± 0.019 | 0.779 ± 0.015 | 0.756 ± 0.029 | 0.779 ± 0.016 | 0.789 ± 0.014 | 0.883 ± 0.019 |
| Vehicle | **0.771** ± 0.016 | 0.537 ± 0.005 | 0.544 ± 0.011 | 0.576 ± 0.016 | 0.589 ± 0.019 | 0.567 ± 0.026 | 0.604 ± 0.022 | 0.611 ± 0.021 | 0.705 ± 0.017 |
| Vowel | **0.869** ± 0.017 | 0.829 ± 0.021 | 0.839 ± 0.017 | 0.836 ± 0.019 | 0.840 ± 0.016 | 0.847 ± 0.019 | 0.843 ± 0.021 | 0.848 ± 0.016 | 0.833 ± 0.023 |
| Wine | **0.999** ± 0.001 | 0.963 ± 0.021 | 0.967 ± 0.017 | 0.975 ± 0.006 | 0.981 ± 0.004 | 0.968 ± 0.007 | **0.999** ± 0.001 | **0.999** ± 0.001 | 0.968 ± 0.019 |
| WineQR | **0.738** ± 0.023 | 0.563 ± 0.014 | 0.578 ± 0.034 | 0.580 ± 0.033 | 0.591 ± 0.031 | 0.602 ± 0.037 | 0.638 ± 0.020 | 0.717 ± 0.019 | 0.688 ± 0.036 |
| WineQW | **0.584** ± 0.028 | 0.487 ± 0.014 | 0.524 ± 0.016 | 0.537 ± 0.024 | 0.533 ± 0.016 | 0.541 ± 0.021 | 0.551 ± 0.028 | 0.561 ± 0.025 | 0.577 ± 0.021 |
| **Average** | **0.854** ± 0.017 | 0.768 ± 0.016 | 0.779 ± 0.019 | 0.795 ± 0.021 | 0.794 ± 0.016 | 0.792 ± 0.024 | 0.813 ± 0.016 | 0.824 ± 0.015 | 0.839 ± 0.018 |

**Table 3.** Comparison of the testing accuracies of MAF-NBC, Dis-NBC, FNB, TAN, AODE, HNB, DFW-NBC, CFW-NBC, and ICA-NBC.

| Data Set | MAF-NBC | Dis-NBC | FNBC | TAN | AODE | HNB | DFW-NBC | CFW-NBC | ICA-NBC |
|---|---|---|---|---|---|---|---|---|---|
| Abalone_3_4 | **0.759** ± 0.030 | 0.685 ± 0.042 | 0.701 ± 0.049 | 0.742 ± 0.032 | 0.733 ± 0.037 | 0.728 ± 0.029 | 0.723 ± 0.031 | 0.751 ± 0.027 | 0.755 ± 0.043 |
| Abalone_9_10 | **0.603** ± 0.028 | 0.587 ± 0.024 | 0.588 ± 0.026 | 0.587 ± 0.027 | 0.589 ± 0.028 | 0.581 ± 0.030 | 0.589 ± 0.031 | 0.582 ± 0.031 | 0.602 ± 0.032 |
| Adult | **0.842** ± 0.014 | 0.795 ± 0.011 | 0.788 ± 0.019 | 0.801 ± 0.025 | 0.817 ± 0.012 | 0.807 ± 0.027 | 0.804 ± 0.012 | 0.811 ± 0.014 | 0.824 ± 0.009 |
| Band | **0.647** ± 0.045 | 0.606 ± 0.028 | 0.601 ± 0.028 | 0.626 ± 0.036 | 0.618 ± 0.038 | 0.609 ± 0.031 | 0.619 ± 0.034 | 0.630 ± 0.036 | 0.619 ± 0.041 |
| Bd | **0.973** ± 0.011 | 0.915 ± 0.017 | 0.921 ± 0.013 | 0.956 ± 0.012 | 0.950 ± 0.010 | 0.946 ± 0.014 | 0.933 ± 0.015 | 0.944 ± 0.016 | 0.948 ± 0.014 |
| Bp | **0.764** ± 0.040 | 0.739 ± 0.047 | 0.746 ± 0.039 | 0.740 ± 0.057 | 0.739 ± 0.052 | 0.737 ± 0.065 | 0.755 ± 0.036 | 0.747 ± 0.041 | 0.746 ± 0.049 |
| Heart | **0.876** ± 0.028 | 0.703 ± 0.022 | 0.711 ± 0.025 | 0.739 ± 0.037 | 0.728 ± 0.034 | 0.789 ± 0.039 | 0.859 ± 0.032 | 0.864 ± 0.035 | 0.851 ± 0.038 |
| Ionosphere | **0.947** ± 0.027 | 0.894 ± 0.024 | 0.906 ± 0.021 | 0.917 ± 0.017 | 0.907 ± 0.025 | 0.901 ± 0.027 | 0.923 ± 0.029 | 0.927 ± 0.020 | 0.931 ± 0.026 |
| Libras | **0.778** ± 0.039 | 0.743 ± 0.047 | 0.756 ± 0.058 | 0.751 ± 0.021 | 0.749 ± 0.048 | 0.748 ± 0.039 | 0.759 ± 0.036 | 0.764 ± 0.039 | 0.759 ± 0.038 |
| Page_small | 0.901 ± 0.019 | 0.896 ± 0.015 | 0.889 ± 0.013 | 0.891 ± 0.016 | 0.881 ± 0.019 | 0.898 ± 0.017 | 0.912 ± 0.015 | **0.913** ± 0.011 | 0.899 ± 0.012 |
| Parkinsons | **0.826** ± 0.042 | 0.756 ± 0.043 | 0.769 ± 0.049 | 0.779 ± 0.059 | 0.771 ± 0.047 | 0.782 ± 0.039 | 0.786 ± 0.034 | 0.807 ± 0.044 | 0.779 ± 0.030 |
| Ring | **0.928** ± 0.017 | 0.781 ± 0.041 | 0.831 ± 0.038 | 0.849 ± 0.049 | 0.834 ± 0.041 | 0.879 ± 0.049 | 0.870 ± 0.028 | 0.910 ± 0.011 | 0.925 ± 0.016 |
| Segment | 0.913 ± 0.016 | 0.869 ± 0.019 | 0.874 ± 0.014 | 0.886 ± 0.018 | 0.895 ± 0.016 | 0.889 ± 0.020 | 0.899 ± 0.015 | 0.906 ± 0.013 | **0.922** ± 0.010 |
| Sonar | 0.816 ± 0.036 | 0.758 ± 0.046 | 0.769 ± 0.053 | 0.779 ± 0.038 | 0.806 ± 0.034 | 0.787 ± 0.023 | 0.796 ± 0.036 | 0.801 ± 0.042 | **0.837** ± 0.029 |
| Spectf | **0.823** ± 0.023 | 0.716 ± 0.037 | 0.727 ± 0.027 | 0.778 ± 0.041 | 0.794 ± 0.030 | 0.757 ± 0.041 | 0.748 ± 0.035 | 0.757 ± 0.030 | 0.801 ± 0.029 |
| Vehicle | **0.611** ± 0.029 | 0.483 ± 0.037 | 0.506 ± 0.039 | 0.529 ± 0.032 | 0.520 ± 0.049 | 0.518 ± 0.054 | 0.537 ± 0.029 | 0.549 ± 0.032 | 0.589 ± 0.028 |
| Vowel | **0.859** ± 0.023 | 0.781 ± 0.057 | 0.781 ± 0.052 | 0.799 ± 0.058 | 0.795 ± 0.057 | 0.801 ± 0.033 | 0.811 ± 0.029 | 0.821 ± 0.024 | 0.816 ± 0.027 |
| Wine | **0.961** ± 0.019 | 0.928 ± 0.024 | 0.926 ± 0.029 | 0.937 ± 0.027 | 0.946 ± 0.021 | 0.939 ± 0.026 | 0.958 ± 0.017 | 0.945 ± 0.022 | 0.931 ± 0.016 |
| WineQR | **0.637** ± 0.018 | 0.489 ± 0.025 | 0.509 ± 0.038 | 0.519 ± 0.037 | 0.528 ± 0.034 | 0.524 ± 0.031 | 0.563 ± 0.028 | 0.587 ± 0.030 | 0.559 ± 0.041 |
| WineQW | **0.534** ± 0.036 | 0.477 ± 0.043 | 0.503 ± 0.033 | 0.508 ± 0.038 | 0.501 ± 0.041 | 0.509 ± 0.029 | 0.508 ± 0.041 | 0.515 ± 0.028 | 0.529 ± 0.039 |
| **Average** | **0.800** ± 0.027 | 0.730 ± 0.032 | 0.740 ± 0.033 | 0.756 ± 0.034 | 0.755 ± 0.034 | 0.756 ± 0.033 | 0.768 ± 0.027 | 0.777 ± 0.027 | 0.781 ± 0.028 |

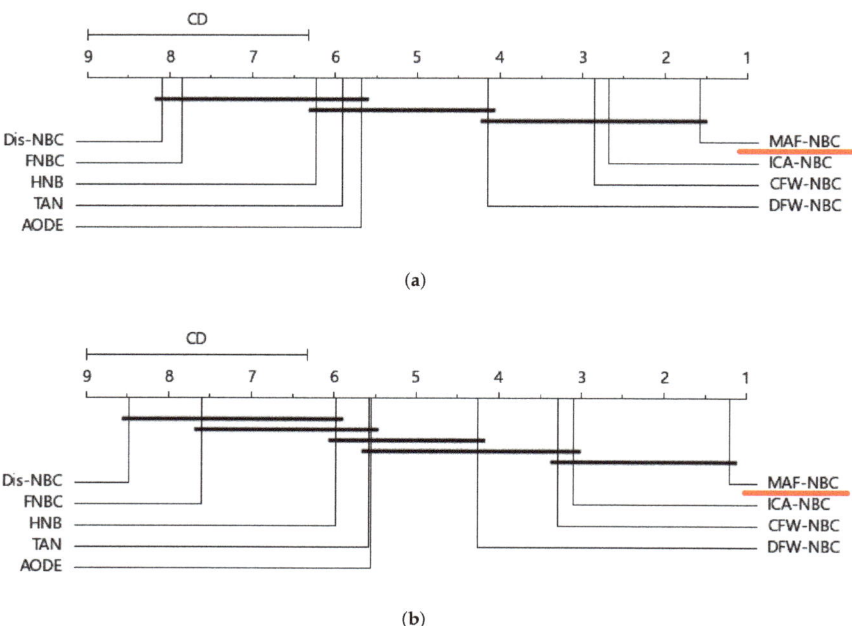


**Figure 5.** CD diagrams corresponding to comparisons in Tables 2 and 3. (**a**) CD diagram of training accuracies. (**b**) CD diagram of testing accuracies.

## 5. Conclusions and Future Works

This paper presented a novel NBC training method for the mixed-attribute data classification problem without continuous attribute discretization and complex Bayesian network structure learning. The original mixed attributes were transformed into a series of continuous attributes with minimum dependence using an autoencoder neural network. To obtain optimal network weights, an effective objective function was designed, and corresponding weight updating rules were derived. The experimental results finally demonstrated improved classification performance for the novel Bayes model in comparison with eight state-of-the-art Bayesian algorithms. The technical advantages of MAF-NBC are four-fold:

- Preserving the information. The autoencoder neural network transforms mixed attributes into independent encoded attributes, which can also be decoded to restore the original mixed attributes. Thus, useful information from the original data set is conserved as much as possible.
- Simple model structure. MAF-NBC effectively handles the attribute dependence problem between discrete attributes and continuous attributes by transforming original attributes into encoded attributes. The simple structure of the NBC was not modified by providing a data transformation-oriented preprocessing strategy.
- Stable training process. The efficient weight updating strategy guarantees the convergence of the autoencoder neural network and thus provided stable data transformation results. In addition, the quasi-normal distribution of independent encoded attributes provided accurate calculations for the joint conditional probability in the NBC.
- Low computation complexity. A single hidden-layer encoder neural network was used to fuse the original mixed attributes. Training to find optimal network weights has a low computational complexity. As mentioned in Section 3, a shallow learning mechanism was able to satisfy the demand of NBC construction. Hence, it is unnec-

essary to explore the availability of deep learning for data transformation-oriented NBC training.

In future studies, we plan (1) to implement MAF-NBC in a distributed environment so that it can be used to deal with large-scale mixed-attribute data-classification problems and (2) to utilize the autoencoder neural network to transform mixed attributes into dependent continuous attributes to construct a non-naive Bayesian classifier [29] based on the joint probability density function estimation technique.

**Author Contributions:** Data curation, formal analysis, and writing—original draft preparation, G.O.; methodology, writing—original draft preparation and review and editing, Y.H.; investigation and writing—review and editing, P.F.-V.; supervision, J.Z.H. All authors have read and agreed to the published version of the manuscript.

**Funding:** This research was funded by the Scientific Research Foundation of National Natural Science Foundation of China (61972261) and Basic Research Foundations of Shenzhen (JCYJ20210324093609026 and JCYJ20200813091134001).

**Institutional Review Board Statement:** None.

**Informed Consent Statement:** None.

**Data Availability Statement:** The data are available in BaiduPan https://pan.baidu.com/s/1741 7zF-IlP6lW_ut7lRTqA (accessed on 14 October 2022) with extraction code *fcju* or GitHub platform https://github.com/ouguiliang110/DataSet_of_MAF-NBC (accessed on 14 October 2022).

**Acknowledgments:** The authors would like to thank the editors and anonymous reviewers who carefully read the paper and provided valuable suggestions that considerably improved the paper. This paper was recommended by the 6th Asian Conference on Artificial Intelligence Technology (ACAIT 2022) which is the 2022 Top Academic Conference recognized by China Association for Science and Technology (CAST).

**Conflicts of Interest:** The authors declare no conflicts of interest.

## References

1. Wu, X.; Kumar, V.; Quinlan, J.R.; Ghosh, J.; Yang, Q.; Motoda, H.; McLachlan, G.J.; Ng, A.; Liu, B.; Yu, P.S.; et al. Top 10 algorithms in data mining. *Knowl. Inf. Syst.* **2008**, *14*, 1–37. [CrossRef]
2. Gil-Begue, S.; Bielza, C.; Larrañaga, P. Multi-dimensional Bayesian network classifiers: A survey. *Artif. Intell. Rev.* **2021**, *54*, 519–559. [CrossRef]
3. John, G.; Langley, P. Estimating continuous distributions in Bayesian classifiers. In Proceedings of the 11th Conference on Uncertainty in Artificial Intelligence, Montreal, QC, Canada, 18–20 August 1995; pp. 1–8.
4. Friedman, N.; Geiger, D.; Goldszmidt, M. Bayesian network classifiers. *Mach. Learn.* **1997**, *29*, 131–163. [CrossRef]
5. Webb, G.I.; Boughton, J.R.; Wang, Z. Not so naive Bayes: aggregating one-dependence estimators. *Mach. Learn.* **2005**, *58*, 5–24. [CrossRef]
6. Jiang, L.; Zhang, H.; Cai, Z. A novel Bayes model: Hidden naive Bayes. *IEEE Trans. Knowl. Data Eng.* **2009**, *21*, 1361–1371. [CrossRef]
7. Jiang, L.; Li, C.; Wang, S.; Zhang, L. Deep feature weighting for naive Bayes and its application to text classification. *Eng. Appl. Artif. Intell.* **2016**, *52*, 26–39. [CrossRef]
8. Jiang, L.; Zhang, L.; Li, C.; Wu, J. A correlation-based feature weighting filter for Naive Bayes. *IEEE Trans. Knowl. Data Eng.* **2018**, *31*, 201–213. [CrossRef]
9. Bressan M.; Vitria J. Improving naive Bayes using class-conditional ICA. In Proceedings of the 2022 Ibero-American Conference on Artificial Intelligence, Seville, Spain, 12–15 November 2002; pp. 1–10.
10. Qin, F.; Ren, S.L.; Cheng, Z.K.; et al. Bayes classification model based on ICA. *Comput. Eng. Des.* **2007**, *28*, 4873–4877.
11. Fan, L.; Poh, K.L. A comparative study of PCA, ICA and class-conditional ICA for naive Bayes classifier. In Proceedings of the 2007 International Work-Conference on Artificial Neural Networks, Porto, Portugal, 9–13 September 2007; pp. 16–22.
12. Jayanthi, S.K.; Sasikala, S. Naive Bayesian classifier and PCA for web link spam detection. *Comput. Sci. Telecommun.* **2014**, *41*, 3–15.
13. Zhang, B.; Liu, Z.; Jia, Y.; Ren, J.; Zhao, X. Network intrusion detection method based on PCA and Bayes algorithm. *Secur. Commun. Netw.* **2018**, *2018*, 1914980. [CrossRef]
14. Kurgan, L.A.; Cios, K.J. CAIM discretization algorithm. *IEEE Trans. Knowl. Data Eng.* **2004**, *16*, 145–153. [CrossRef]

15. Li, M.; Zhao, S.; Chen, S.; Huang, D. A new method of online fault diagnosis based on incremental continuous attribute naive Bayesian. In Proceedings of the 2018 IEEE International Conference on Prognostics and Health Management, Seattle, WA, USA, 11–13 June 2018; pp. 1–7.
16. Scanagatta, M.; Salmerón, A.; Stella, F. A survey on Bayesian network structure learning from data. *Prog. Artif. Intell.* **2019**, *8*, 425–439. [CrossRef]
17. Parzen, E. On estimation of a probability density function and mode. *Ann. Math. Stat.* **1962**, *33*, 1065–1076. [CrossRef]
18. Yang, Y.; Webb, G.I. A comparative study of discretization methods for naive-Bayes classifiers. In Proceedings of the 2002 Pacific Rim Knowledge Acquisition Workshop, Tokyo, Japan, 18–22 August 2002; pp. 159–173.
19. Yang, Y.; Webb, G.I. On why discretization works for naive-Bayes classifiers. In Proceedings of the 2003 Australasian Joint Conference on Artificial Intelligence, Perth, Australia, 3–5 December 2003; pp. 440–452.
20. Yang, Y.; Webb, G.I. Discretization for naive-Bayes learning: managing discretization bias and variance. *Mach. Learn.* **2009**, *74*, 39–74. [CrossRef]
21. He, Y.L.; Ye, X.; Haung, D.F.; Fournier-Viger, P.; Huang, J.Z. A hybrid method to measure distribution consistency of mixed attribute data sets. *IEEE Trans. Artif. Intell.* **2022**. [CrossRef]
22. Qi, Z.; Wang, B.; Tian, Y.; Zhang, P. When ensemble learning meets deep learning: A new deep support vector machine for classification. *Knowl.-Based Syst.* **2016**, *107*, 54–60. [CrossRef]
23. Feng, J.; Yu, Y.; Zhou, Z.H. Multi-layered gradient boosting decision trees. In Proceedings of the 32nd Conference on Neural Information Processing Systems, Montreal, QC, Canada, 3–8 December 2018; pp. 1–11.
24. Guo, M.; Fei, J.; Meng, Y. Deep nearest neighbor website fingerprinting attack technology. *Secur. Commun. Netw.* **2021**, *2021*, 5399816. [CrossRef]
25. De Freitas, J.F.; Niranjan, M.; Gee, A.H.; Doucet, A. Sequential Monte Carlo methods to train neural network models. *Neural Comput.* **2000**, *12*, 955–993. [CrossRef] [PubMed]
26. Alcalá-Fdez, J.; Fernández, A.; Luengo, J.; Derrac, J.; García, S.; Sánchez, L.; Herrera, F. KEEL data-mining software tool: data set repository, integration of algorithms and experimental analysis framework. *J.-Mult.-Valued Log. Soft Comput.* **2011**, *17*, 255–287.
27. Helal, A.; Otero, F.E. A mixed attribute approach in ant-miner classification rule discovery algorithm. In Proceedings of the Genetic and Evolutionary Computation Conference, Denver, CO, USA, 20–24 July 2016; pp. 13–20.
28. Demšar, J. Statistical comparisons of classifiers over multiple data sets. *J. Mach. Learn. Res.* **2006**, *7*, 1–30.
29. Wang, X.Z.; He, Y.L.; Wang, D.D. Non-naive Bayesian classifiers for classification problems with continuous attributes. *IEEE Trans. Cybern.* **2013**, *44*, 21–39. [CrossRef] [PubMed]

*Article*

# A Distributed Bi-Behaviors Crow Search Algorithm for Dynamic Multi-Objective Optimization and Many-Objective Optimization Problems

**Ahlem Aboud [1,2], Nizar Rokbani [2,3], Bilel Neji [4,*], Zaher Al Barakeh [4], Seyedali Mirjalili [5,6] and Adel M. Alimi [2,7]**

1    University of Sousse, ISITCom, Sousse 4011, Tunisia; aboud.ahlem.tn@ieee.org
2    REGIM Lab: REsearch Groups in Intelligent Machines, University of Sfax, National Engineering School of Sfax (ENIS), BP 1173, Sfax 3038, Tunisia; nizar.rokbani@ieee.org (N.R.); adel.alimi@ieee.org (A.M.A.)
3    High Institute of Applied Science and Technology of Sousse, University of Sousse, Sousse 4003, Tunisia
4    College of Engineering and Technology, American University of the Middle East, Egaila 54200, Kuwait; zaher.al-barakeh@aum.edu.kw
5    Yonsei Frontier Lab, Yonsei University, Seoul 03722, Korea; seyedali.mirjalili@alumni.griffithuni.edu.au
6    Centre for Artificial Intelligence Research and Optimisation, Torrens University Australia, Brisbane 2607, Australia
7    Department of Electrical and Electronic Engineering Science, Faculty of Engineering and the Built Environment, University of Johannesburg, Johannesburg 2006, South Africa
*    Correspondence: bilel.neji@aum.edu.kw

**Citation:** Aboud, A.; Rokbani, N.; Neji, B.; Al Barakeh, Z.; Mirjalili, S.; Alimi, A.M. A Distributed Bi-Behaviors Crow Search Algorithm for Dynamic Multi-Objective Optimization and Many-Objective Optimization Problems. *Appl. Sci.* **2022**, *12*, 9627. https://doi.org/10.3390/app12199627

Academic Editors: Yue Wu, Xinglong Zhang and Pengfei Jia

Received: 4 August 2022
Accepted: 17 September 2022
Published: 25 September 2022

**Publisher's Note:** MDPI stays neutral with regard to jurisdictional claims in published maps and institutional affiliations.

**Abstract:** Dynamic Multi-Objective Optimization Problems (DMOPs) and Many-Objective Optimization Problems (MaOPs) are two classes of the optimization field that have potential applications in engineering. Modified Multi-Objective Evolutionary Algorithms hybrid approaches seem to be suitable to effectively deal with such problems. However, the standard Crow Search Algorithm has not been considered for either DMOPs or MaOPs to date. This paper proposes a Distributed Bi-behaviors Crow Search Algorithm (DB-CSA) with two different mechanisms, one corresponding to the search behavior and another to the exploitative behavior with a dynamic switch mechanism. The bi-behaviors CSA chasing profile is defined based on a large Gaussian-like Beta-1 function, which ensures diversity enhancement, while the narrow Gaussian Beta-2 function is used to improve the solution tuning and convergence behavior. Two variants of the proposed DB-CSA approach are developed: the first variant is used to solve a set of MaOPs with 2, 3, 5, 7, 8, 10,15 objectives, and the second aims to solve several types of DMOPs with different time-varying Pareto optimal sets and a Pareto optimal front. The second variant of DB-CSA algorithm (DB-CSA-II) is proposed to solve DMOPs, including a dynamic optimization process to effectively detect and react to the dynamic change. The Inverted General Distance, the Mean Inverted General Distance and the Hypervolume Difference are the main measurement metrics used to compare the DB-CSA approach to the state-of-the-art MOEAs. The Taguchi method has been used to manage the meta-parameters of the DB-CSA algorithm. All quantitative results are analyzed using the non-parametric Wilcoxon signed rank test with 0.05 significance level, which validated the efficiency of the proposed method for solving 44 test beds (21 DMOPs and 23 MaOPS).

**Keywords:** beta function; crow search algorithm; dynamic multi-objective optimization problems; evolutionary algorithm; many-objective optimization problems

## 1. Introduction

During the last decade, a wide range of meta-heuristics have been designed to solve many complex problems based on Evolutionary Algorithms (EA), such as the Genetic Algorithm (GA) [1], and Swarm Intelligence (SI) approaches, such as the Particle Swarm Optimization (PSO) algorithm [2–5]. Different Multi-Objective Evolutionary Algorithms (MOEAs) have been employed to solve static Single Objective Optimization Problems

(SOPs) and static Multi-Objective Optimization Problems (MOPs), where the main challenge is to find the best solution for SOP and a set of optimal solutions when solving MOP to balance the convergence and diversity in the search space. However, this process becomes more challenging when solving Dynamic Multi-Objective Optimization Problems (DMOPs), characterized by several types of time-varying Pareto Optimal Sets (POSs) and Pareto Optimal Fronts (POFs) [6]. Equations (1) and (2) investigate the main differences between the static Multi-Objective Optimization Problem (MOP), static Many-Objective Optimization Problem (MaOP) and dynamic Multi-Objective Optimization Problem (DMOP). First, the common characteristics are as follows:

- Each problem (MOP, MaOP, DMOP) has a fitness function $F(X)$, which can be minimized or maximized with simultaneously performed $M$ objectives.
- Each problem has a set of solutions, $X$, presented by a set of bounded decision variables, $X$, with a d-dimensional search space generated between the minimum ($X_{min}$) and the maximum ($X_{max}$) boundaries.
- The search space of each problem can be limited by a set of inequality and equality constraint.

$$F(X) = \begin{cases} F(X) = (f_1(X), \ldots, f_M(X)) \\ \text{Subject to}: g(X_i) \leq 0 \text{ or } g(X_i) \geq 0, h(X_i) = 0 \\ \forall i = 1, \ldots, d, x \in [X_{min}, X_{max}] \\ 1 < M \leq 3 \end{cases} \tag{1}$$

where $F(X)$ is the fitness function, $M$ is the number of objectives, and $g(X)$ and $h(X)$ are the inequality and the equality constraints, respectively.

The main difference between the static MOP, MaOP and DMOP is investigated by looking at the number of objectives, nature of decision space, objective space, and constraints.

- For the MOP and the DMOP, the number of objectives ($M$) is fixed to 2 or 3 ($1 < M \leq 3$) and ($M$) is upper than 3 for MaOP ($M > 3$).
- However, the DMOP has a time-varying decision variables, objectives and/or constraints, and is presented in Equation (2).
- The DMOP has M dynamic objective functions $F(X, t)$ and a set of time-varying variables $X(t) = X_1(t), \ldots, X_n(t)$, which are subject to different bounded constraints, limiting the search space.

$$F(X,t) = \begin{cases} F(X,t) = (f_1(X,t), f_2(X,t), \ldots, f_M(X,t)) \\ \text{Subject to}: g_i(X,t) \leq 0 \text{ and } 0h_j(X,t) = 0 \\ \forall i = 1, \ldots, d_g(t) \text{ and }, j = 1, \ldots, d_h(t)] \\ X \in [X_{min}, X_{max}], t \in [t_{begin}, t_{end}] \\ X \in \Omega_X, t \in \Omega_t \end{cases} \tag{2}$$

where $M$ is the number of conflicting objective functions, $d_g(t)$ and $d_h(t)$ are the number of inequality and quality constraints at time $t$, respectively, and $X$ is a set of bounded decision variables with a d-dimensional search space generated between minimum boundary ($X_{min}$) and maximum boundary ($X_{max}$). $F(X,t)$ is the objective vector that optimizes solution $X$ at time $t$. $\Omega_X \subseteq \mathbb{R}^n$ is the decision space, and $\Omega_t \in \mathbb{R}$ is the time space, bounded between the starting time $t_{begin}$ and the ending time $t_{end}$. The objective vector is denoted by $F(X,t) : \Omega_X \times \Omega_t \rightarrow \mathbb{R}^{M_t}$, presenting the resulting values for each solution $X$ at time $t$.

Generally speaking, Dynamic Multi-Objective Evolutionary Algorithms (DMOEAs) are designed to effectively detect and react to the changes that may affect the POS and POF, while conserving both convergence and diversity concepts [7,8]. However, Evolutionary Dynamic Optimization (EDO) approaches include an explicit or implicit mechanisms to detect and correctly react to the dynamic change. A change detection mechanism can be maintained through detectors using a feasible search population, such as the current best solutions, memory of optimal solutions or some predefined sub-population. Furthermore, this can be assumed separately to the search space using a set of random selected solutions, a fixed point, a regular grid of solutions or a set of determined points. In addition, the algorithm behaviors have considered a robust detection strategy based on the average of best-found solutions, time-varying observation of different sub-swarms, diversity of the solutions compared to the success rate, time-varying distributions and statistical methods. Increasing the mutation rate (hyper-mutation) or adding a randomly new member and relocating some useful solutions are the main mechanisms used to manage the loss of diversity in a dynamic search space which may fall within undetected regions of potential solutions.

The efficiency of standard MOEAs significantly decreases when dealing with Many-Objective Optimization Problems (MaOPs) where the number of objectives that need to be satisfied is higher than 3. Furthermore, three main issues are introduced when solving MaOPs, including: (i) the inutility of the dominance operator when dealing with a large number of objectives, (ii) the loss of diversity and premature convergence, and (iii) the exponentially increase of the population size. The Crow Search Algorithm (CSA) [9] is a meta-heuristic that simulates the social behavior of crows for food-searching. Crows are characterized by their ability to memorize food sources, as well as sources that other member of the flock may hold or hide. The CSA algorithm was first proposed as a mono-objective optimization technique and then extended to solve static Multi-Objective Problem (MOP) and constrained engineering optimization problems, in which the algorithm showed a relative effectiveness in comparison with techniques such as the harmony search (HS) [10], GA [1] and PSO approach. The main contribution is as follows:

- This contribution presents a novel Distributed Bi-behaviours Crow Search Algorithm (DB-CSA) to solve both Dynamic Multi-Objective Optimization Problems (DMOPs) and static Many-Objective Optimization Problems (MaOPs), which have not yet been performed using the standard CSA algorithm.
- The main difference between the original CSA algorithm and the new DB-CSA algorithm is as follows: the proposed DB-CSA approach presents two new chasing profiles, denoted by Beta Distribution profiles over the large Gaussian Beta-1 function for diversity enhancement and the narrow Gaussian Beta-2 function for convergence improvements.
- The proposed approach tends to achieve a dynamic balance between exploitation and exploration at each iteration during the optimization process, which makes it more suitable for both dynamic multi-objective optimization and many-objective optimization.
- A dynamic optimization mechanism is considered within the proposed DB-CSA algorithm when solving DMOPs to detect the time-varying POS and POF, effectively reacts to the dynamic changes and is denoted by the second variant (DB-CSA-II). This process aims to manage and control the dynamic change in a time-varying search space.

The reminder of this manuscript is organized as follows: Section 2 presents an overview of the best-known Dynamic Multi-Objective Optimization methods, Many-Objective Optimization Approaches and some existing Crow Search Algorithms based-methods. Section 3 presents the proposed Distributed Bi-behaviours Crow Search Algorithm (DB-CSA). Section 4 details the experimental evaluation, which is based on two comparative studies: one for DMOPs and the second for MaOPs. The results are presented in terms of their mean and standard deviation. Then, a statistical comparison between the proposed DB-CSA algorithm and the state-of-art methods is carried out using the non-parametric Wilcoxon signed rank test. Finally, Section 5 concludes this paper and presents some future work.

## 2. State-of-the-Art on Evolutionary Multi-Objective Optimization

This section presents a set of comparable DMOEAs and MaOEAs, designed for both Dynamic Multi-Objective Optimization and Many-Objective Optimization, presented in Sections 2.1 and 2.2, respectively. In addition, the crow search-based methods are given in Section 2.3.

### 2.1. Dynamic Multi-Objective Optimization Methods

Several Dynamic Multi-Objective Evolutionary Algorithms (DMOEAs) have been designed in the literature to solve DMOPs with time-varying objective, variables or constraints. A set of these are visible in Table 1. Five groups of DMOEAs are available in the literature to solve DMOPs: diversity-based techniques, memory-based approaches, prediction methods, parallel systems and transfer learning-based algorithms. The diversity-based approach [1] has shown the ability to solve dynamic problems with continuous and small time-varying parameters, and showed its limit when faced a severe environmental changes. Furthermore, the DMOPs have presented some periodical or recurrent changes, making storing the historical experience of solutions useful in preserving diversity. The dynamic non-dominated sorting genetic algorithm II (DNSGA-II) [1] is proposed to enhance the diversity of solutions when solving DMOPs. In DNSGA-II, a set of solutions was randomly selected for use as detectors and re-evaluated after each change. Then, if a change was detected, all selected solutions were re-initialized or hyper-mutated.

Memory-based approaches use redundant representations of an evolutionary algorithm, using extra-memory components to detect future changes [11]. These approaches are very effective to solve DMOPs with periodically time-varying properties. However, such mechanisms slow the convergence and strengthen diversity in the EDO approaches. The main disadvantage of memory-based algorithms is the ineffectiveness of the redundant solutions stored in the archive. However, prediction-based methods tend to predict changes based-on limited patterns. Such a system can quickly detect the best global solution, but fail when the changes are stochastic, which increases the relative training error rates. The Steady-State and Generational Evolutionary Algorithm (SGEA) [11] was designed to effectively detect and react to the change in a steady-state manner. If a change is detected, a number of good solutions are re-used in the next processing step; then, a combination of previous and new solutions are used to approximate the new Pareto optimal front.

The parallel approaches present an optimization process over multiple sub-swarms, which can handle the problem over a separate search space and are recommended for multi-modal problems, while being computationally expensive. A key challenge for these methods is finding an appropriate number of sub-swarms and their sizes. The Competitive-Cooperative Co-evolutionary Algorithm (dCOEA) in [12] aims to track the time-varying POF based on the decomposition of the optimization process. However, only the winners of each sub-population are used to manage the optimal solutions. The MOEA/D [13] is a decomposition-based approach, which aims to subdivide the population into several sub-populations and solve many sub-problems separately and simultaneously, making the MOEA/D system lower and more time-consuming.

The prediction-based methods were developed based on machine-learning algorithms and can efficiently determine and optimize the initial population based on previous experience. However, the main limit is the insufficiency of useful knowledge at the beginning of the optimization process, and they are time-consuming. The population prediction strategy (PPS) [14] is a prediction-based method, which divides the non-dominated solutions into a centerpoint and a manifold; then, both are used to predict the future center point and manifold, respectively. When a change is detected based a population, re-initialization is operated. Transfer-learning-based techniques are reliable alternatives for DMOPs, based on the use of MOEA/D [13] as a baseline system. In 2020, the new memory-driven manifold transfer learning was proposed, based on the evolutionary algorithm (MMTL-MOEA/D) [15]. This approach combined the memory mechanism to preserve the previous best solutions and the manifold transfer learning feature to estimate the best solutions, so that the best solutions are conserved and set as the initial population of the next generation.

**Table 1.** Classification of the MOEAs for DMOPs.

| | |
|---|---|
| Diversity-Based Approaches | DNSGA-II [1] |
| Memory-Based Approaches | SGEA [11] |
| Prediction-Based methods | PPS [14] |
| Parallel Approaches | MOEA/D [13] |
| | dCOEA [12] |
| Transfer Learning-Based Methods | MMTL-MOEA/D [15] |
| | RI-MOEA/D [15] |
| | SVR-MOEA/D [16] |
| | Tr-MOEA/D [17] |
| | KF-MOEA/D [18] |

The random re-initialization mechanism in (RI-MOEA/D) [15] selects 10% of the selected populations after each change to maintain the diversity. A combination of PPS [14] and the MOEA/D is considered in the PPS-MOEA/D algorithm to solve the DMOP. The support vector regression (SVR) based on evolutionary algorithm (SVR-MOEA/D) is proposed in [16], designed to solve the nonlinear correlation between two historical optimization processes. The SVR is used to predict a new population after each change in the search space. A transfer-learning-based dynamic multi-objective evolutionary algorithm (Tr-MOEA/D) is proposed in [17], aiming to solve the issue of non-independent and identically distributed data in a dynamic environment. The Tr-MOEA/D system implements a transfer learning mechanism to reuse the previous historical population after each change, which speeds up the optimization process. In the KF-MOEA/D [18] system, a Kalman filter (KF) is used to predict a new population prior to performing the convergence concept. The transfer-learning-based method adapts a set of machine-learning techniques to improve the performance of heuristics when solving DMOPs. The transfer-learning-based method aims to re-use previous computational experience to improve the efficiency of the newly generated populations after each detected change. However, this category of methods presents a major limit in the parameter-tuning procedure, which is a time-consuming process and requires trial and error [17,18].

### 2.2. Many-Objective Optimization Methods

Many-objective optimization algorithms are designed to manage the issues related to the exploitation and exploration concepts to preserve convergence and diversity in the search space. Table 2, presents several Many-Objective Evolutionary Algorithms (MaOEAs) and is classified into five classes, namely, decomposition-based methods, indicator-based approaches, diversity-based selection criterion, modified dominance relation-based approaches and preference-based approaches. Many Pareto-based approaches show their limits to rank and determine the set non-dominated solutions using the dominance operator since a high number of solutions have a large number of objectives leading to the poor convergence implied by the Active Diversity Promotion (ADP) phenomenon [19]. As a solution, a variety of enhancements have been adopted in the original MOEAs when solving MaOPs, including the decomposition-based and indicator-based approaches. Decomposition mechanisms combine multiple objectives into a single problem or a series of sub-problems. Some popular techniques of this type are Pareto sampling [20], improved Pareto sampling (MSOPS-II) [21] and Multi-Objective Evolutionary Algorithm based on Decomposition (MOEA/D) [13]. The decomposition-based approach becomes more effective with a set of sub-MOPs, such as those presented in the Reference Vector-Guided Evolutionary Algorithm (RVEA) [22], MOEA/D-M2M [23], NSGA-III [24], MOEA/DD [25] and the MOEA/D-ROD [26]. In addition, a set of performance metrics are considered to guide the optimization process over different indicator-based approaches, such as the fast hypervolume-based evolutionary algorithm (HypE) [27], S-metric-selection-based evolutionary multi-objective algorithm (SMS-EMOA) [28], indicator-based evolutionary algorithm (IBEA) [29], and Evolutionary Many-Objective Optimization Algorithm based on an IGD Indicator with Region Decomposition [30] and MaOEA/IGD [31].

A set of dominance operators are proposed to deal with the issue of the ineffectiveness of sorting non-dominated solutions using the dominance operator proposed in the Pareto-based methods. The most know novel dominance operators as the following; L-optimality [32], $\varepsilon$-dominance [33], fuzzy dominance [34], the Grid-based Evolutionary Algorithm (GrEA) [35], $\theta$ Dominance-based Evolutionary Algorithm ($\theta$-DEA) [36] and the preference order ranking procedure [37]. Diversity management techniques are proposed to arrange a good balance between the convergence and the diversity when solving MaOPs. In [35], a three grid-based criterion was proposed to maintain diversity, including the grid crowding distance, the grid coordinate point distance and the grid ranking. A diversity promotion mechanism, DM, is introduced in [38] to activate or disactivate the diversity of the population based on the spread and the crowding distance of solutions. In NSGA-III algorithm [24], the reference point-based strategy is used to solve MaOPs. The shift-based density estimation (SDE) strategy [39] was utilized to replace the dominance operators of MOEAs. The knee point-driven evolutionary algorithm (KnEA) [40] was developed using both knee point-based selection and dominance-based selection. Three groups of preference-based approaches, including prior algorithms, interactive algorithms and posterior algorithms, were employed to deal with the issue of population size limitations regarding the large dimension of the objective space. The best-known posterior approaches are the Preference-Inspired Co-evolutionary Algorithms (PICEA-g) [41], the novel, two-archive algorithm (TAA) [42], and its improved version (Two_Arch2) [43].

In addition, the Particle Swarm Optimization (PSO) algorithm received a great attention in MaOPs. The Control Dominance Area of Solutions (CDAS) [44] has been used with SMPSO and SigmaMOPSO for MaOPs. Indicator-based PSO systems were proposed to maintain the leader's selection using the R2 indicator, as presented in H-MOPSO [45] or the hypervolume metric in S-MOPSO [46]. A two-stage strategy and a parallel cell coordinate system were adopted in MaOPSO/2s-pccs [47]. A preference-based method was proposed, using a PSO system, focusing on solutions around the knee point and called knee-driven particle swarm optimization (KnPSO) [48]. In [49], the MaPSO method uses the leader's selection from a certain number of historical solutions using scalar projection. In addition, the HGLSS-MOPSO algorithm [50] adopted the Hybrid Global Leader Selection (HGLSS) using two global leader selection mechanisms: the first for exploration and the second for exploitation. A recently published paper [51] presented an adaptive localized decision variable analysis approach under the decomposition-based framework to solve the Large-Scale Multi-Objective Optimization problems and Multi-Tasking Optimization Problems in MaOPs. As a conclusion, all the abovementioned Many-Objective Evolutionary Algorithms (MaOEAs) are presented as highly complex and time-consuming systems, especially when using decomposition-based mechanisms and/or quality indicators to separately deal with convergence and diversity.

The high number of objectives is managed through a decomposition-based mechanism to obtain a set of single or multi-objectives problems. The multiple single-objective Pareto sampling (MSOPS) [20] algorithm generates a set of target vectors. Then, it undergoes multiple single-objective optimization processes to solve MaOP. Such a strategy may return moderated results, since it does not correctly manage the multi-objectives of MaOPs. The enhanced MSOPS-II [19] uses a set of target vectors based on the current population to guide the optimization process at each iteration. Then, the aggregation of fitness functions is used to evaluate the performances of the proposed solutions. The MOEA/D [13] algorithm proceeds with a decomposition of the many-objectives into a set of single objectives using a uniformly distributed weight vector. Similar to the weight vectors of the MOEA/D algorithm, the NSGA-III [24] used a number of well-spread reference points to approximate non-dominated solutions; then, it enhances the diversity of the population. Based on the main idea of the NSGA-II, the reference vector-guided evolutionary algorithm (RVEA) [22] adopted two reference vectors: one for selection and the second for adaptation. In the RVEA system, the concept of convergence and diversity are dynamically managed using the Angle Penalized Distance (APD).

A vector angle-based evolutionary algorithm (VaEA) [52] is proposed for Unconstrained MaOPs. The VaEA algorithm uses the maximum-vector-angle as a selection mechanism to guarantee a good distribution to approximate to the true Pareto front, while the worse solutions are replaced with newly generated random solutions. The $\theta$-DEA [36] system is based on NSGA-III with a new, $\theta$-non-dominated concept, which is different to the original dominance operator, as used on the Pareto-based methods. It employs a set of reference points to cluster the solutions set and enhance the exploration phase. The NSGA-II/SDR is a modified version of the NSGA-II, with a Strengthened Dominance Relation (SDR), presented in [53] to solve MaOP. The NSGA-II/SDR adopts the angle and the niching mechanism to select the best converged solutions. MOEA/DD, MOEA dominance and decomposition [25] form a hybridization of the MOEA/D [13] and the NSGA-III [24], where the multiple objectives are decomposed into sub-problems and a dominance criterion is used to aggregate the global solution. Different grid-based criteria, such as the grid crowding distance (GCD), the grid ranking (GR) and the grid coordinate point distance (GCPD) are integrated in MOEAs to evaluate the fitness function of the MaOP. In addition, the GrEA system [35], is designed to maintain a good balance between convergence and diversity over both the grid dominance and grid difference to evaluate the fitness function and push the system toward the most optimal solutions. Two variants of the Pareto-based evolutionary algorithm using the penalty mechanism (PMEA) are presented in [54]: the MPEA-MP and the MPEA*-MA. The PMEA-MA is developed using the Manhattan-distance and the cosine distance as the convergence and distribution metrics; it includes a population-prepossessing to enhance the diversity. The second variant, PMEA*-MA, is a simplified one, which does not adopt the prepossessing step.

The Angle-based Selection algorithm (AnD) [55] is a non-Pareto-based method that aims to maintain the diversity of the population using an angle-based selection technique, then picks optimized members in the same search direction as a sorting solution. A hybridization of the Strength Pareto Evolutionary Algorithm (SPEA) and the shift-based density estimation (SDE) strategy in [39] is denoted by (SPEA/SDE). This estimates the density of the population; then, individuals who do not converge are eliminated to enhance diversity among the divergent solutions. In [56], the SPEAR leverages a reference direction-based density estimator using the standard SPEA algorithm for multi/many-objective optimization problems. The knee point-driven evolutionary algorithm (KnEA), proposed in [40], evolves a population and selects non-dominated solutions based on knee point criterion, which may be assumed to be a Pareto strategy. Furthermore, the two-stage evolutionary algorithm (TSEA) is developed in [57]. In the first stage, several sub-populations are optimized to converge to different regions of the Pareto front; then, the non-dominated solutions of each sub-population are considered as individuals that could be optimized in the second stage.

In indicator-based methods, several quality metrics are used to perform the optimization process; for example, the Monto Carlo simulation is used in the HypE algorithm [27] to minimize the computation cost and approximate the results. The preference-based approaches use different adaptation mechanisms like an external memory for convergence and diversity preservation as well as to make a decision regarding the true Pareto front. In [41], the PICEA-g algorithm integrates the co-evolution as a posterior adaptation mechanism with a set of candidate solutions to help with decision-making and approximate the entire of the POF. Two archives are used in the Two_Arch2 [43] system: the first is considered for convergence (CA) and the second is used to maintain diversity (DA). A crossover operator is used as a selector mechanism between the CA and DA and a mutation operator is used in CA memory.

**Table 2.** Classification of the MOEAs for MaOPs.

| | |
|---|---|
| Decomposition-based approaches | MSOPS [20] and MSOPS-II [19] |
| | MOEA/D [13] |
| | MOEA/DD [25] |
| | TSEA [57] |
| | MPEA-MP and MPEA*-MA [54] |
| Indicator-based approaches | HypE [27] |
| Diversity-based selection criterion | NSGA-III [24] |
| | SPEA/SDE [39] |
| | KnEA [40] |
| | SPEAR [56] |
| Modified dominance relation-based approaches | GrEA [35] |
| | VaEA [52] |
| | $\theta$-DEA [36] |
| | NSGA-II/SDR [53] |
| | AnD [55] |
| Preference-based approaches | RVEA [22] |
| | PICEA-g [41] |
| | Two_Arch2 [43] |

*2.3. Existing Crow Search-Based Methods*

The Crow Search Algorithm (CSA) [9] was first proposed in 2016 to solve constrained engineering optimization problems. In [58], Meriahi et al. published a new overview paper to present all modified versions of the CSA algorithm. A set of novel CSA algorithms are extended for solving MOPs. A Multi-Objective Crow Search Algorithm (MOCSA) is proposed in [59], in which chaos and orthogonal opposition-based operators are used to hybridize CSA, (M2O-CSA) with a focus on solving MOPs. Also, a hybridization of the CSA algorithm with a clustering model is published and denoted by the Multi-objective Taylor Crow Optimization algorithm (MOTCO) considered for solving the clustering-aware wireless sensor network [60]. Furthermore, two binary versions of the CSA algorithm are proposed in [61,62]. The first one is the BCSA [61], which is used as a V-shaped transfer function to obtain a binary representation of continuous data with applications in feature selection. The second binary CSA algorithm [62] consists of applying a sigmoid transformation for solving the 2D bin packing problem. Several modified versions of the CSA algorithm tend to manage the loss of diversity in the search space based on the Gaussian distribution for diversity enhancement as in [63]. A priority-based technique is used to determine the sufficient flight length for each crow and update their position based on a followed crow. This technique is considered for electromagnetic optimization of the usability factors' hierarchical model in a prediction model [64] as well as for the economic load dispatch problem [65]. Also, a modified crow search algorithm is proposed for solving a power system problems and aims to modify the standard CSA parameters such as the awareness probability and the random perturbation by a dynamic awareness probability (DAP) and a Lévy flights to simulate the evasion behavior of each crow [66]. Furthermore, Huang et al. [67] proposed a hybrid CSA algorithm (HCSA) based on the Variable Neighborhood Search (VNS) and the standard SCA method. The HCSA algorithm is proposed for dealing with an NP-hard combinatorial optimization problem, such as the permutation flow shop scheduling problem (PFSP), aiming to retrieve an actionable permutation order and handle a large number of jobs. An Improved Crow Search Algorithm was proposed by Primitivo et al. [68], to solve complex energy problems. The authors aim to modify the awareness probability (AP) and the random perturbation of the standard CSA algorithm with a new dynamic awareness probability (DAP) and Lévy flight distribution to balance exploration and exploitation in the search space. The robustness of the CSA algorithm is proven when solving different complex problems. Meddeb et al. [69] proposed a novel meta-heuristic approach based on the crow search algorithm (CSA) to solve the optimal reactive power dispatch (ORPD) problem.

A set of mechanisms has been used to improve the CSA algorithm, including a search bounds limits management strategy [70], adding an archive component [71], and restructuring awareness probabilities [72] to enhance the random perturbation and the dynamic probability of the CSA system. Several operators were added to achieve a good balance between convergence and diversity, such as the Roulette wheel selection tool and the inertia weight, the Levy flight and the adaptive adjustment factors. In addition, a crossover and a mutation operator were proposed to intrinsically hybridize CSA in [73], with applications in hybrid renewable energy PV/wind/battery system. Many hybridization methods have been developed to combine the CSA algorithm with the Grey Wolf Optimizer (GWO) [74], the Cat Swarm Optimization (CSO), the Crow PSO [75] and the Crow Search Mating-based Lion Algorithm [76].

### 3. The Proposed Distributed Bi-Behaviors Crow Search Algorithm

Different MOEAs were designed to solve the DMOP, and should be able to detect and respond to problem pattern changes. However, many modified evolutionary approaches have been designed for MaOP to deal with a high number of objective functions. The state-of-the-art optimization approaches designed for DMOPs and MaOPs are characterized by their complexity in terms of time and resources. This work proposes a new Distributed Bi-behaviours Crow Search Algorithm (DB-CSA) to dynamically manage both convergence and diversity concepts when solving both DMOPs and MaOPs. The new DB-CSA is classified as a diversity-based approach, combining the simplicity of the CSA algorithm and the flexibility of the Beta function [77] to produce several forms and configurations of distributions, including the normal Gaussian one. More details about the proposed DB-CSA are presented in the next subsection.

### 3.1. The Standard Crow Search Algorithm

The Crow Search Algorithm (CSA) was proposed by Askarzadeh in 2016 [9] as a metaheuristic to solve constrained engineering optimization problems. Crows are known to be a social bird with the ability to memorize and use food source positions when needed; these sources may be the result of a personal search or from the crows' social activities. The CSA algorithm mimics the crow flock's search mechanisms and uses them for optimization purposes. The search process is detailed in Algorithm 1, and begins with a random initialization of N crow's positions with d dimensional search space.

---

**Algorithm 1** The Standard Crow Search Algorithm (CSA)

---

Begin
Randomly initialize the position (X)of N crows
Evaluate the position of the crows
Initialize the memory of each crow
While iteration < Max-Iteration
    For $i = 1$ to $i \leq$ N do
        Randomly choose one ($i$) crow to follow
        Define the awareness probability $AP_i(t)$
        Define the flight length $Fl_i(t)$
        Update the crow position using Equation (1)
    End For
Check the feasibility of new positions
Evaluate the new position of the crows
Update the memory (*Mem*) of crows using Equation (2)
End While
End

---

Each crow $i$ is characterized by a position vector $X_i$, defined by: $X_i = X_1^i, X_2^i, \ldots, X_d^i$ and their memory position $Mem_i$ used to achieve the best food positions. All crows are flying in the search space and at each iteration we aim to optimize the fitness function

$Fit(X_i)$ of each crow based on the updated position and their memories. While exploring the search space for new food positions, a crow needs to remember the best location in which it hid its own food, and should remain aware if other crows discover this location. Assuming that the *i*-th crow decides to visit a previously memorized position at iteration *(t)* $(Mem_i, t)$, and assuming that congener (*i*) follows the crow (*i*), two controversial behaviors may occur, with each one represented by a particular state:

- The first state is when the crow *i* ignores being followed, and simply continues its search, considering what it previously found $(Mem_{i,t})$
- The second state is when the crow is aware of being followed; in this case, the crow will simply hide its food source and undergo a completely random search.
- These two position updates are detailed in Equation (3).

$$X_i(t-1) \begin{cases} //\text{State 1} \\ \text{If } R_j(t) \geq AP_j(t) \text{then:} \\ X_i(t) + R_i(t)Fl_i(t)(M_j(t) - X_i(t)) \\ \text{Else:} \\ //\text{State2:} \\ \text{random update} \end{cases} \tag{3}$$

where $R_i(t)$ is a random number with uniform distribution between the interval [0, 1] at iteration $t$, $Fl_i(t)$ is the flight length of the crow *i*, and $AP_i(t)$ is the awareness probability of the crow *i*. In the CSA algorithm, the balance between exploration and exploitation during the optimization process is achieved by the flight length (Fl) of the *i*th crow. However, the memory $Mem_i(t+1)$ of each crow *i* is updated using Equation (4). All the optimization processes are executed until a predefined maximum number of iterations is reached.

$$Mem_i(t-1) = \begin{cases} X_i(t+1) \text{ If } Fit(X_i(t+1)) \geq Fit(Mem_i(t)) \\ Mem_i(t) \text{ Otherwise} \end{cases} \tag{4}$$

### 3.2. The Distributed Bi-Behaviours Crow Search Algorithm (DB-CSA)

The Distributed Bi-behaviours Crow Search Algorithm (DB-CSA) is based on a Beta distribution profiles inspired from the Beta distributed PSO [78] developed for solving the Inverse Kinematics problem and considered to enhance the exploitation and exploration of solutions in the search space. The DB-CSA algorithm is developed to solve both static Many-Objective Optimization Problems (MaOPs) and Dynamic Multi-Objective Optimization Problems (DMOPs). Considering the state-of-the-art methods, the main motivation of the proposed DB-CSA is as follows:

- For MaOPs, the original meta-heuristics, such as the CSA algorithm [9] suffer from a robust mechanism to optimize a problem with a high number of objectives (more than three), which are optimized at the same time.
- For DMOPs, most of the existing approaches cannot manage the dynamic change in decision variables and objective values in the POS and the POF, respectively.

To overcome both limits, two optimization processes are separately proposed to solve static MaOPs and DMOPs. The detailed description of the proposed DB-CSA algorithms are as follows:

- The first variant (DB-CSA) has the same optimization process as the standard CSA algorithm [9] where the main difference is provided in the convergence and the diversity enhancement during the optimization process. A modified rules are considered in the first CSA algorithm to update the position of each crow. The first version of the proposed DB-CSA algorithm is developed to solve static MaOPs. The general flowchart is shown in Figure 1, and in the pseudo-code is given in Algorithm 2. More details are presented in Section 3.2.1.

- The second variant (DB-CSA-II) is proposed based on the first DB-CSA algorithm. The main difference is that this investigates the dynamic optimization mechanism to efficiently detect and react to the change when solving DMOPs with a time-varying Pareto Optimal Set (POS) and dynamic Pareto Optimal Front (POF). The general flowchart of the second version is shown in Figure 2 and the pseudo-code is presented by Algorithm 3. More details are presented in Section 3.2.2.

---

**Algorithm 2** The pseudo-code of the proposed Distributed Bi-behaviours Crow Search Algorithm (DB-CSA)

---

Begin
Randomly initialize the position (X) of the flock of N crows with d-dimensional search space;
Initialize the memory (*Mem*) of each crow;
Initialize the archive (A) to store the non-dominated solutions;
Evaluate the position of the crows;
while iteration < $Max_{Iterations}$ do
  for $i = 1$ to $i \leq$ N do
    Evaluate the fitness function of the crow $i$: $Fit(X_i(t))$
    Choose the followed crow ($i$) randomly;
    Determine the average crow $i$: Mean ( $Fit(X_i(t))$ )
      if $Fit(X_i(t)) \geqslant$ Mean( $Fit(X_i(t))$ )then
        Update the crow position using Equation (8) on Beta-1 exploitation profile;
      else
        Update the crow position using Equation (8) on Beta-2 exploration profile;
      end if
    Update the memory using Equation (4);
  end For
  Apply the mutation operators using Equations (9) and (10);
  Update the archive (A) of non-dominated solutions;
end while
Return the archive (A) of the non-dominated solutions;
End

---

### 3.2.1. First Variant: DB-CSA for Static MaOPs

The key processing steps in the first variant of the proposed DB-CSA approach are shown in Figure 1 and detailed as follows:

1. Initialization of population positions and their memories: In DB-CSA algorithm, each crow $i$ is presented as a potential solution in the search space. The DB-CSA starts with a random initialization of position (X) and the memory (*Mem*) of the flock of $N$ crows when each crow $i$ has presented a potential solution in the search space.
2. Initialization of the archive of the non-dominated solutions: the archive (A) is initially created to store all the non-dominated solutions during the optimization process. After that, all the following steps are executed until a predefined number of iterations is reached.
3. Fitness Function Evaluation: for each crow $i$, the fitness function $Fit(X_i)$ is evaluated.
4. Determine the followed crow $i$: at each iteration, one of the main behaviors of crow $i$ is to determine one crow $i$ to follow by selecting a random position value between zero and the size of the flock of best crows.
5. Determine the average crow $i$: the aggregated value of $K$ objectives are computed as the fitness function $Fit(X_i(t)$ of each crow $i$; then, the average value of all fitness functions is selected to determine the mean solution.
6. Update the crow position using the bi-behaviours' beta-distribution profiles
7. Update the memory (*Mem*): the memory of each crow $i$ is updated using Equation (4).
8. Apply the mutation operators

9. Update the archive of non-dominated solutions: at each time $t$ of the optimization procedure, all the non-dominated solutions are stored in the archive (A) based on the $\varepsilon - dominance$ operator.

10. Generate OUTPUT = the best Pareto solutions from the archive (A).

---

**Algorithm 3** The DB-CSA-II Algorithm with Dynamic Optimization Process

---

Begin
Randomly initialize the position $X$ of N crows with d-dimensional search space;
Initialize the memory *(Mem)* of each crow;
Initialize the archive (A) of non-dominated solutions;
Evaluate the position of the crows;
Initialize the flock of crows at iteration ($t$): $POS(t) \leftarrow \varnothing$;
Initialize the previous flock of crows at iteration ($t - 1$): $POS(t - 1) \leftarrow \varnothing$;
Initialize non-dominated solutions counter $n_p = 0$;
Initialize dominated solutions counter $n_q = 0$;
while iteration < $Max_{Iterations}$ do
for $i = 1$ to $i \leq$ N do
  Evaluate the fitness function of crow $i$: $Fit(X_i(t))$;
  Set $POS(t) \leftarrow POS(t) \cup solution_i(t)$;
  Set $POS(t - 1) \leftarrow POS(t - 1) \cup solution_i(t - 1)$
  if Dynamic Change = True then
    for $p = 1$ to $|POS(t)|$ do
      for $q = 1$ to $|POS(t - 1)|$ do
      Compare objective values $POS(t)$ and $POS(t - 1)$ using dominance operator;
        if $solution_p$ dominates $solution_q$ then
          $n_p = n_p + 1$;
          New population $POS(t + 1) \leftarrow POS(t + 1) \cup solution_p$;
        else if $solution_q$ dominates $solution_p$ then
          $n_q = n_q + 1$;
          Re-initialize $solution_p$;
        end if
      end for
    end for
    Update the archive (A) $\leftarrow non - dominated solutions$ $(POF(t))$;
  end if
Choose the followed crow ($i$) randomly;
Determine the average crow $i$: $Mean(Fit(X_i(t)))$
if $Fit(X_i(t)) \geqslant Mean(Fit(X_i(t))$ then
  Update the crow's position using Equation (8) on Beta-1 exploitation profile
else
  Update the crow's position using Equation (8) on Beta-2 exploration profile
end if
Update the memory using Equation (4)
end For
Apply the mutation operators using Equations (9) and (10)
Update the archive (A) of non-dominated solutions
end while
Return the archive (A) of the non-dominated solutions
End

---

In the standard CSA algorithm, the crow position is updated according to the Equation (3), while the convergence and the diversity stages are treated separately, causing premature convergence. However, this issue was treated by the new DB-CSA algorithm using a bi-behaviours beta-distribution profile to assume a dynamic and good balance between both stages. The two beta-distribution profiles are presented in Equation (8) and Figure 3, denoted by *Beta1_rand* and *Beta2_rand*, respectively, for exploitation and exploration enhancement. The couple of beta-profiles are used to modify the original Equation (3) of the standard CSA algorithm, presenting the update process for each crow $i$. The two profiles were proposed based on the beta-function of Alimi [77], as presented in Equations (5)–(7). The main advantage in using the beta function presented here is their capacity to produce several forms and configurations of distributions, including the normal Gaussian one. The one-dimensional Beta function is defined in Equation (5).

$$\beta(x, p, q, x_0, x_1) = \begin{cases} \left(\dfrac{x - x_0}{x_c - x_0}\right)^p \left(\dfrac{x_1 - x}{x_1 - x_c}\right)^q & \text{If } x \in [x_0, x_1] \\ \text{Otherwise} \end{cases} \tag{5}$$

where $p$, $q$, $x_0$ and $x_1$ are real values, with $(x_0 < x_1) \in I\!R$ , and $x_c$ is detailed in Equation (6).

$$x_c = \frac{(p \times x_1) + (q \times x_0)}{p + q} \tag{6}$$

However, the multi-dimensional beta function provided in the mathematical definition (7), presenting m product of the one-dimensional beta function in (5).

$$\beta(x) = \prod_{k=1}^{m} \beta(x_k, p_k, q_k, x_{0,k}, x_{1,k}) \tag{7}$$

The dynamic switch mechanism between the bi-behaviors' Beta-1 and Beta-2 profiles are assumed using a comparison between the fitness function $Fit(X_i(t))$ of each crow $i$ and the average solution (crow). If the fitness function $Fit(X_i(t)) = \sum_{k=1}^{k} f_k$ is greater than the mean value, we assume an exploration stage for the crow optimization process using the Beta-1 behaviour in Equation (8), which is used to update the crow position. Otherwise, the second Beta-2 behaviour in Equation (8) is considered, pushing each solution to the exploitation stage. As illustrated in Figure 3, the two beta-distribution profiles are detailed as follows:

- The first large Gaussian Beta-1 exploitation profile is used to update the crows positions which is characterized by a large standard deviation pushing the population to achieve good diversity in the search space with $p$ and $q$ variables of the beta-function in Equation (8), which are equal to 5.
- The second narrow Gaussian Beta-2 exploration profile adapts a limited standard deviation to update the crows positions with $p$ and $q$ in Equation (8), which are equal to 50 allowing for a good convergence to the optimal solution over time.

$$X_i(t+1) = \begin{cases} //\textbf{Beta 1 Behaviour for exploitation profile:} \\ \text{IF } Fit(X_i(t)) \geq Mean(Fit(X_i(t)) \text{ then:} \\ X_i(t) + Beta1\_rand(i) \times (Mem_i(t) - X_i(t)) \\ \text{Else:} \\ //\textbf{Beta 2 Behaviour for exploration profile:} \\ Beta2\_rand() \end{cases} \tag{8}$$

where Beta-1 is a beta random distribution over [0, 1], which assimilates to a fine search step around the optimal solution, while Beta-2 is more like a random exploration mechanism performed away from the previous optimal solution, $Mem_i(t)$. Both Beta-1 and Beta-2 values are determined using Equation (5) with different configurations of the two properties $p$ and $q$.

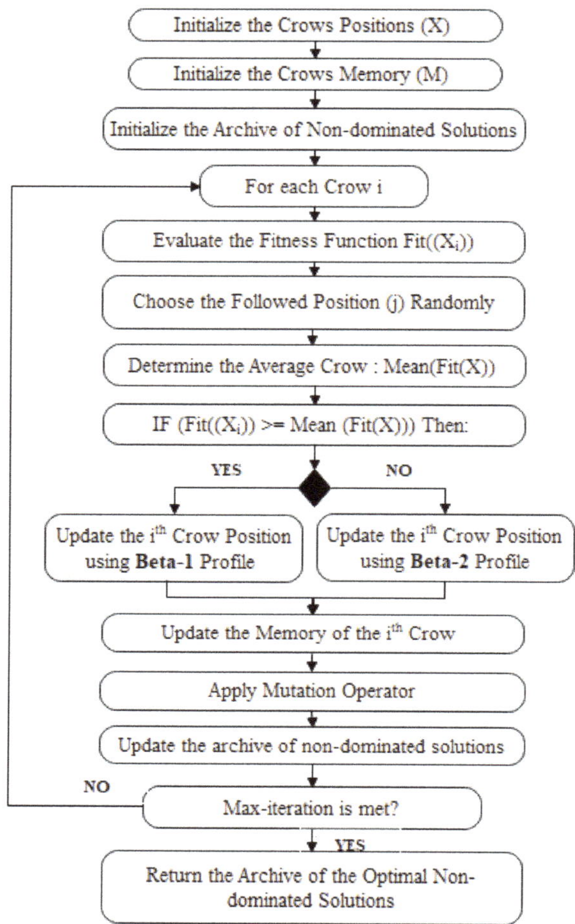

**Figure 1.** A flowchart of the proposed Distributed Bi-behaviours Crow Search Algorithm (DB-CSA).

The mutation operators in [79] are added to maintain more diversity in the flock of $N$ crows. The nonuniform and the boundary mutation operators in Equations (9) and (10) are applied to modify the variables $X_i = X_1^i, X_2^i, \ldots, X_d^i$ of each crow $i$, according to the probability mutation $P_m$ equal to $1/d$, where $d$ is the dimensional search space and $X_i \in [a_i, b_i]$ where $a_i$ and $b_i$ are the lower and upper bounds, respectively. The nonuniform mutation in Equation (9) is applied when the modulo value (mod) that divides the crow position $i$ by three is equal to zero. However, if the remainder is equal to one, the boundary mutation in Equation (10) is used. Otherwise, all variables are considered without mutation operators.

$$X_i' = \begin{cases} X_i + (b_i - X_i) \times \left( r_1 \times \left( 1 - \dfrac{iteration}{Max_{iterations}} \right)^b \right), \text{if } r_1 \geq 0.5,\ i \bmod 3 = 0 \\ X_i + (X_i - a_i) \times \left( r_2 \times \left( 1 - \dfrac{iteration}{Max_{iterations}} \right)^b \right), \text{if } r_2 > 0.5,\ i \bmod 3 = 0 \\ X_i, \text{otherwise} \end{cases} \tag{9}$$

where $r_1$ and $r_2$ are a random value of between 0 and 1.

$$X_i' = \begin{cases} a_i, if\, X_i + (r - 0.5 \times P_m) < a_i, i \bmod 3\ = 1 \\ b_i, if\, X_i + (r - 0.5 \times P_m) \geq b_i, i \bmod 3\ = 1 \\ X_i + (r - 0.5 \times P_m); \text{otherwise, where } r = U(0,1) \end{cases} \quad (10)$$

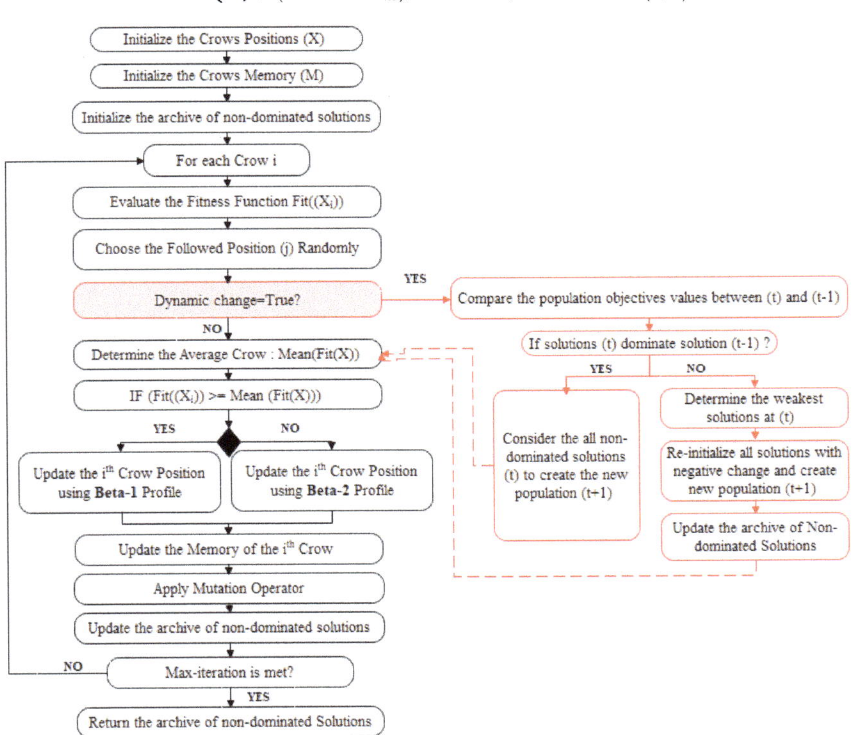


**Figure 2.** A modified flowchart of the proposed Distributed Bi-behaviours Crow Search Algorithm (DB-CSA-II) with Dynamic Optimization Process.

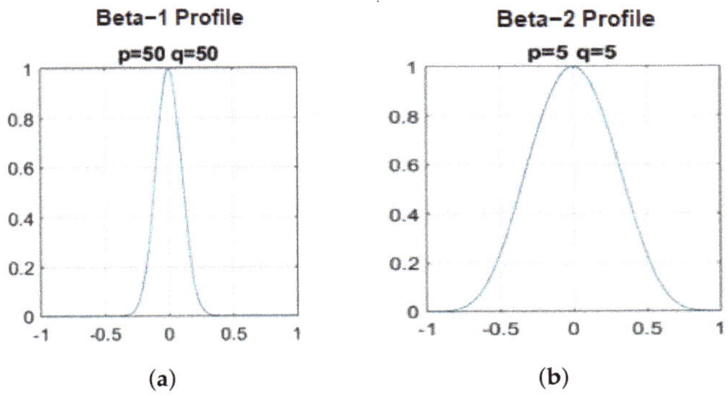


**Figure 3.** Beta profiles , respectively (**a**) Beta-1 and (**b**) Beta-2 distributions.

### 3.2.2. The Second Variant: DB-CSA for DMOPs, DB-CSA-II

The second variant (DB-CSA-II) of the proposed DB-CSA algorithm has the same optimization steps as the first variant (DB-CSA) and the main difference is in its investigating the dynamic optimization process to manage the time-varying change that occurs when solving DMOPs. The dynamic handling mechanism starts with the extraction of the population of crows of both the current iteration *(t)* and previous iteration *(t − 1)*, denoted, respectively, as *POS(t)* and *POS(t − 1)*. Then, the objective values are compared using the Pareto dominance operator [80]. Pareto dominance is a useful mechanism in multi-objective optimization to compare the evolution and  deterioration of two solutions $Solution_p$ *(t)* in *POS(t)* and $Solution_q$ *(t − 1)* from *POS(t − 1)* based on their objective function vectors: $F(Solution_p$ *(t))* $= f_1$ *(t)*, $\ldots, f_M$ *(t)* and $F(Solution_q$ *(t − 1))* $= f_1$ *(t − 1)*, $\ldots, f_M$ *(t − 1)*. During the optimization process, solution $Solution_p$ *(t)* dominates $Solution_q$ *(t − 1)*  if both dominance conditions are verified:

- Check dominance condition (2): the $Solution_p(t)$ is strictly better than $Solution_q(t-1)$, $F(Solution_p(t)) > F(Solution_q(t-1))$ for at least one objective.
- Check the dynamic change: if the current solution $Solution_p(t)$ dominates the previous $Solution_q(t-1)$ based on both conditions (1) and (2), so the dynamic change is successfully detected. The next step aims to effectively react to the dynamic change and to enhance all deterioration in the time-varying search space.
- React to the dynamic change: all non-dominated $Solution_p(t)$ are considered to create the next population at iteration $(t + 1)$, and all deteriorated solutions $Solution_p(t)$ are randomly re-initialized. Then, all non-dominated solutions in the archive (A) are updated. This process aims to enhance the convergence and diversity of crows in a dynamic search space. The general flowchart of the second version of DB-CSA algorithm is shown in Figure 2.

### 3.3. The Complexity Analysis of the Proposed DB-CSA Approach

The dynamic beta-distributed profiles are the main properties of the DB-CSA algorithm, investigating the use of beta function that provide a high flexibility to produce several forms of data distributions. Using both large Beta-1 and narrow Beta-2 functions provided the standard CSA with a new mechanism to assume a good population distribution toward the best approximated results. The advantage of the proposed DB-CSA algorithm is proved by their simplicity and robustness to balance between the convergence and the diversity in a static and dynamic search space. The time complexity is independent of the number of objective functions and the nature of the search space. The proposed DB-CSA algorithm is developed to optimize both static Many-Objective Optimization Problems (MaOPs) and Dynamic Multi- Objective Optimization Problems (DMOPs). In the worst case scenario, the time complexity is equal to $O(M \times Nlog(T)) + O(M \times N^2)$ where $M{=}2$ or $O(M \times Nlog(T)) + O(Nlog(M − 2)N)$ where $M{>}2$ and is obtained as follows: The initialization of the position for N crows with a d-dimensional search space takes $O(N \times d)$. The initialization of the memory of $N$ crows take a complexity time $O(N)$. The evaluation of the fitness function with $M$ objectives for N crows takes $O(M \times N)$. The process to rank and compare all solutions to determine the set of non-dominated solutions takes $O(M \times N^2)$, where $M{=}2$ is the number of objectives and it takes $O(Nlog(M − 2)N)$ where M>2. The optimization process of the proposed DB-CSA algorithm is executed until the maximum number of iteration $(Max_{Iterations})$ is reached. At each iteration *(t)*, the following steps are iterative repeated, including updating the positions, evaluating fitness functions, and updating the archive (A). In sum, the overall optimization process of the DB-CSA-I algorithm is equal to $O(M \times Nlog(T)) + (O(M \times N^2)) = O(N \times d) + O(M \times N) + O(M \times N^2)$. However, the second version of the proposed DB-CSA-II algorithm includes a dynamic optimization process for solving DMOPs. The main difference between the first (DB-CSA-I) and the second version (DB-CSA-II) of the proposed DB-CSA algorithm is presented in the additional process to manage the dynamic change that occurs when solving DMOPs and takes a time complrexity equal to $O(M \times N^2)$.

In sum, the overall complexity of the second version designed for DMOPs is equal to $O(M \times N log(T)) + O(M \times N^2) = O(N \times d) + O(M \times N) + O(M \times N) + O(M \times N^2)$.

## 4. Experimental Study

This section presents the experimental study and all the presented results are for two comparative studies, as detailed in Tables 3 and 4:

- The first was to compare the new proposed DB-CSA-II to a set of DMOEAs designed for Dynamic Multi-Objective Optimization Problems (DMOPs).
- The second was for Many-Objective Optimization Problems (MaOPs).
- Algorithms configurations and parameters are listed in Tables 3 and 4 for DMOPs and MaOPs, respectively.

### 4.1. Quality Indicators

The performance measurements of all tested systems were carried out using the minimum values of the three quality indicators (QI), including the Inverted General Distance (IGD), the Mean Inverted General Distance (MIGD) and the Hypervolume Difference (HVD), which are presented, respectively, in Equations (11)–(13) respectively. All these metrics are used to measure both the convergence and diversity of the tested DMOEAs.

- The Inverted General Distance (IGD) [11] in Equation (11) measures a Euclidean distance $d(i,POF)$ between the ith points in the non-dominated solutions $POF^*$ and the nearest approximated Pareto front (POF).

$$IGD(POF_t^*, POF_t) = \frac{\sum_{i \in POF^*} d(i, POF)}{|POF|} \tag{11}$$

- The Mean Inverted General Distance (MIGD) [11] is presented in Equation (12), presenting the average of the IGD values at each iteration $t \in T$.

$$MIGD((POF_t^*, POF_t) = 1/T \times \sum_{t \in T} IGD(POF_t^*, POF_t) \tag{12}$$

- The Hypervolume Difference (HVD) [11] detailed in Equation (13) aims to compute the difference between the Hypervolume (HV) of the true $POF^*$ and the approximated $POF$

$$HVD = HV(POF_t^*) - HV(POF_t) \tag{13}$$

### 4.2. Tested Benchmarks

Forty-four benchmarks were used to evaluate the relative performances of the proposed method upon the two scenarios. The twenty-one DMOPs test beds are as follows: five FDA [6], three dMOP [12], seven UDF [81] and six F(ZJZ) [14] functions. The twenty-three MaOP problems are composed of seven MaF test suite MaF1-7, seven DTLZ1-7 functions and nine WFG1-9 problems. The test configurations are detailed in Table 5 according to the number of variables (D) and objectives (M). For dynamic multi-objective optimization, Farina et al. [6] presented three types of DMOPs, which are classified into three categories according to the time-varying POF and POS. The DMOPs in type I have a dynamic change in the POS and the POF remains the stable. Both POS and POF are changed for the DMOPs in type II. However, type III of DMOP presents a time-varying POF, where the POS is unchanged. The main properties of all tested problems are reported in Table 5, presenting a variation in both POS and POF.

### 4.3. Experimental Settings

This section presents the experimental settings for all the compared state-of-the-art methods. To assume that the experimental studies have a fairness configuration, all the parameter settings for experimental studies (1) and (2) were fixed, referring to the original papers [11,15,54,55]. The experimental study was conducted using a personal computer with 8 Go of Ram and an i7 Intel processor. A Java implementation of the proposed method was performed on the jMetal framework [82]. Furthermore, the Taguchi method was used to select the meta-parameters of the beta-behaved profiles.

### 4.3.1. Comparative Study (1) for DMOPs

The first comparative test was performed for DMOPs using FDA, dMOP, UDF and F(ZJZ) benchmarks, with two and three objectives. Five standard MOEAs [11] and the six-transfer learning-based methods [15] were compared to the new proposal DB-CSA-II system. All the compared algorithms have the same parameters settings, referring to the original publications [11,15] . However, all DMOPs were characterized by a dynamic POS or/and POF according to the time-varying property $t$, which changes at each instance, as in Equation (14).

$$t = \frac{1}{n_t} \times \left| \frac{\tau}{\tau_t} \right| \tag{14}$$

where $n_t, \tau$, and $\tau_t$ are the severity of the change, the iteration counter and the frequency of the change, respectively. Three categories of environmental change were considered in this study and differentiated according to the values of $n_t$, fixed to 10, and the variation in the frequency $\tau_t$. The property $\tau_t$ was equal to 5, 10 and 20 for severe, moderate and slight environmental changes, respectively, as fixed in the original papers [11,15] .

As resumed in Table 3, the swarm and the archive size were equal to 100, as fixed in [11,15]. All DMOEAs were independently executed 30 times, and each run was stopped when the maximum number of iterations was reached and computed as follows: $Max_{iter} = 3 \times n_t \times \tau_t + 50$. For each DMOP, the number of variables (D) and objectives (M) are fixed in Table 5.

**Table 3.** Parameters Settings and DMOEAs Solvers used for the comparative study (1).

| Parameters | Comparative Study (1) for DMOPs | |
|---|---|---|
| | Five MOEAs [11] | Six transfer learning-based methods [15] |
| Comparable DMOEAs | DNSGA-II [1]<br>SGEA [11]<br>dCOEA [12]<br>PPS [14]<br>MOEA/D [13] | MMTL-MOEA/D [15]<br>RI-MOEA/D [15]<br>PPS-MOEA/D [15]<br>SVR-MOEA/D [16]<br>Tr-MOEA/D [17]<br>KF-MOEA/D [18] |
| Quality indicators | IGD and HVD | MIGD |
| Testbeds (DMOPs) | FDA, dMOP, UDF, F | FDA, dMOP |
| Number of Objectives (M) | 2 and 3 | 2 and 3 |
| Population size | 100 | |
| Max-Iteration (Max$_{iter}$) | $3 \times n_t \times \tau_t + 50$ | |
| Independent runs | 30 | |

### 4.3.2. Comparative Study (2) for MaOPs

The second experimental test was carried out for many-objective optimization, referring to the contributions [54,55], to compare the proposed DB-CSA approach to seven and thirteen Many Objective Evolutionary Algorithms (MaOEAs), respectively. As mentioned in Table 4, the population size was fixed according to the number of objectives (M). The seven and thirteen MaOEAs were executed during 30 and 31 independent runs, respectively. Each run was stopped when the maximum number of iterations $(Max_{iter})$ was reached. As per the recommendations in [55] , the number of objectives (M) for both MaF and WFG test suites was set to 2, 3 and 7 and the number of variables (D) were computed as follows: D = M + K − 1, where k is set to 10 for MaF1-MaF6 and 20 for MaF7. However, the WFG test suite present the following configurations including; the number of decision variables (D) are equal to D=M+9, the number of position-related variables (K) are equal to K = M − 1, and the number of distance-related variables (L) set as L = D − k. Furthermore, in [54] both WGF and DTLZ functions were tested with 3, 5, 8, 10 and 15 objectives.

### 4.3.3. Taguchi Method for Orthogonal Experimental Design

Comparative studies were conducted using the Taguchi method [83] to analyse the sensitivity of the parameter design of the proposed DB-CSA and DB-CSA-II algorithms. The Taguchi method was provided by Genichi Taguchi in 2004 and aims to analyse the sensitivities of user-defined parameters and propose a reasonable combination of parameter designs, using the orthogonal arrays (OAs) mechanism. The Taguchi method aims to minimize the number of runs that are needed for the experimental study. OAs are denoted by $L_a(b^c)$, where $a$ is the number of experimental runs, $b$ is the number of levels of each factor, $c$ is the number of columns in the array, and $L$ denotes Latin square design. In this study, the DB-CSA algorithm consists of six key parameters with an array of two-level factors.

- The swarm size is fixed with two-level factors ($N$), for DMOPs $N \in \{100, 200\}$ and MaOPs $N \in \{100, 300\}$.
- The maximum number of iterations is fixed with two-level factors, for DMOPs $T_{max} \in \{250, 350\}$ for severe and moderate change and for MaOPs $T_{max} \in \{100, 25{,}000\}$.
- Both parameters of the Beta-1 function (p1) and (q1) are fixed to 5 or 50.
- Both parameters of the Beta-2 function (p2) and (q2) are fixed to 5 or 50.

**Table 4.** Parameters Settings and MaOEAs Solvers used for the comparative study (2).

| Parameters | Comparative Study (2) for DMOPs | |
|---|---|---|
| | Thirteen MOEAs [55] | Seven MaOEAs [54] |
| Comparable MaOEAs | MSOPS-II [21]<br>MOEA/D [13]<br>HypE [27]<br>picea-G [41]<br>SPEA/SDE [39]<br>GrEA [35]<br>NSGA-III [24]<br>KnEA [40]<br>RVEA [22]<br>Two_Arch2 [41]<br>$\theta$-dea [36]<br>MOEA/DD ] [25]<br>AnD [55] | PMEA-MA [54]<br>PMEA*-MA [54]<br>SPEA2/SDE [39] NSGA-II/SDR [53]<br>MaOEA/IDG [31]<br>VaEA [52]<br>spea [56] |
| Quality indicators | IGD | IGD |
| Benchmarks | WFG, MaF | WFG, DTLZ |
| Number of Objectives (M) | 2, 3, and 7 | 3, 5, 8, 10, and 15 |
| Population size | 100 | 92, 224, 164, 280, and 152 |
| Max-Iteration (Max$_{iter}$) | 25.000 | 300: WFG1-9 and DTLZ 2,4,5,6,<br>1000: DTLZ 1,3,6 |
| Independent runs | 31 | 30 |

**Table 5.** Properties of the tested benchmarks: DMOPs and MaOPs.

| DMOPs | | D | M | Properties |
|---|---|---|---|---|
| Dynamic Multi Objective Optimization Problems (DMOPs) | FDA1 | 20 | 2 | Type I, convex, POS: sinusoidal and vertical shift |
| | FDA2 | 15 | 2 | Type II, POF: convex to concave, dynamic density, POS: sinusoidal and vertical shift |
| | FDA3 | 30 | 2 | Type II, POF: convex, dynamic spread, POS: sinusoidal and vertical shift |
| | FDA4 | 12 | 3 | Type I, POF: concave, dynamic spread, POS: sinusoidal and vertical shift |
| | FDA5 | 12 | 3 | Type II, POF: concave, dynamic spread, POS: sinusoidal and vertical shift |
| | dMOP1 | 10 | 2 | Type III, POF: convex to concave, POS: no change |
| | dMOP2 | 10 | 2 | Type II, POF: convex to concave, POS: sinusoidal and vertical shift |
| | dMOP3 | 10 | 2 | Type I, POF: convex, dynamic spread, POS: sinusoidal and vertical shift |
| | F5, F6, F7 | 20 | 2 | Type II, POF: convex to concave, POS: |
| | F9, F10 | 20 | 2 | Type II, POF: convex to concave, POS: |
| | F8 | 20 | 3 | trigonometric and vertical |
| | UDF1 | 10 | 2 | Type I, POF: linear continuous, POS: trigonometric and vertical shift |
| | UDF2 | 10 | 2 | Type I, POF: linear continuous, POS: polynomial and vertical shift |
| | UDF3 | 10 | 2 | Type III, POF: discontinuous, POS: trigonometric and no variation |
| | UDF4 | 10 | 2 | Type II, convex to concave, POS: trigonometric and horizontal shift |
| | UDF5 | 10 | 2 | Type II, convex to concave, POS: polynomial + vertical shift |
| | UDF6 | 10 | 2 | Type III, discontinuous, POS: trigonometric and no variation |
| | UDF7 | 10 | 3 | Type III, POF :3D radius concave, POS: trigonometric and no variation |
| Many Objective Optimization Problems (MaOPs) | MaF1 | | | Linear |
| | MaF2 | 11 | 2 | Concave |
| | MaF3 | 12 | 3 | Convex, multimodal |
| | MaF4 | 16 | 7 | Concave, multimodal |
| | MaF5 | | | Convex, biased |
| | MaF6 | | | Concave, degenerate |
| | MaF7 | 21 | 2 | Mixed |
| | | 22 | 3 | Disconnected |
| | | 26 | 7 | Multimodal |

**Table 5.** *Cont.*

| DMOPs | | D | M | Properties |
|---|---|---|---|---|
| Many Objective Optimization Problems (MaOPs) | WFG1 | 11 | 2 | Convex, unimodal |
| | WFG2 | 12 | 3 | Convex, disconnected |
| | WFG3 | 16 | 7 | Linear, unimodal |
| | WFG4 | | | Concave, multimodal |
| | WFG5 | 12 | 3 | Concave, deceptive |
| | WFG6 | 14 | 5 | Concave, unimodal |
| | WFG7 | 17 | 8 | Concave, unimodal |
| | WFG8 | 19 | 10 | Concave, unimodal |
| | WFG9 | 24 | 15 | Concave, multimodal |
| | DTLZ1 | 7 | 3 | Linear |
| | | 9 | 5 | |
| | | 12 | 8 | |
| | | 14 | 10 | |
| | | 19 | 15 | |
| | DTLZ2 | 12 | 3 | Concave |
| | DTLZ3 | 14 | 5 | Concave |
| | DTLZ4 | 17 | 8 | Concave |
| | DTLZ5 | 19 | 10 | Degenerate |
| | DTLZ7 | 22 | 3 | Disconnected |

Figure 4 presents a different data distribution according to the configurations of parameters $p$ and $q$ of the Beta function, which are fixed to 5 or 50. As mentioned in Table 6, the application of the Taguchi orthogonal arrays identified an array of $L_8(2^6)$ with only eight best runs for a different combination of parameter designs, with a two-level design for six control factors. The sensitivity of parameter design of the proposed DB-CSA algorithm was confirmed using the Taguchi method. As shown in Table 7, the parameters of Beta1 function with p1 and q1 are equal to 5, and those of Beta2 function with p2 and q2 are equal to 50, to achieve the smallest mean values for MIGD, IGD and HVD metrics and solve DMOPs with severe and moderate changes and MaOPs with seven objectives. However, the experimental study was carried out using the best configuration, obtained using the Taguchi method.

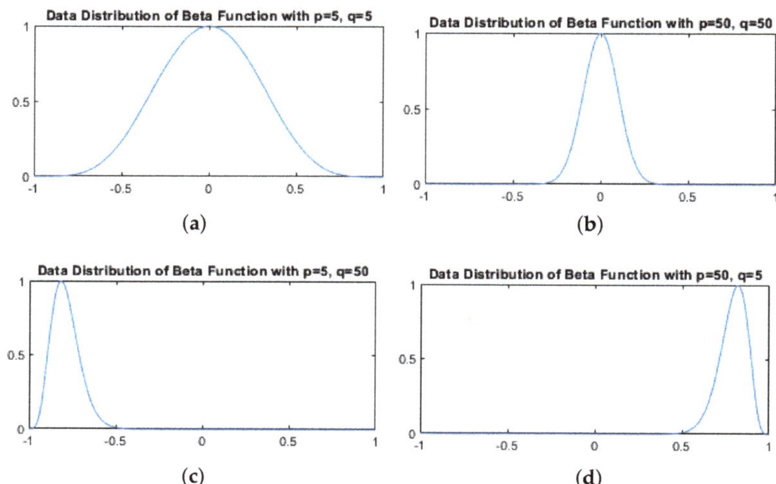

**Figure 4.** Different Data Distributions of Beta Function with Different Configurations of $p$ and $q$ : (**a**) large Gaussian data distribution, (**b**) narrow Gaussian data distribution, (**c**) exponential decrease data distribution, (**d**) exponential increase data distribution.

**Table 6.** Taguchi Array Design $L_8(2^6)$ of DB−CSA−II for DMOPs and of DB−CSA for MaOPs.

| Problems | ID Run | Population Size | Max-Iteration | Parameters of Beta Function | | | |
|---|---|---|---|---|---|---|---|
| | | | | p1 | p2 | q1 | q2 |
| DMOPs | 1 | 100 | 250 | 5 | 5 | 5 | 5 |
| | 2 | 100 | 250 | 5 | 50 | 5 | 50 |
| | 3 | 200 | 250 | 50 | 5 | 50 | 5 |
| | 4 | 200 | 250 | 50 | 50 | 5 | 50 |
| | 5 | 100 | 350 | 50 | 5 | 5 | 50 |
| | 6 | 100 | 350 | 50 | 50 | 50 | 5 |
| | 7 | 200 | 350 | 5 | 50 | 5 | 50 |
| | 8 | 200 | 350 | 5 | 50 | 5 | 5 |
| MaOPs | 1 | 100 | 100 | 5 | 5 | 5 | 5 |
| | 2 | 100 | 100 | 5 | 50 | 50 | 50 |
| | 3 | 100 | 25.000 | 50 | 5 | 5 | 50 |
| | 4 | 100 | 25.000 | 5 | 50 | 5 | 50 |
| | 5 | 300 | 100 | 50 | 5 | 50 | 5 |
| | 6 | 300 | 100 | 50 | 50 | 5 | 50 |
| | 7 | 300 | 25.000 | 5 | 5 | 50 | 50 |
| | 8 | 300 | 25.000 | 5 | 50 | 5 | 5 |

**Table 7.** Best Experimental Design using Taguchi Method for MIGD, IGD, HVD Metric for DMOPs (FDA, dMOP) with Severe and Moderate Dynamic Changes and MaOPs with 7 Objectives (WFG).

| Problems | | Population Size | Max-Iteration | Parameters of Beta Function | | | | Best Mean Values | | |
|---|---|---|---|---|---|---|---|---|---|---|
| | | | | p1 | p2 | q1 | q2 | MIGD | IGD | HVD |
| DMOPs with Severe Change | FDA1 | 100 | 250 | 5 | 50 | 5 | 50 | $6.37 \times 10^{-7}$ | $5.73 \times 10^{-4}$ | $2.22 \times 10^{-2}$ |
| | FDA2 | 100 | 250 | 5 | 50 | 5 | 50 | $3.33 \times 10^{-6}$ | $2.99 \times 10^{-3}$ | $7.96 \times 10^{-1}$ |
| | FDA3 | 100 | 250 | 5 | 50 | 5 | 50 | $2.63 \times 10^{-4}$ | $2.36 \times 10^{-1}$ | $4.89 \times 10^{-1}$ |
| | FDA4 | 100 | 250 | 5 | 50 | 5 | 50 | $1.60 \times 10^{-6}$ | $1.44 \times 10^{-3}$ | $7.92 \times 10^{-2}$ |
| | FDA5 | 200 | 250 | 50 | 5 | 50 | 5 | $3.14 \times 10^{-6}$ | $2.83 \times 10^{-3}$ | $1.48 \times 10^{-1}$ |
| | dMOP1 | 100 | 250 | 5 | 50 | 5 | 50 | $7.28 \times 10^{-7}$ | $6.55 \times 10^{-4}$ | $4.14 \times 10^{-3}$ |
| | dMOP2 | 200 | 250 | 50 | 50 | 5 | 50 | $1.26 \times 10^{-6}$ | $1.13 \times 10^{-3}$ | $3.80 \times 10^{-2}$ |
| | dMOP3 | 200 | 250 | 50 | 50 | 5 | 50 | $5.21 \times 10^{-5}$ | $4.69 \times 10^{-2}$ | $4.23 \times 10^{+0}$ |
| DMOPs with Moderate change | FDA1 | 100 | 350 | 5 | 50 | 5 | 50 | $6.35 \times 10^{-7}$ | $5.71 \times 10^{-4}$ | $1.96 \times 10^{-2}$ |
| | FDA2 | 100 | 350 | 5 | 50 | 5 | 50 | $4.19 \times 10^{-6}$ | $3.77 \times 10^{-3}$ | $4.33 \times 10^{+0}$ |
| | FDA3 | 100 | 350 | 5 | 50 | 5 | 50 | $4.69 \times 10^{-4}$ | $4.22 \times 10^{-1}$ | $6.24 \times 10^{-1}$ |
| | FDA4 | 100 | 350 | 5 | 50 | 5 | 50 | $1.43 \times 10^{-6}$ | $1.29 \times 10^{-3}$ | $3.67 \times 10^{-2}$ |
| | FDA5 | 100 | 350 | 5 | 50 | 5 | 50 | $3.90 \times 10^{-6}$ | $3.51 \times 10^{-3}$ | $1.61 \times 10^{-1}$ |
| | dMOP1 | 100 | 350 | 5 | 50 | 5 | 50 | $5.95 \times 10^{-7}$ | $5.35 \times 10^{-4}$ | $1.99 \times 10^{-2}$ |
| | dMOP2 | 100 | 350 | 50 | 5 | 5 | 50 | $1.02 \times 10^{-6}$ | $9.18 \times 10^{-4}$ | $3.13 \times 10^{-2}$ |
| | dMOP3 | 200 | 350 | 5 | 50 | 5 | 50 | $3.36 \times 10^{-5}$ | $3.02 \times 10^{-2}$ | $1.68 \times 10^{-1}$ |
| MaOPs with 7 objectives | WFG1 | 100 | 25.000 | 5 | 50 | 5 | 50 | - | $2.42 \times 10^{-4}$ | - |
| | WFG2 | 100 | 25.000 | 50 | 5 | 5 | 50 | - | $2.53 \times 10^{-4}$ | - |
| | WFG3 | 100 | 25.000 | 5 | 50 | 5 | 50 | - | $5.87 \times 10^{-5}$ | - |
| | WFG4 | 100 | 25.000 | 50 | 5 | 5 | 50 | - | $4.09 \times 10^{-4}$ | - |
| | WFG5 | 100 | 25.000 | 50 | 5 | 5 | 50 | - | $1.54 \times 10^{-4}$ | - |
| | WFG6 | 100 | 25.000 | 50 | 5 | 5 | 50 | - | $3.57 \times 10^{-4}$ | - |
| | WFG7 | 100 | 25.000 | 50 | 5 | 5 | 50 | - | $3.87 \times 10^{-4}$ | - |
| | WFG8 | 100 | 25.000 | 5 | 50 | 5 | 50 | - | $1.91 \times 10^{-3}$ | - |
| | WFG9 | 100 | 25.000 | 5 | 50 | 5 | 50 | - | $7.93 \times 10^{-4}$ | - |

*4.4. Results Analysis and Discussion*

This subsection presents the analysis of comparative studies which is conducted through the non-parametric Wilcoxon sign rank test [84] and the box-plots of the one-way ANOVA test [85]. The statistical analysis methods were used to estimate the *p*-value property to determine the statistically significant difference between the compared methods.In this study case, the Wilcoxon sign rank test is used to compare two paired approaches and executed according to the following steps:

- Step 1: define the null hypothesis ($H_0$): the means of the paired approaches are the same.
- Step 2: define ix, the alternative hypothesis ($H_1$): the means of the paired approaches are different.
- Step 3: compute the difference between the mean values.
- Step 4: assign a rank for the results obtained in step 3.
- Step 5: compute the number of ranks with negative values ($R_-$) and positive values ($R_+$).
- Step 6: compute the probability (*p*-value) that the null hypothesis is true and is computed based on the test statistic value (z-score) produced by the Wilcoxon Rank–Sum test.

If the *p*-value is less than or equal to 0.05, the statistical results are considered significantly important and we can conclude that the difference between the mean value of the paired approaches is statistically significant; otherwise, this difference is not statistically significant. All quantitative results are presented in the Supplementary file, Tables S1–S9.

4.4.1. Analysis of the Comparative Study (1) for FDA and dMOP Problems

The comparative study (1) is first considered to compare the proposed DB-CSA-II algorithm to six transfer learning-based methods, the standard CSA algorithm and the first variant DB-CSA and solve FDA and dMOP test beds with severe ($\tau_t = 5, n_t = 10$), moderate ($\tau_t = n_t = 10$) and slight ($\tau_t = 20, n_t = 10$) environmental changes. The MIGD metric aims to measure the mean value of the obtained IGD values and is used to measure the convergence and diversity of the obtained POF compared to the true one. Table S1 (see the Supplementary File) presents the mean and standard deviation values through the MIGD metric; the efficiency of the new DB-CSA-II system shows the best mean and standard deviation values for all test suites with different environmental changes compared to the six transfer learning-based approaches and the standard CSA algorithm. Based on the statistical results obtained using the Wilcoxon signed rank test on the following Table 8, we can determine the importance of the new DB-CSA-II with a $p$-value of less than 0.05, defining a significant difference in the mean values compared to MMTL-MOEA/D, KF-MOEA/D, PPS-MOEA/D, SVR-MOEA/D, Tr-MOEA/D, and RI-MOEA/D approaches. The one-way ANOVA box-plots in Figure 5 determines the importance of the DB-CSA-II algorithm compared to six transfer-learning-based methods. Compared with the DB-CSA-II algorithm, which aims to control the evolution and deterioration of optimal solutions, most TL-based approaches were designed to predict the new population after each dynamic change, but tend to be inefficient when managing dynamic change in both POS and POF and balancing convergence and diversity within dynamic search spaces.

**Table 8.** Non-parametric statistical analysis based on a Wilcoxon signed rank test of DB-CSA-II vs. six peer transfer-learning based approaches over MIGD metric for FDA and dMOP functions.

| DB-CSA-II (with Dynamic Process) vs. | QI | Prob. | R− | R+ | $p$-Value | Best Method |
|---|---|---|---|---|---|---|
| MMTL-MOEA/D | | | 300 | 0 | 0.000018 | |
| KF-MOEA/D | | | 300 | 0 | 0.000018 | |
| PPS-MOEA/D | | | 300 | 0 | 0.000018 | |
| SVR-MOEA/D | MIGD | FDA & dMOP | 300 | 0 | 0.000018 | DB-CSA-II |
| Tr-MOEA/D | | | 300 | 0 | 0.000018 | |
| RI-MOEA/D | | | 300 | 0 | 0.000018 | |
| CSA | | | 300 | 0 | 0.000018 | |
| DB-CSA | | | 275 | 25 | 0.000355 | |

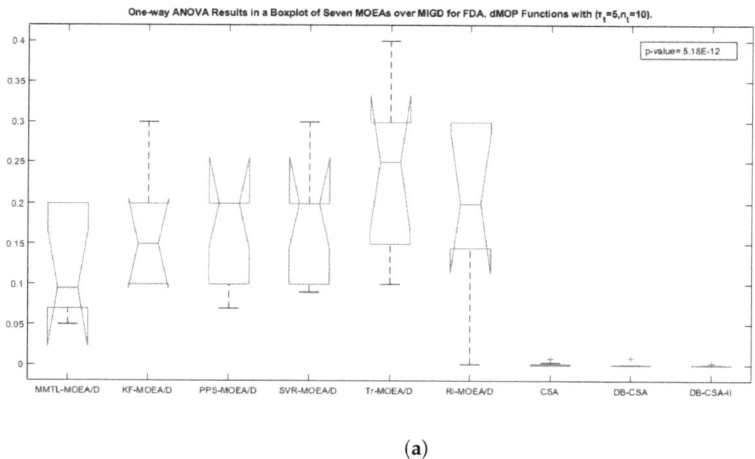


(a)

**Figure 5.** *Cont.*

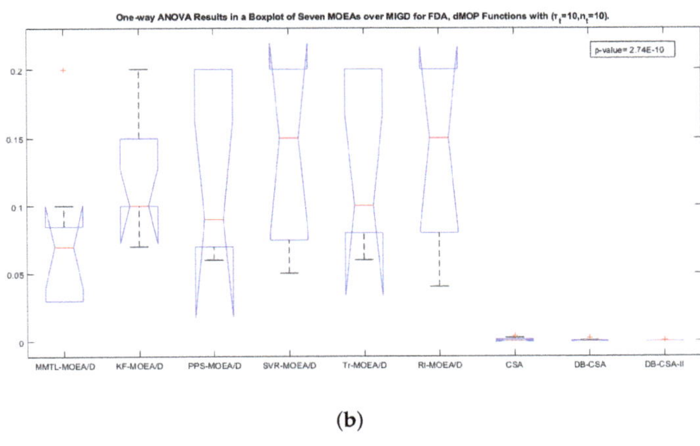



**(b)**

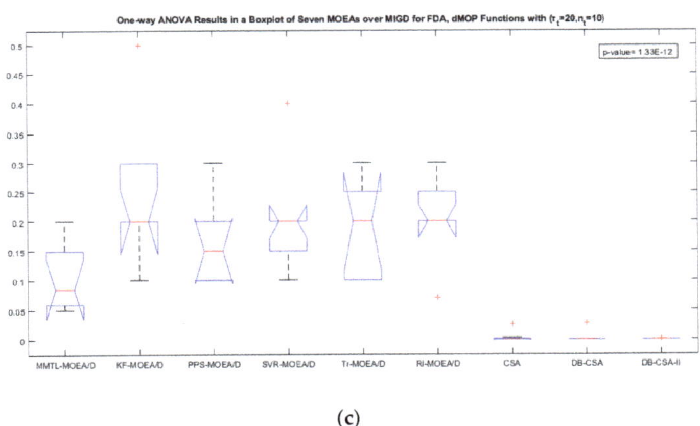



**(c)**

**Figure 5.** One-way ANOVA Box-plots of 7 MOEAs over MIGD of FDA, dMOP for (**a**) severe with ($\tau_t = 5$, $n_t = 10$), (**b**) moderate ($\tau_t = 10$, $n_t = 10$), and (**c**) slight with ($\tau_t = 20$, $n_t = 10$) environmental changes respectively.

Second, the five standard MOEAs (DNSGA-II, dCOEA, PPS, MOEA/D, and SGEA) were compared to the new DB-CSA-II algorithm. The average and the standard deviation values for both FDA and dMOP test suites over the IGD and HVD metrics, respectively, can be seen in Tables S2 and S3 of the Supplementary File. Based on the IGD metric on Table S2, we can argue the superiority of DB-CSA-II method compared to five standard MOEAs designed for dynamic multi-objective optimization. The results based on Wilcoxon signed rank test are presented in Table 9, indicating that DB-CSA-II is the best method when compared to IGD at a statistically significant level of 0.05 compared to other MOEAs. The same conclusion is confirmed using the box plot over a one-way ANOVA test in Figure 6.

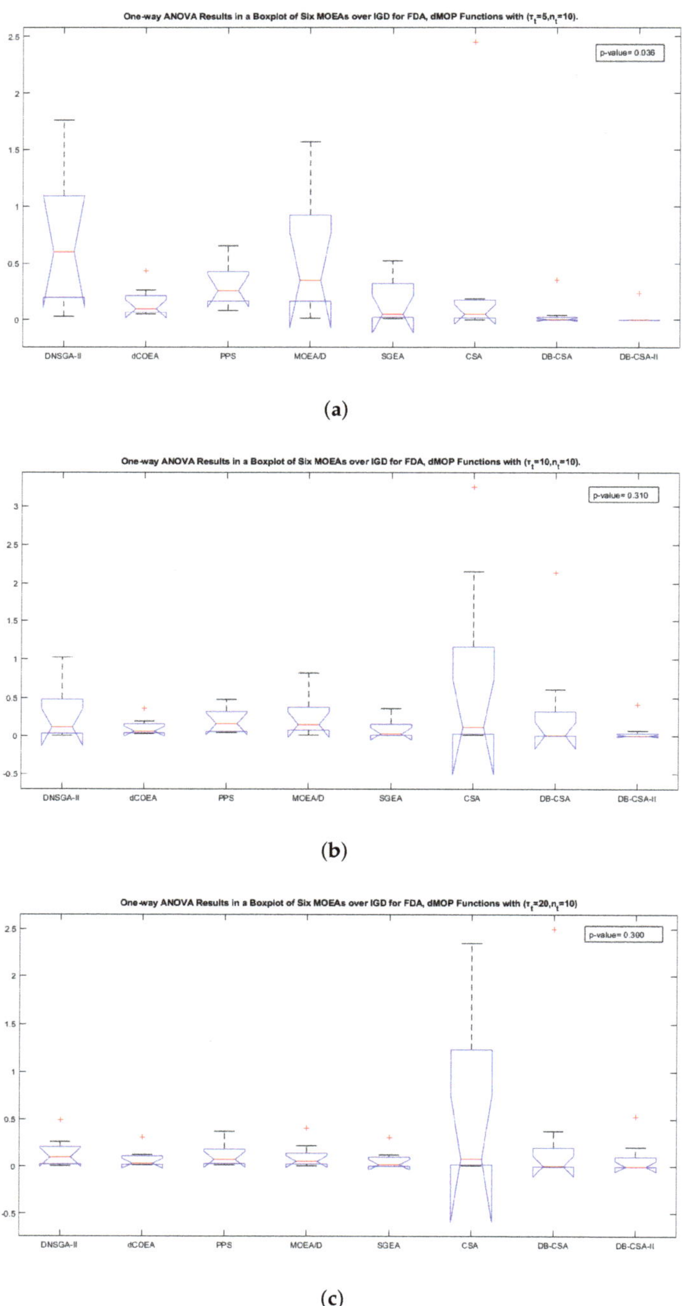



(a)

(b)

(c)

**Figure 6.** One-way ANOVA Box-plots of 6 MOEAs over IGD of FDA, dMOP for (**a**) severe with ($\tau_t = 5$, $n_t = 10$), (**b**) moderate ($\tau_t = 10$, $n_t = 10$), and (**c**) slight ($\tau_t = 20$, $n_t = 10$) environmental changes, respectively.

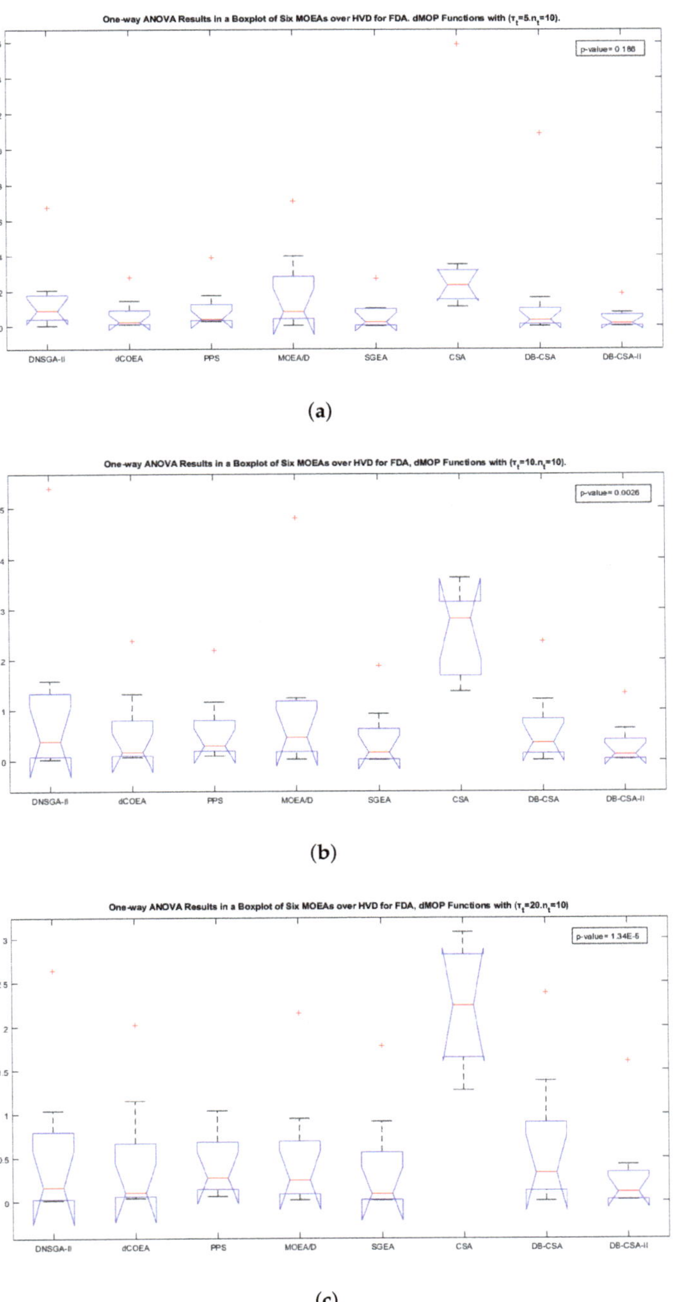

**Figure 7.** One-way ANOVA  Box-plots of 6 MOEAs over HVD of FDA, dMOP for (**a**) severe with ($\tau_t = 5$, $n_t = 10$), (**b**) moderate ($\tau_t$=10, $n_t = 10$), and (**c**) slight ($\tau_t = 20$, $n_t = 10$) environmental changes, respectively.

Table S3 reports the quantitative results using an HVD quality indicator. We can conclude that the proposed DB-CSA is the winner when solving different types of DMOPs, including FDA1 in type I, with dynamic POS and static POF, FDA3, FDA5, and dMOP2 in type II, with time-varying POS and POF and dMOP1, and in type III, with unchangeable POS and dynamic POF, regarding all categories of environmental changes. Meanwhile the DB-CSA-II obtained similar results to the SGEA system when solving FDA2 function in type II, characterized by a dynamic density of the solutions and a cyclic change of the POF from convex to concave as well as the FDA4 problem in type I characterized by a time-varying spread of solutions in a severe dynamic change. In addition, the dCOEA algorithm has a closed mean value for solving the dMOP3 function, characterized by the static curvature of the estimated POF and dynamic spread of the solution set compared to the proposed DB-CSA-II.

Table 9 shown the negative and positive Wilcoxon ranks, we can conclude that the DB-CSA-II algorithm is the best method compared to other comparable approaches based on the HVD quality indicator. This importance does not determine statistically significance with a $p$-value greater than 0.05. The one-way ANOVA results in Figure 7 assuming the competitive importance of DNSGA-II, dCOEA, PPS, MOEA/D, and SGEA to solve FDA and dMOPs test functions with 2 and 3 objectives, including the different environmental changes found using the HVD metric.

**Table 9.** Non-parametric statistical analysis based on a Wilcoxon signed rank test of DB-CSA-II vs. five peer MOEAs over IGD, HVD metrics for FDA, dMOP, UDF and F functions.

| DB-CSA-II (with Dynamic Process) vs. | Prob. | QI | R− | R+ | $p$-Value | Best Method |
|---|---|---|---|---|---|---|
| DNSGA-II | | | 235 | 65 | 0.015158 | DB-CSA-II |
| dCOEA | | | 213 | 87 | 0.071861 | ≅ |
| PPS | FDA | | 241 | 59 | 0.009322 | DB-CSA-II |
| MOEA/D | & | IGD | 231 | 69 | 0.020652 | DB-CSA-II |
| SGEA | dMOP | | 204 | 96 | 0.122865 | ≅ |
| CSA | | | 300 | 0 | 0.000018 | DB-CSA-II |
| DB-CSA | | | 275 | 25 | 0.000355 | DB-CSA-II |
| DNSGA-II | | | 222 | 78 | 0.039672 | DB-CSA-II |
| dCOEA | | | 206 | 94 | 0.109599 | ≅ |
| PPS | FDA | | 219 | 81 | 0.048675 | DB-CSA-II |
| MOEA/D | & | HVD | 223 | 77 | 0.037005 | DB-CSA-II |
| SGEA | dMOP | | 205 | 95 | 0.116083 | ≅ |
| CSA | | | 289 | 11 | 0.000071 | DB-CSA-II |
| DB-CSA | | | 240 | 60 | 0.010128 | DB-CSA-II |
| DNSGA-II | | | 91 | 0 | 0.001474 | DB-CSA-II |
| dCOEA | | | 90 | 0 | 0.001871 | DB-CSA-II |
| PPS | UDF | | 86 | 5 | 0.004649 | DB-CSA-II |
| MOEA/D | & | IGD | 91 | 0 | 0.001474 | DB-CSA-II |
| SGEA | F | | 85 | 6 | 0.005772 | DB-CSA-II |
| CSA | | | 90 | 1 | 0.001871 | DB-CSA-II |
| DB-CSA | | | 91 | 0 | 0.001474 | DB-CSA-II |
| DNSGA-II | | | 36 | 55 | 0.506746 | ≅ |
| dCOEA | | | 34 | 57 | 0.421579 | ≅ |
| PPS | UDF | | 35 | 56 | 0.463071 | ≅ |
| MOEA/D | & | HVD | 36 | 55 | 0.506746 | ≅ |
| SGEA | F | | 35 | 56 | 0.463071 | ≅ |
| CSA | | | 69 | 22 | 0.100525 | ≅ |
| DB-CSA | | | 71 | 20 | 0.074735 | ≅ |

### 4.4.2. Analysis of the Comparative Study (1) for UDF and F Problems

Considering the quantitative results of the Unconstrained Dynamic Functions (UDF1-UDF7) in Table S4 of the Supplementary File, it appears that DB-CSA-II has the greatest values for all UDF functions. Furthermore, we can resume the stability of the new DB-CSA algorithm when solving the tri-objective problem (F8) and the bi-objectives function (F10) over IGD metrics compared to the Population Prediction Strategy (PPS) approach, which is only performed to solve the F5, F6, F7 and F9 test functions. However, the F(ZJZ) problems provide a complex benchmark, including a time-varying POF and POS with a nonlinear correlation between the decision variables. Based on the Wilcoxon sign rank in Table 9, we assume that the DB-CSA-II is the best method; however, this importance does not present a high statistical significance, with $p$-values greater than 0.05 compared to the five MOEAs over the IGD metric.

Based on the HVD results reported in Table S5 of the Supplementary File, the DB-CSA-II obtains good results for the majority of UDF benchmarks, and only fails when solving the disconnected UDF6 compared to the DNSGA-II system. However, we can assume that the PPS system is important for solving F5, F7 and F10 and the SGEA for F6 and F9. The Wilcoxon signed rank test presented in Table 9 shows the same, statistically significant results between all compared algorithms with a $p$-value exceeding a 0.05 significance level. Figure 8 reported the one-way ANOVA results in a box plot of the six MOEAs over IGD and HVD metrics. Figures 9–11 present the plot of the MIGD, IGD and HVD values of the proposed DB-CSA-II algorithm to solve FDA and dMOP problems with severe, moderate and slight changes.

### 4.4.3. Analysis of the Comparative Study (2) for MaF and WFG Problems with 2, 3 and 7 objectives

For the second comparative study (2), thirteen multiple-objective evolutionary approaches (MSOPS-II, MOEA/D, HypE, PICEA-g, SPEA/SDE, GrEA, NSGA-III, KnEA, RVEA, two_Arch2, $\theta$-DEA, MOEA/DD, AnD) are first compared to the new proposed DB-CSA system based on a set of multiple-objective optimization problems, as denoted by the MaF and WFG test suites, with 2, 3 and 7 objectives, including different numbers of decision variables, as detailed in Table 5. The results are reported in Table S6 of the Supplementary File, showing the IGD results of the 14 compared Many-Objective Evolutionary Algorithms regarding their ability to solve nine MaOPs (WFG1-WFG9), characterized by the dynamic shape of the POF, which changes from convex to concave.

The DB-CSA algorithm was first ranked to solve seven WFG test suites from nine (7/9), including WFG1, WFG3, WFG4, WFG5, WFG6, WFG8 and WFG9 and failed only for solving WFG2 compared to HypE and $\theta$-DEA, which have almost the same mean values as the IGD metric for WFG7 when the number of objectives is equal to 2. By increasing the number of objectives to 3 and 7, the WFG becomes more complex and the issue of a lack of convergence and diversity presents a challenging task. Based on the reported IGD values of the tri-objective WFG functions in Table S6 of the Supplementary File, we can conclude the efficiency of the newly proposed DB-CSA approach to deal with the increasing number of objectives. Table S6, showed the best values for MaOPS, with seven objectives.

In addition, Table S7 of the Supplementary File shows the mean and the standard deviation values over the IGD metric to solve the MaF test suite (MaF1-MaF7) with 2, 3 and 7 objective functions. Figure 12 presents the approximated POF for the MaF test suite. The new DB-CSA is presented a good method for solving the MaF test suite compared to the thirteen state-of-the-art MaOEAs. Table 10 shows the importance of DB-CSA over the Wilcoxon signed rank test, while all the computed $p$-values are less than 0.05, assuming the statistically significant difference in DB-CSA compared to the thirteen MaOEAs, including MSOPS-II, MOEA/D, HypE, PICEA-g, SPEA/SDE, GrEA, NSGA-III, KnEA, RVEA, two_Arch2, $\theta$-DEA, MOEA/DD, AnD, to solve the MaF test suite with the 2, 3 and 7 compared objectives. The dynamic treatment of both convergence and diversity concepts is very useful when solving a set of complex MaOPs with a high number of objectives.

**Table 10.** Non-parametric statistical analysis based on Wilcoxon signed rank test of DB-CSA vs. thirteen peer MAOEAs over the IGD metric for WFG, MaF and DTLZ functions.

| DB-CSA vs. | QI | Prob. | R− | R+ | $p$-Value | Best Method |
|---|---|---|---|---|---|---|
| MSOPS-II | | | 378 | 0 | 0.000006 | |
| MOEA/D | | | 378 | 0 | 0.000006 | |
| HypE | | | 374 | 4 | 0.000009 | |
| PICEA-g | | | 378 | 0 | 0.000006 | |
| SPEA/SDE | | | 376 | 2 | 0.000007 | |
| GrEA | | | 378 | 0 | 0.000006 | |
| NSGA-III | IGD | WFG | 373 | 5 | 0.000010 | DB-CSA |
| KnEA | | | 378 | 0 | 0.000006 | |
| RVEA | | | 378 | 0 | 0.000006 | |
| Two_Arch2 | | | 374 | 4 | 0.000009 | |
| $\theta$-DEA | | | 376.5 | 1.5 | 0.000007 | |
| MOEA/DD | | | 377 | 1 | 0.000006 | |
| AnD | | | 378 | 0 | 0.000006 | |
| CSA | | | 378 | 0 | 0.000006 | |
| MSOPS-II | | | 231 | 0 | 0.000060 | |
| PMEA*-MA | | | 1035 | 0 | 0.000006 | |
| MOEA/D | | | 231 | 0 | 0.000060 | |
| HypE | | | 231 | 0 | 0.000060 | |
| PICEA-g | | | 231 | 0 | 0.000060 | |
| SPEA/SDE | | | 231 | 0 | 0.000060 | |
| GrEA | IGD | MaF | 231 | 0 | 0.000060 | DB-CSA |
| NSGA-III | | | 231 | 0 | 0.000060 | |
| KnEA | | | 231 | 0 | 0.000060 | |
| RVEA | | | 231 | 0 | 0.000060 | |
| Two_Arch2 | | | 231 | 0 | 0.000060 | |
| $\theta$-DEA | | | 231 | 0 | 0.000060 | |
| MOEA/DD | | | 231 | 0 | 0.000060 | |
| AnD | | | 231 | 0 | 0.000060 | |
| CSA | | | 231 | 0 | 0.000060 | |
| PMEA-MA | | | 1035 | 0 | $5.179 \times 10^{-9}$ | |
| PMEA*-MA | | | 1035 | 0 | $5.179 \times 10^{-9}$ | |
| SPEA2/SDE | | | 1035 | 0 | $5.179 \times 10^{-9}$ | |
| NSGA-II/SDR | IGD | WFG | 1035 | 0 | $5.179 \times 10^{-9}$ | DB-CSA |
| MaOEA/IGD | | | 1035 | 0 | $5.179 \times 10^{-9}$ | |
| VaEA | | | 1035 | 0 | $5.179 \times 10^{-9}$ | |
| SPEA | | | 1035 | 0 | $5.179 \times 10^{-9}$ | |
| CSA | | | 1035 | 0 | $5.179 \times 10^{-9}$ | |
| PMEA-MA | | | 629 | 1 | $2.70 \times 10^{-7}$ | |
| PMEA*-MA | | | 1035 | 0 | $2.70 \times 10^{-7}$ | |
| PMEA*-MA | | | 629 | 1 | $2.70 \times 10^{-7}$ | |
| SPEA2/SDE | IGD | DTLZ | 630 | 0 | $2.48 \times 10^{-7}$ | DB-CSA |
| NSGA-II/SDR | | | 630 | 0 | $2.48 \times 10^{-7}$ | |
| MaOEA/IGD | | | 630 | 0 | $2.48 \times 10^{-7}$ | |
| VaEA | | | 629 | 1 | $2.70 \times 10^{-7}$ | |
| SPEA | | | 630 | 0 | $2.48 \times 10^{-7}$ | |

### 4.4.4. Analysis of the Comparative Study (2) for DTLZ and WFG Problems with 3, 5, 8, 10 and 15 Objectives

In the second part of the comparative study (2), seven MaOEAs (PMEA-MA, PMEA*-MA, SPEA2/SDE, NSGA-II/SDR, MaOEA/IGD, VaEA, SPEA) were compared to the new

DB-CSA approach to solve a set of complex DTLZ and WFG test suites with 3, 5, 8, 10 and 15 objectives. Figure 13 presents the box-plots of all comparable approaches based on the one-way ANOVA test and the obtained figures show the importance of the DB-CSA algorithm for solving the WFG tests with 3, 5 and 15 objectives. Some qualitative results are presented in Figures 14 and 15 to present the estimated POF of the true optimal solutions for both WFG and DTLZ, with 10 and 15 objectives, respectively. However, all quantitative results are given in Tables S8 and S9 of the Supplementary File, presenting the efficiency of the new DB-CSA approach compared to the IGD metric to solve the complex set of the nine tested WFG1-9 problems and seven DTLZ1-7 functions, respectively. However, this difference is reported as being particularly statistically significant when using the Wilcoxon signed rank test with a 0.05 significance level, as detailed in Table 10, where all computed *p*-values are less than 0.05.

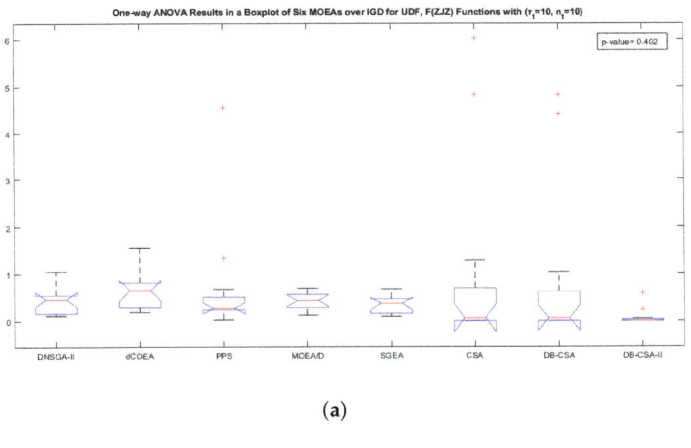

(**a**)

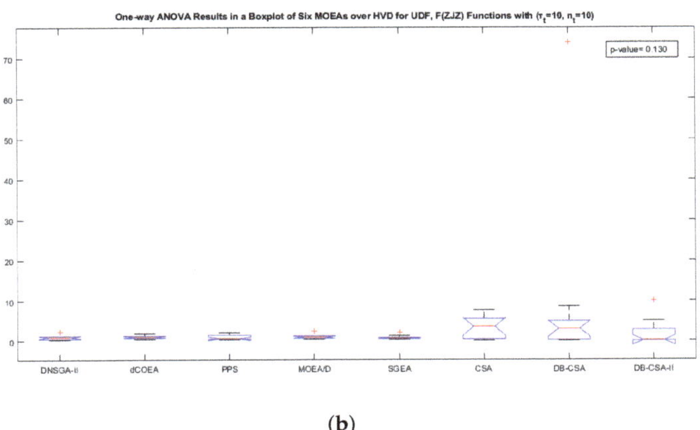

(**b**)

**Figure 8.** One-way ANOVA Box-plots of 6 MOEAs over (**a**) IGD and (**b**) HVD of UDF, F function for moderate ($\tau_t = 10$, $n_t = 10$) environmental changes.

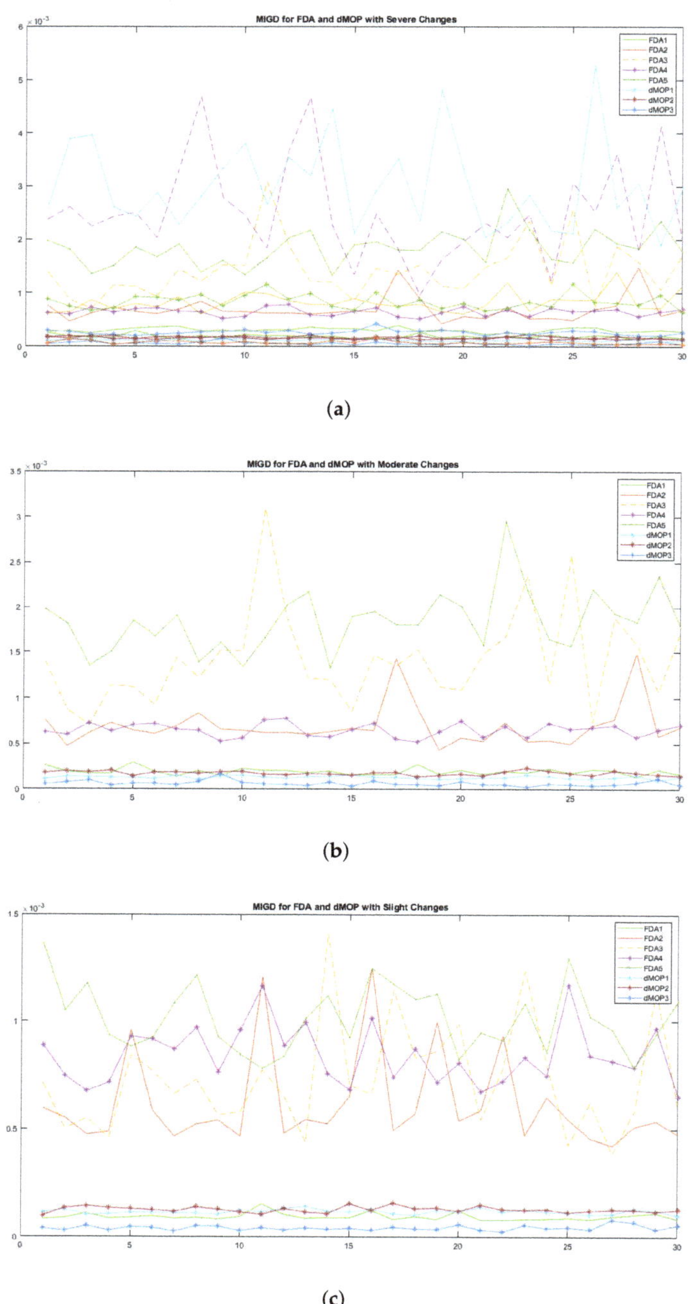

(a)

(b)

(c)

**Figure 9.** The plots of MIGD values for FDA, dMOP functions with (**a**) severe, (**b**) moderate and (**c**) slight environmental changes using the DB-CSA algorithm.

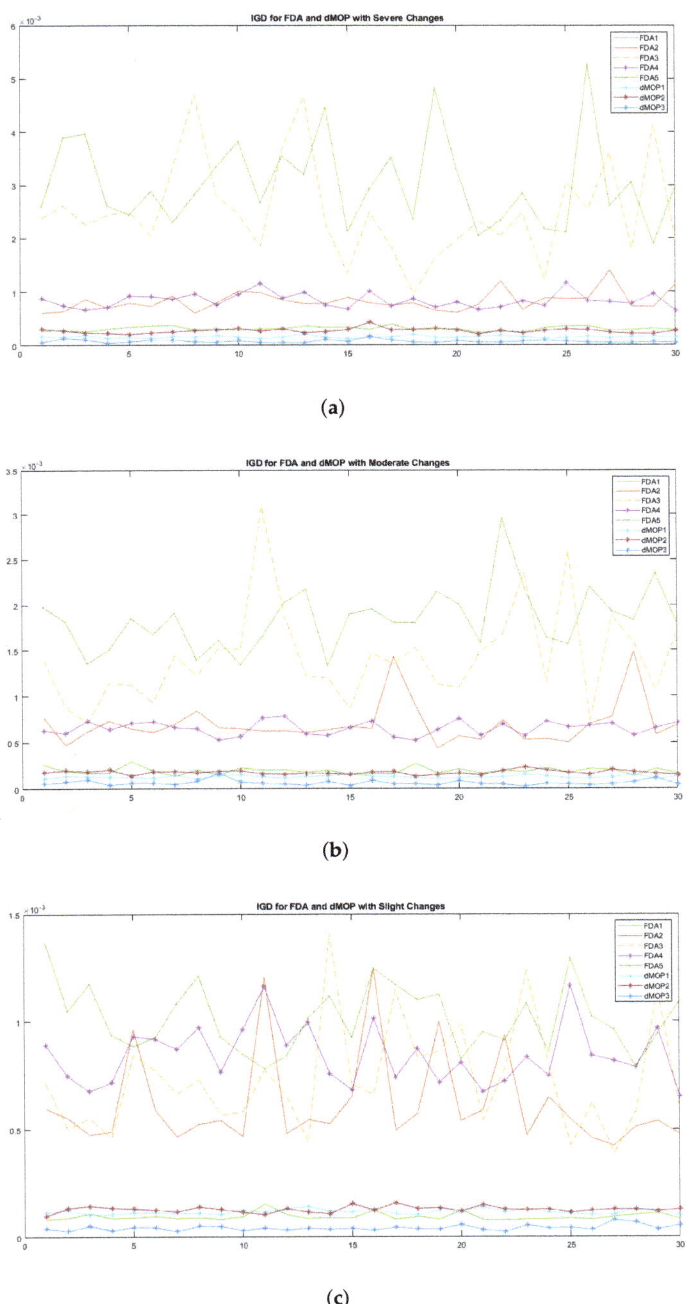

(**a**)

(**b**)

(**c**)

**Figure 10.** The plots of IGD values for FDA, dMOP functions with (**a**) severe, (**b**) moderate and (**c**) slight environmental changes using the DB-CSA algorithm.

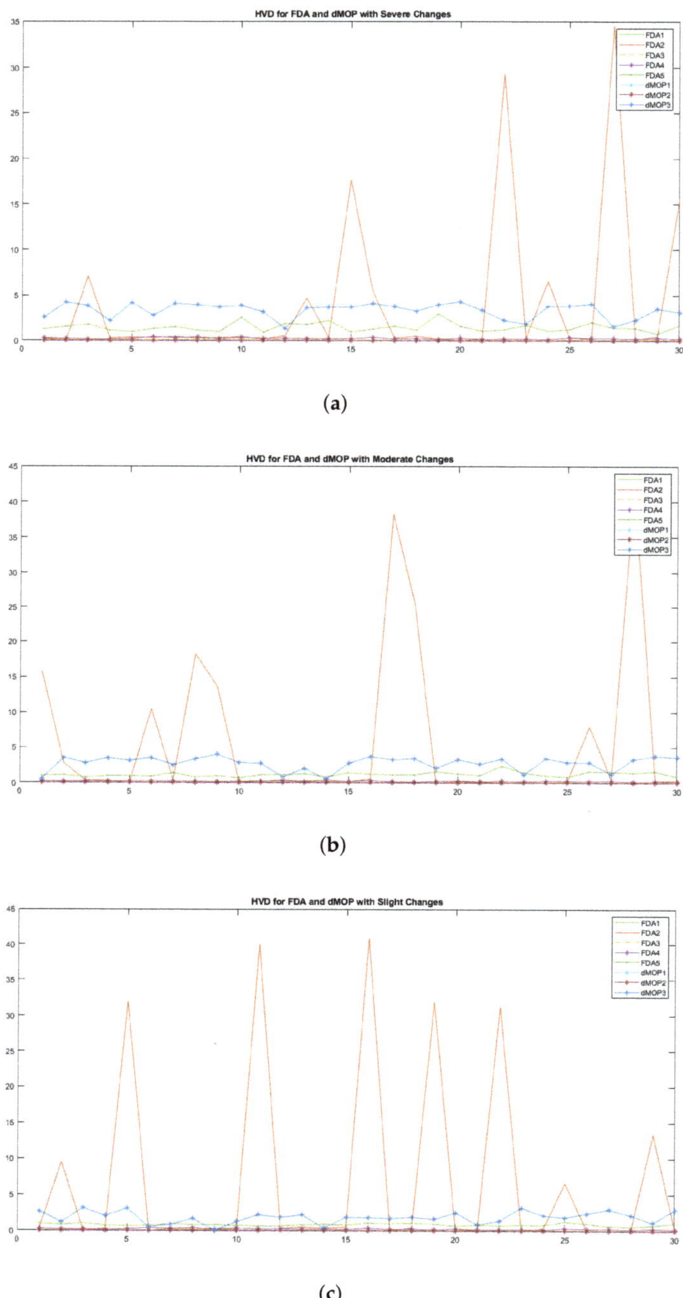

(a)

(b)

(c)

**Figure 11.** The plots of HVD values for FDA, dMOP functions with (**a**) severe, (**b**) moderate and (**c**) slight environmental changes using the DB-CSA algorithm.

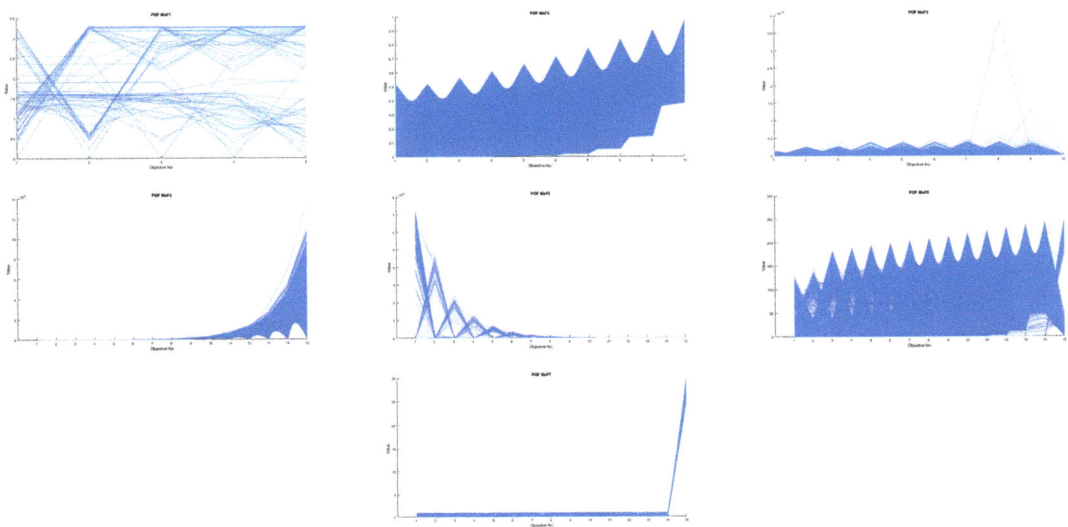

**Figure 12.** The plots of POF for MaF1-7 functions with 7 objectives using the DB-CSA algorithm.

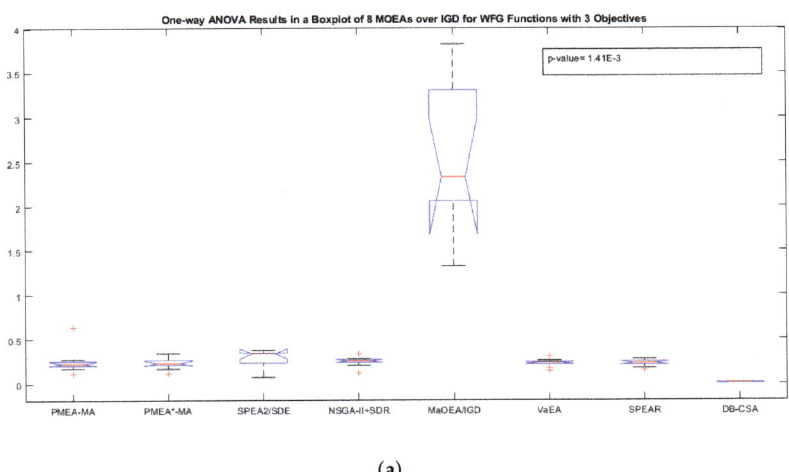


(**a**)

**Figure 13.** *Cont.*

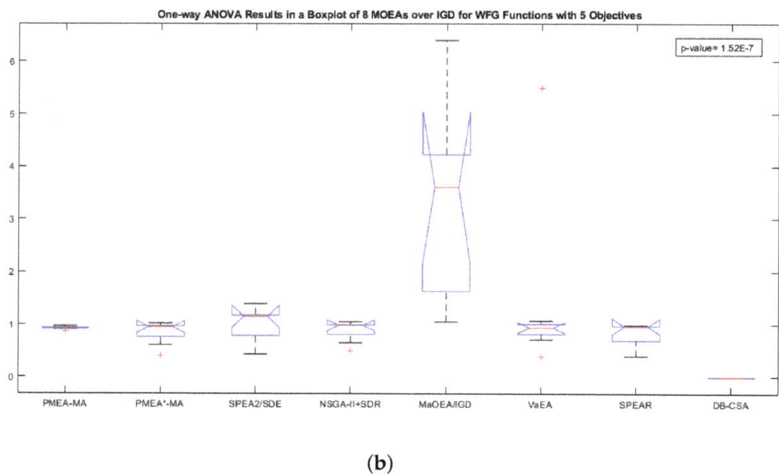


(**b**)

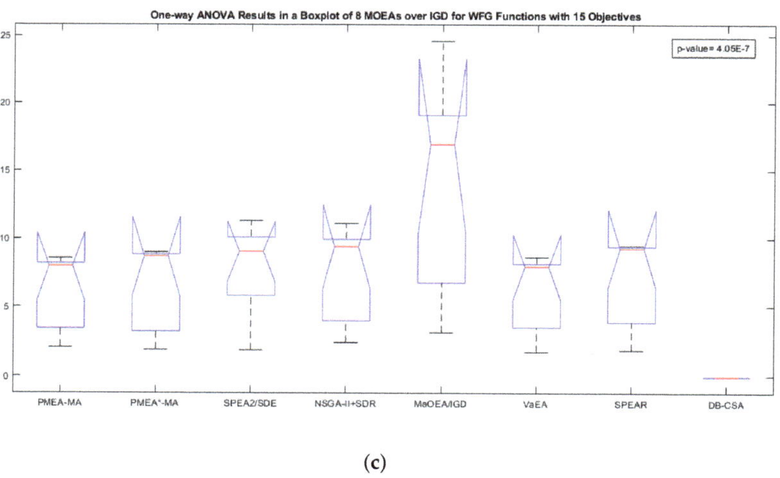


(**c**)

**Figure 13.** One-way ANOVA box-plots of 8 MOEAs over IGD for WFG functions with (**a**) 3, (**b**) 5 and (**c**) 15 objectives.

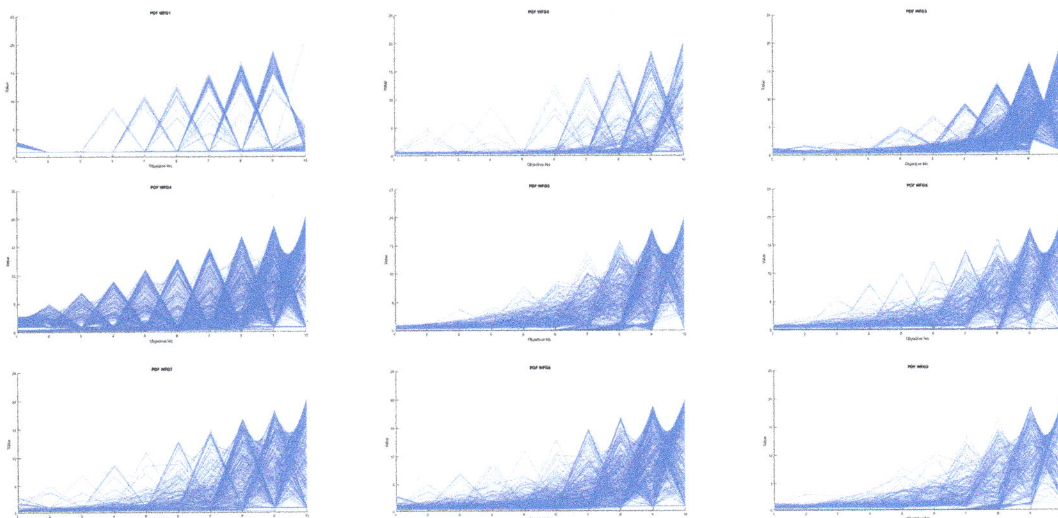

**Figure 14.** The plots of POF for WFG1-9 functions with 10 objectives using DB-CSA algorithm.

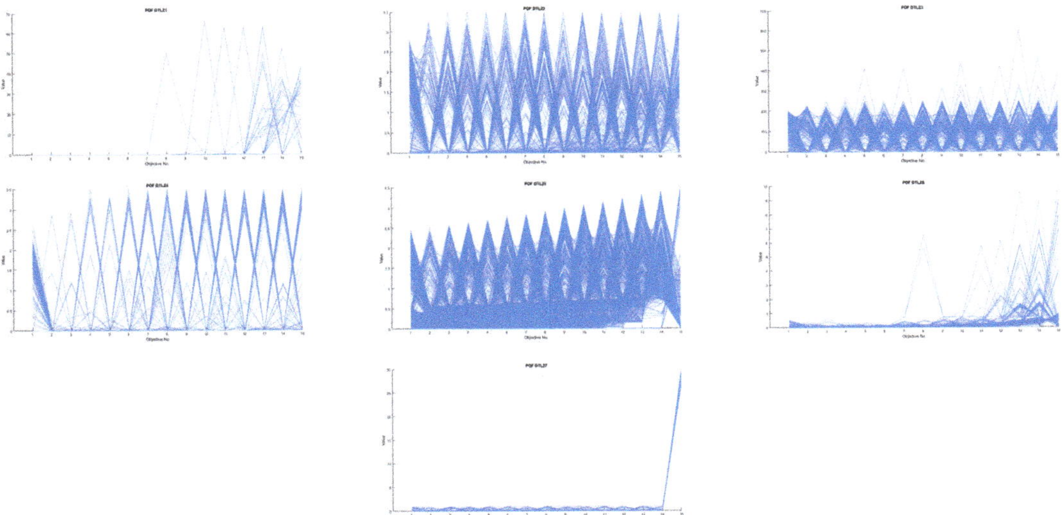

**Figure 15.** The plots of POF for DTLZ1-7 functions with 15 objectives using DB-CSA algorithm.

As a global conclusion and based on comparative studies (1) and (2), all quantitative results showed the efficiency of DB-CSA and DB-CSA-II variants and their flexibility in solving eight DMOPs (FDA and dMOP) with 2 and 3 objectives, including several types of time-varying POF and POS, compared to the seven transfer-learning based methods (MMTL-MOEA/D, KF-MOEA/D, PPS-MOEA/D, SVR-MOEA/D, Tr-MOEA/D, and RI-MOEA/D) using the MIGD metric. By considering the plot of the MIGD, IGD and HVD values in Figures 9–11 during 30 independent runs, we can determine the importance of DB-CSA for solving DMOPs in types I (FDA1, FDA4, dMOP3), II (dMOP2) and III (dMOP1). By analyzing the perturbation of MIGD, IGD and HVD plots, we can see the challenging results obtained when solving FDA1, FDA4, dMOP1, dMOP2 and dMOP3 compared to FDA5 and FDA3 in type II with time-varying POF and POS in both severe and moderate search spaces, and FDA5, FDA3 and FDA2 with a slight change.

The efficiency of DB-CSA-II is demonstrated when solving a dynamic tri-objective FDA4 with dynamic POS. However, the proposed DB-CSA-II algorithm assumed a competitive importance compared to the five standard MOEAs (DNSGA-II, dCOEA, PPS, MOEA/D and SGEA) when solving five FDA functions and three dMOP problems over the IGD metric, including different types of environmental changes. This contradicts the HVD metric when all results are not statistically significant at the level of 0.05. Furthermore, the importance of DB-CSA does not assume a high significance level compared to the five standard MOEAs when solving seven UDF and six F problems in type II with a time-varying POF and POS in a moderate environmental change.

Finally, we can assume the importance of the DB-CSA algorithm compared to 13 MaOEAs including; MSOPS-II, MOEA/D, HypE, PICEA-g, SPEA/SDE, GrEA, NSGA-III, KnEA, RVEA, Two_Arch2, $\theta$-DEA, MOEA/DD, AnD for solving a set of many-objective optimization problems (9 WFG and 7 MaF) with 2, 3 and 7 objectives. Also the proposal has achieved the best results for solving the more complex DTLZ and WFG test suites with 3, 5, 8, 10 and 15 objectives compared to the seven MaOEAs (PMEA-MA, PMEA*-MA, SPEA2/SDE, NSGA-II/SDR, MaOEA/IGD, VaEA, SPEA). The main weakness of the proposed DB-CSA-II algorithm is presented when solving DMOPs in type I and II (FDA1, FDA3 and FDA4) characterized by a time-varying POS and POF, a dynamic spread or the dynamic density of the approximated solution set, with a nonlinear correlation between the decision variables. The high number of objectives also leads to the need for additional computational resources and increases the execution time.

### 4.4.5. Time Processing Cost

The time needed to process the proposed DB-CSA-II variant was computed to solve the 5 FDA and 3 dMOP test suites, as shown in Table 11. The DB-CSA-II algorithm was compared to six state-of-the-art transfer-learning-based approaches: (MMTL-MOEA/D, KF-MOEA/D, PPS-MOEA/D, SVR-MOEA/D, Tr-MOEA/D, DB-CSA-II). The run-time of the six transfer learning approaches was obtained from the original paper [15] where the important values are in bold. The PPS-MOEA/D algorithm was found to be fast for solving FDA1, FDA2, and FDA3 test beds. The SVR-MOEA/D approach has the fastest running time for FDA4, the MMTL-MOEA/D algorithm for both FDA5, dMOP1, and the KF-MOEA/D method for solving dMOP2 and dMOP3. We can conclude that the novel DB-CSA-II algorithm is not very fast in terms of computation time compared to the state of the art methods. However, the robust performance of the proposed DB-CSA-II algorithm is proved by the obtained means and standard deviation values based on the MIGD metric. The time comparison should be moderated, since it depends on the processor capacities as well as the hardware configurations used to conduct the tests.

**Table 11.** Run-times of DMOEAs for Solving FDA and dMOP Test Suites (Unit: Seconds).

| DMOPs | MMTL-MOEA/D | KF-MOEA/D | PPS-MOEA/D | SVR-MOEA/D | Tr-MOEA/D | DB-CSA-II |
|---|---|---|---|---|---|---|
| FDA1 | 5.83 | 3.55 | **3.11** | 4.85 | 42.54 | 10.32 |
| FDA2 | 5.40 | 3.46 | **3.08** | 4.63 | 45.27 | 12.86 |
| FDA3 | 5.09 | 3.92 | **3.64** | 4.78 | 57.71 | 14.96 |
| FDA4 | 10.06 | 9.78 | 13.32 | **9.66** | 132.59 | 21.92 |
| FDA5 | **9.94** | 10.24 | 14.83 | 11.52 | 115.52 | 19.85 |
| dMOP1 | **7.05** | 7.25 | 8.96 | 8.34 | 80.23 | 9.05 |
| dMOP2 | 7.74 | **4.02** | 4.93 | 4.47 | 73.53 | 9.29 |
| dMOP3 | 5.63 | **3.52** | 4.15 | 4.61 | 75.55 | 9.73 |

## 5. Conclusions and Perspectives

In this paper, a new Distributed Bi-behaviors Crow Search Algorithm (DB-CSA) is proposed for the dynamic treatment of both convergence and diversity concepts, based on two new mechanisms: distributed bi-behavior profiles, characterized by a large Gaussian

Beta-1 and narrow Gaussian Beta-2 functions for exploitation and exploration enhancement, respectively. All quantitative results were analyzed using the non-parametric Wilcoxon signed rank test with a 0.05 significance level. The experimental studies showed that the proposed DB-CSA is significantly better than the state-of-the-art methods. The novel DB-CSA-II algorithm achieved good results for solving dynamic multi-objective problems characterized by different types of dynamic change in the POS and the POF including 2 or 3 conflicting objective functions. The comparative study (1) included seven transfer-learning based methods (MMTL-MOEA/D, KF-MOEA/D, PPS-MOEA/D, SVR-MOEA/D, Tr-MOEA/D, and RI-MOEA/D) used the MIGD metric and the five popular DMOEAs (DNSGA-II, dCOEA, PPS, MOEA/D and SGEA) to solve twenty-one DMOPs with different types of changes on both POF and POS usign the IGD and HVD quality indicators and it is proved that the proposal relative results are better for all test beds. Based on the comparative study (2), we can resume the efficiency of DB-CSA system compared to thirteen MaOEAs (MSOPS-II, MOEA/D, HypE, PICEA-g, SPEA/SDE, GrEA, NSGA-III, KnEA, RVEA, Two_Arch2, $\theta$-DEA, MOEA/DD, AnD) for solving sixteen many-objective optimization problems (9 WFG and 7 MaF) with 2, 3 and 7 objectives, as well as the more complex DTLZ and WFG test suites with 3, 5, 8, 10 and 15 objectives compared to the seven MaOEAs (PMEA-MA, PMEA*-MA, SPEA2/SDE, NSGA-II/SDR, MaOEA/IGD, VaEA, SPEA). All results confirmed the relevance of the proposed DB-CSA approach and its capacity to correctly manage convergence and diversity concepts when solving DMOPs and MaOPS. For future works, it is recommended to investigate the impact that the beta-profiles have on performances when solving a DMOP characterized by a time-varying POS and POF, a dynamic spread or the dynamic density of the approximated solution set with a nonlinear correlation between the decision variables. Both variants of the DB-CSA method are worthy of consideration when solving a set of Evolutionary Transfer Multi/Many-objective Optimization Problems.

**Supplementary Materials:** The following supporting information can be downloaded at: https://www.mdpi.com/article/10.3390/app12199627/s1, Table S1: MIGD results (Mean and Standard Deviation) of the 6 DMOEAs [15] compared to the standard CSA and the proposed DB-CSA algorithm for solving FDA and dMOP functions; Table S2: IGD results (Mean and Standard Deviation) of the 5 DMOEAs [11] compared to the standard CSA and the proposed DB-CSA algorithm for solving FDA and dMOP functions; Table S3: HVD results (Mean and Standard Deviation) of the 5 DMOEAs [11] compared to the standard CSA and the proposed DB-CSA algorithm for solving FDA and dMOP functions; Table S4: IGD results (Mean and Standard Deviation) of the 5 DMOEAs [11] compared to the standard CSA and the proposed DB-CSA algorithm for solving UDF and F(ZJZ) functions with ($\tau_t = n_t = 10$); Table S5: HVD results (Mean and Standard Deviation) of the 5 DMOEAs [11] compared to the standard CSA and the proposed DB-CSA algorithm for solving UDF and F(ZJZ) functions with ($\tau_t = n_t = 10$); Table S6: IGD results (Mean and Standard Deviation) of the 13 MOEAs [55] compared to the standard CSA and the proposed DB-CSA algorithm on the 2, 3 and 7 objective WFG problems; Table S7: IGD results (Mean and Standard Deviation) of the 13 MOEAs [55] compared to the standard CSA and the proposed DB-CSA algorithm on the 2, 3 and 7 objective MaF problems; Table S8: IGD results (Mean and Standard Deviation) of the 7 MOEAs [54] compared to DB-CSA on the WFG test suite; Table S9: IGD results (Mean and Standard Deviation) of the 7 MOEAs [54] compared to the standard CSA and the proposed DB-CSA algorithm on the DTLZ test suite.

**Author Contributions:** Conceptualization, A.A.; Formal analysis, A.A.; Funding acquisition, B.N. and Z.A.B.; Investigation, A.A. and Z.A.B.; Methodology, A.A., N.R. and S.M.; Project administration, N.R., B.N. and A.M.A.; Resources, Z.A.B.; Supervision, N.R., B.N. and A.M.A.; Validation, N.R., B.N. and Z.A.B.; Visualization, A.A. and S.M.; Writing—original draft, A.A.; Writing—review & editing, A.A., N.R., B.N., Z.A.B., S.M. and A.M.A. All authors have read and agreed to the published version of the manuscript.

**Funding:** The research leading to these results has received funding from the Ministry of Higher Education and Scientific Research of Tunisia under the grant agreement number LR11ES48.

**Institutional Review Board Statement:** Not applicable.

**Informed Consent Statement:** Not applicable.

**Data Availability Statement:** As an alternative, the data has been shared on the Mendeley Data Repository and will be public to the community at the following DOI Link: 10.17632/hydzpsv4tp.2.

**Conflicts of Interest:** The authors declare no conflict of interest.

# References


1. Deb, K.; N., U.; Karthik, S. Dynamic multi-objective optimization and decision-making using modified NSGA-II: A case study on hydro-thermal power scheduling. In *Lecture Notes in Computer Science (Including Subseries Lecture Notes in Artificial Intelligence and Lecture Notes in Bioinformatics)*; Springer: Berlin/Heidelberg, Germany, 2007; pp. 803–817. [CrossRef]
2. Aboud, A.; Fdhila, R.; Alimi, A. Dynamic Multi Objective Particle Swarm Optimization Based on a New Environment Change Detection Strategy. In *Lecture Notes in Computer Science (Including Subseries Lecture Notes in Artificial Intelligence and Lecture Notes in Bioinformatics)*; Springer: Cham, Switzerland, 2017; Volume 10637, pp. 258–268. [CrossRef]
3. Aboud, A.; Fdhila, R.; Alimi, A. MOPSO for dynamic feature selection problem based big data fusion. In Proceedings of the 2016 IEEE International Conference on Systems, Man, and Cybernetics, SMC 2016—Conference Proceedings, Budapest, Hungary, 9–12 October 2016; pp. 3918–3923. [CrossRef]
4. Aboud A.; Rokbani N.; Fdhila R.; Qahtani A.M.; Almutiry O.; Dhahri H.; Hussain A.; Alimi A.M. DPb-MOPSO: A dynamic Pareto bi-level Multi-objective Particle Swarm Optimization Algorithm. *App. Soft Comput.* **2022**, 109622. [CrossRef]
5. Kennedy, J.; Eberhart, R. Particle swarm optimization. In Proceedings of ICNN'95—International Conference on Neural Networks, Perth, WA, Australia, 27 November–1 December 1995; Volume 4, pp. 1942–1948. [CrossRef]
6. Farina, M.; Deb, K.; Amato, P. Dynamic multiobjective optimization problems: Test cases, approximations, and applications. *IEEE Trans. Evol. Comput.* **2004**, *8*, 425–442. [CrossRef]
7. Ou, J. A pareto-based evolutionary algorithm using decomposition and truncation for dynamic multi-objective optimization. *Appl. Soft Comput.* **2019**, *85*, 105673. [CrossRef]
8. Zou, J.; Li, Q.; Yang, S.; Bai, H.; Zheng, J. A prediction strategy based on center points and knee points for evolutionary dynamic multi-objective optimization. *Appl. Soft Comput.* **2017**, *61*, 806–818. [CrossRef]
9. Askarzadeh, A. A novel metaheuristic method for solving constrained engineering optimization problems: Crow search algorithm. *Comput. Struct.* **2016**, *169*, 1–12. [CrossRef]
10. Geem, Z.; Kim, J.; Loganathan, G. A New Heuristic Optimization Algorithm: Harmony Search. *Simulation* **2001**, *76*, 60–68. [CrossRef]
11. Jiang, S.; Yang, S. A Steady-State and Generational Evolutionary Algorithm for Dynamic Multiobjective Optimization. *IEEE Trans. Evol. Comput.* **2017**, *21*, 65–82. [CrossRef]
12. Goh, C.; Tan, K. A competitive-cooperative coevolutionary paradigm for dynamic multiobjective optimization. *IEEE Trans. Evol. Comput.* **2009**, *13*, 103–127. [CrossRef]
13. Zhang, Q.; Li, H. MOEA/D: A multiobjective evolutionary algorithm based on decomposition. *IEEE Trans. Evol. Comput.* **2007**, *11*, 712–731. [CrossRef]
14. Zhou, A.; Jin, Y.; Zhang, Q. A Population prediction strategy for evolutionary dynamic multiobjective optimization. *IEEE Trans. Cybern.* **2014**, *44*, 40–53. [CrossRef]
15. Jiang, M.; Wang, Z.; Qiu, L.; Guo, S.; Gao, X.; Tan, K. A Fast Dynamic Evolutionary Multiobjective Algorithm via Manifold Transfer Learning. *IEEE Trans. Cybern.* **2020**, *51*, 3417–3428. [CrossRef] [PubMed]
16. Cao, L.; Xu, L.; Goodman, E.; Bao, C.; Zhu, S. Evolutionary Dynamic Multiobjective Optimization Assisted by a Support Vector Regression Predictor. *IEEE Trans. Evol. Comput.* **2020**, *24*, 305–319. [CrossRef]
17. Jiang, M.; Huang, Z.; Qiu, L.; Huang, W.; Yen, G. Transfer Learning-Based Dynamic Multiobjective Optimization Algorithms. *IEEE Trans. Evol. Comput.* **2018**, *22*, 501–514. [CrossRef]
18. Muruganantham, A.; Tan, K.; Vadakkepat, P. Evolutionary Dynamic Multiobjective Optimization Via Kalman Filter Prediction. *IEEE Trans. Cybern.* **2016**, *46*, 2862–2873. [CrossRef]
19. Purshouse, R.; Fleming, P. On the evolutionary optimization of many conflicting objectives. *IEEE Trans. Evol. Comput.* **2007**, *11*, 770–784. [CrossRef]
20. Hughes, E.J. Multiple single objective Pareto sampling. In Proceedings of the 2003 Congress on Evolutionary Computation, Canberra, Australia, 8–12 December 2003; Volume 4, p. 2678–2684. [CrossRef]
21. Hughes, E.J. MSOPS-II: A general-purpose many-objective optimiser. In Proceedings of the 2007 IEEE Congress on Evolutionary Computation, CEC 2007, Singapore, 25–28 September 2007; pp. 3944–3951. [CrossRef]
22. Cheng, R.; Jin, Y.; Olhofer, M.; Sendhoff, B. A Reference Vector Guided Evolutionary Algorithm for Many-Objective Optimization. *IEEE Trans. Evol. Comput.* **2016**, *20*, 773–791. [CrossRef]
23. Liu, H.; Gu, F.; Zhang, Q. Decomposition of a multiobjective optimization problem into a number of simple multiobjective subproblems. *IEEE Trans. Evol. Comput.* **2014**, *18*, 450–455. [CrossRef]
24. Deb, K.; Jain, H. An evolutionary many-objective optimization algorithm using reference-point-based nondominated sorting approach, Part I: Solving problems with box constraints. *IEEE Trans. Evol. Comput.* **2014**, *18*, 577–601. [CrossRef]

25. Li, K.; Deb, K.; Zhang, Q.; Kwong, S. An evolutionary many-objective optimization algorithm based on dominance and decomposition. *IEEE Trans. Evol. Comput.* **2015**, *19*, 694–716. [CrossRef]
26. Lu, X.; Tan, Y.; Zheng, W.; Meng, L. A Decomposition Method Based on Random Objective Division for MOEA/D in Many-Objective Optimization. *IEEE Access* **2020**, *8*, 103550–103564. [CrossRef]
27. Bader, J.; Zitzler, E. HypE: An algorithm for fast hypervolume-based many-objective optimization. *Evol. Comput.* **2011**, *19*, 45–76. [CrossRef] [PubMed]
28. Beume, N.; Naujoks, B.; Emmerich, M. SMS-EMOA: Multiobjective selection based on dominated hypervolume. *Eur. J. Oper. Res.* **2007**, *181*, 1653–1669. [CrossRef]
29. Zitzler, E.; Künzli, S. Indicator-based selection in multiobjective search. In *Lecture Notes in Computer Science (Including Subseries Lecture Notes in Artificial Intelligence and Lecture Notes in Bioinformatics*; Springer: Berlin/ Heidelberg, Germany, 2004; Volume 3242, pp. 832–842. [CrossRef]
30. Feng, S.; Wen, J. An Evolutionary Many-Objective Optimization Algorithm Based on IGD Indicator and Region Decomposition. In Proceedings of the 2019 15th International Conference on Computational Intelligence and Security, CIS 2019, Macau, China, 13–16 December 2019; pp. 206–210. [CrossRef]
31. Sun, Y.; Yen, G.; Yi, Z. IGD Indicator-Based Evolutionary Algorithm for Many-Objective Optimization Problems. *IEEE Trans. Evol. Comput.* **2019**, *23*, 173–187. [CrossRef]
32. Zou, X.; Chen, Y.; Liu, M.; Kang, L. A new evolutionary algorithm for solving many-objective optimization problems. *IEEE Trans. Syst. Man, Cybern. Part B Cybern.* **2008**, *38*, 1402–1412. [CrossRef]
33. Hadka, D.; Reed, P. Borg: An auto-adaptive many-objective evolutionary computing framework. *Evol. Comput.* **2013**, *21*, 231–259. [CrossRef]
34. Gaoping, W.; Huawei, J. Fuzzy-dominance and its application in evolutionary many objective optimization. In Proceedings of the CIS Workshops 2007, 2007 International Conference on Computational Intelligence and Security Workshops, Harbin, China, 15–19 December 2007; pp. 195–198. [CrossRef]
35. Yang, S.; Li, M.; Liu, X.; Zheng, J. A grid-based evolutionary algorithm for many-objective optimization. *IEEE Trans. Evol. Comput.* **2013**, *17*, 721–736. [CrossRef]
36. Yuan, Y.; Xu, H.; Wang, B.; Yao, X. A New Dominance Relation-Based Evolutionary Algorithm for Many-Objective Optimization. *IEEE Trans. Evol. Comput.* **2016**, *20*, 16–37. [CrossRef]
37. Pierro, F.; Khu, S.; Savić, D. An investigation on preference order ranking scheme for multiobjective evolutionary optimization. *IEEE Trans. Evol. Comput.* **2007**, *11*, 17–45. [CrossRef]
38. Adra, S.; Fleming, P. Diversity management in evolutionary many-objective optimization. *IEEE Trans. Evol. Comput.* **2011**, *15*, 183–195. [CrossRef]
39. Li, M.; Yang, S.; Liu, X. Shift-based density estimation for pareto-based algorithms in many-objective optimization. *IEEE Trans. Evol. Comput.* **2014**, *18*, 348–365. [CrossRef]
40. Zhang, X.; Tian, Y.; Jin, Y. A knee point-driven evolutionary algorithm for many-objective optimization. *IEEE Trans. Evol. Comput.* **2015**, *19*, 761–776. [CrossRef]
41. Wang, R.; Purshouse, R.; Fleming, P. Preference-inspired coevolutionary algorithms for many-objective optimization. *IEEE Trans. Evol. Comput.* **2013**, *17*, 474–494. [CrossRef]
42. Praditwong, K.; Yao, X. A new multi-objective evolutionary optimisation algorithm: The two-archive algorithm. In Proceedings of the 2006 International Conference on Computational Intelligence and Security, ICCIAS 2006, Guangzhou, China, 3–6 November 2006; Volume 1, p. 286–291. [CrossRef]
43. Wang, H.; Jiao, L.; Yao, X. Two Arch2: An Improved Two-Archive Algorithm for Many-Objective Optimization. *IEEE Trans. Evol. Comput.* **2015**, *19*, 524–541. [CrossRef]
44. Carvalho, A.; Pozo, A. Measuring the convergence and diversity of CDAS Multi-Objective Particle Swarm Optimization Algorithms: A study of many-objective problems. *Neurocomputing* **2012**, *75*, 43–51. [CrossRef]
45. Castro, O.; Pozo, A. A MOPSO based on hyper-heuristic to optimize many-objective problems. In Proceedings of the 2014 IEEE Symposium on Swarm Intelligence, Proceedings, Orlando, FL, USA, 9–12 December 2014; pp. 251–258. [CrossRef]
46. Sun, X.; Chen, Y.; Liu, Y.; Gong, D. Indicator-based set evolution particle swarm optimization for many-objective problems. *Soft Comput.* **2016**, *20*, 2219–2232. [CrossRef]
47. Hu, W.; Yen, G.; Luo, G. Many-Objective Particle Swarm Optimization Using Two-Stage Strategy and Parallel Cell Coordinate System. *IEEE Trans. Cybern.* **2017**, *47*, 1446–1459. [CrossRef] [PubMed]
48. Maltese, J.; Ombuki-Berman, B.; Engelbrecht, A. Pareto-based many-objective optimization using knee points. In Proceedings of the 2016 IEEE Congress on Evolutionary Computation, CEC 2016, Vancouver, BC, Canada, 24–29 July 2016; pp. 3678–3686. [CrossRef]
49. Xiang, Y.; Zhou, Y.; Chen, Z.; Zhang, J. A Many-Objective Particle Swarm Optimizer with Leaders Selected from Historical Solutions by Using Scalar Projections. *IEEE Trans. Cybern.* **2020**, *50*, 2209–2222. [CrossRef]
50. Leung, M.; Coello, C.; Cheung, C.; Ng, S.; Lui, A. A hybrid leader selection strategy for many-objective particle swarm optimization. *IEEE Access* **2020**, *8*, 189527–189545. [CrossRef]
51. Ma, L.; Huang, M.; Yang, S.; Wang, R.; Wang, X. An Adaptive Localized Decision Variable Analysis Approach to Large-Scale Multiobjective and Many-Objective Optimization. *IEEE Trans. Cybern.* **2021**. [CrossRef] [PubMed]

52. Xiang, Y.; Zhou, Y.; Li, M.; Chen, Z. A Vector Angle-Based Evolutionary Algorithm for Unconstrained Many-Objective Optimization. *IEEE Trans. Evol. Comput.* **2017**, *21*, 131–152. [CrossRef]
53. Tian, Y.; Cheng, R.; Zhang, X.; Su, Y.; Jin, Y. A Strengthened Dominance Relation Considering Convergence and Diversity for Evolutionary Many-Objective Optimization. *IEEE Trans. Evol. Comput.* **2019**, *23*, 331–345. [CrossRef]
54. Liu, Y.; Zhu, N.; Li, M. Solving Many-Objective Optimization Problems by a Pareto-Based Evolutionary Algorithm with Preprocessing and a Penalty Mechanism. *IEEE Trans. Cybern.* **2020**, *51*, 5585–5594. [CrossRef]
55. Li, K.; Wang, R.; Zhang, T.; Ishibuchi, H. Evolutionary Many-Objective Optimization: A Comparative Study of the State-of-The-Art. *IEEE Access* **2018**, *6*, 26194–26214. [CrossRef]
56. Jiang, S.; Yang, S. A strength pareto evolutionary algorithm based on reference direction for multiobjective and many-objective optimization. *IEEE Trans. Evol. Comput.* **2017**, *21*, 329–346. [CrossRef]
57. Chen, H.; Cheng, R.; Pedrycz, W.; Jin, Y. Solving Many-Objective Optimization Problems via Multistage Evolutionary Search. *IEEE Trans. Syst. Man Cybern. Syst.* **2019**, *51*, 3552–3564. [CrossRef]
58. Meraihi, Y.; Gabis, A.; Ramdane-Cherif, A.; Acheli, D. A comprehensive survey of Crow Search Algorithm and its applications. *Artif. Intell. Rev.* **2021**, *54*, 2669–2716. [CrossRef]
59. Nobahari, H.; Bighashdel, A. MOCSA: A Multi-Objective Crow Search Algorithm for Multi-Objective optimization. In Proceedings of the 2nd Conference on Swarm Intelligence and Evolutionary Computation, CSIEC 2017–Proceedings, Kerman, Iran, 7–9 March 2017; pp. 60–65. [CrossRef]
60. John, J.; Rodrigues, P. MOTCO: Multi-objective Taylor Crow Optimization Algorithm for Cluster Head Selection in Energy Aware Wireless Sensor Network. *Mob. Netw. Appl.* **2019**, *24*, 1509–1525. [CrossRef]
61. Souza, R.; Coelho, L.; MacEdo, C.; Pierezan, J. A V-Shaped Binary Crow Search Algorithm for Feature Selection. In Proceedings of the 2018 IEEE Congress on Evolutionary Computation (CEC), Rio de Janeiro, Brazil, 8–13 July 2018. [CrossRef]
62. Laabadi, S.; Naimi, M.; Amri, H.; Achchab, B. A Binary Crow Search Algorithm for Solving Two-dimensional Bin Packing Problem with Fixed Orientation. *Procedia Comput. Sci.* **2020**, *167*, 809–818. [CrossRef]
63. Coelho, L.S.; Richter, C.; Mariani, V.; Askarzadeh, A. Modified crow search approach applied to electromagnetic optimization. In Proceedings of the 2016 IEEE Conference on Electromagnetic Field Computation (CEFC), Miami, FL, USA, 13–16 November 2016. [CrossRef]
64. Gupta, D.; Rodrigues, J.; Sundaram, S.; Khanna, A.; Korotaev, V.; Albuquerque, V. Usability feature extraction using modified crow search algorithm: A novel approach. *Neural Comput. Appl.* **2020**, *32*, 10915–10925. [CrossRef]
65. Mohammadi, F.; Abdi, H. A modified crow search algorithm (MCSA) for solving economic load dispatch problem. *Appl. Soft Comput. J.* **2018**, *71*, 51–65. [CrossRef]
66. Cuevas, E.; Espejo, E.; Enríquez, A. A modified crow search algorithm with applications to power system problems. In *Studies in Computational Intelligence*; Springe: Cham, Switzerland, 2019; Volume 822, pp. 137–166. [CrossRef]
67. Huang, K.W.; Girsang, A.S.; Wu, Z.X.; Chuang, Y.W. A Hybrid Crow Search Algorithm for Solving Permutation Flow Shop Scheduling Problems. *Appl. Sci.* **2019**, *9*, 1353. [CrossRef]
68. Díaz, P.; Pérez-Cisneros, M.; Cuevas, E.; Avalos, O.; Gálvez, J.; Hinojosa, S.; Zaldivar, D. An Improved Crow Search Algorithm Applied to Energy Problems. *Energies* **2018**, *11*, 571. [CrossRef]
69. Meddeb, A.; Amor, N.; Abbes, M.; Chebbi, S. A Novel Approach Based on Crow Search Algorithm for Solving Reactive Power Dispatch Problem. *Energies* **2018**, *11*, 3321. [CrossRef]
70. Javidi, A.; Salajegheh, E.; Salajegheh, J. Enhanced crow search algorithm for optimum design of structures. *Appl. Soft Comput. J.* **2019**, *77*, 274–289. [CrossRef]
71. Bhullar, A.; Kaur, R.; Sondhi, S. Enhanced crow search algorithm for AVR optimization. *Soft Comput.* **2020**, *24*, 11957–11987. [CrossRef]
72. Cuevas, E.; Gálvez, J.; Avalos, O. An Enhanced Crow Search Algorithm Applied to Energy Approaches. In *Studies in Computational Intelligence*; Springer: Cham, Switzerland, 2020; Volume 854, pp. 27–49. [CrossRef]
73. Moghaddam, S.; Bigdeli, M.; Moradlou, M.; Siano, P. Designing of stand-alone hybrid PV/wind/battery system using improved crow search algorithm considering reliability index. *Int. J. Energy Environ. Eng.* **2019**, *10*, 429–449. [CrossRef]
74. Arora, S.; Singh, H.; Sharma, M.; Sharma, S.; Anand, P. A New Hybrid Algorithm Based on Grey Wolf Optimization and Crow Search Algorithm for Unconstrained Function Optimization and Feature Selection. *IEEE Access* **2019**, *7*, 26343–26361. [CrossRef]
75. Huang, K.W.; Wu, Z.X. CPO: A Crow Particle Optimization Algorithm. *Int. J. Comput. Intell. Syst.* **2019**, *12*. [CrossRef]
76. Gaddala, K.; Raju, P. Merging Lion with Crow Search Algorithm for Optimal Location and Sizing of UPQC in Distribution Network. *J. Control. Autom. Electr. Syst.* **2020**, *31*, 377–392. [CrossRef]
77. Alimi, A. Beta Neuro-Fuzzy Systems. *Task Q.* **2003**, *7*, 23–41.
78. Rokbani N.; Slim M.; Alimi A.M. The Beta distributed PSO, $\beta$-PSO, with application to Inverse Kinematics In Proceedings of National Computing Colleges Conference (NCCC), 2021; pp. 1–6. [CrossRef]
79. Garzelli, A.; Capobianco, L.; Nencini, F. Fusion of multispectral and panchromatic images as an optimisation problem. In *Image Fusion*; Elsevier: Amsterdam, The Netherlands, 2008; pp. 223–250. [CrossRef]
80. Deb, K.; Pratap, A.; Agarwal, S.; Meyarivan, T. A fast and elitist multiobjective genetic algorithm: NSGA-II. *IEEE Trans. Evol. Comput.* **2002**, *6*, 182–197. [CrossRef]

81.   Biswas, S.; Das, S.; Suganthan, P.; Coello, C. Evolutionary multiobjective optimization in dynamic environments: A set of novel benchmark functions.   In Proceedings of the 2014 IEEE Congress on Evolutionary Computation, CEC 2014, Beijing, China, 6–11 July 2014; pp. 3192–3199. [CrossRef]
82.   Durillo, J.; Nebro, A. JMetal: A Java framework for multi-objective optimization. *Adv. Eng. Softw.* **2011**, *42*, 760–771. [CrossRef]
83.   Genichi, T.; Rajesh, J.; Shin, T. *Computer-based Robust Engineering: Essentials for DFSS*; ASQ Quality Press: Milwaukee, WI, USA, 2004; p. 217.
84.   Rokbani, N.   Bi-heuristic ant colony optimization-based approaches for traveling salesman problem.   *Soft Comput.* **2021**, *25*, 3775–3794. [CrossRef]
85.   Dordevic, M. Statistical analysis of various hybridization of evolutionary algorithm for traveling salesman problem. In Proceedings of the IEEE International Conference on Industrial Technology, Melbourne, Australia, 13–15 February 2019; pp. 899–904. [CrossRef]

*applied*
*sciences*

*Article*

# Combining UNet 3+ and Transformer for Left Ventricle Segmentation via Signed Distance and Focal Loss

**Zhi Liu** *,†, **Xuelin He** † **and Yunhua Lu**

School of Artificial Intelligence, Chongqing University of Technology, Chongqing 401120, China
* Correspondence: liuzhi@cqut.edu.cn; Tel.: +86-139-8398-2460
† These authors contributed equally to this work.

**Abstract:** Left ventricle (LV) segmentation of cardiac magnetic resonance (MR) images is essential for evaluating cardiac function parameters and diagnosing cardiovascular diseases (CVDs). Accurate LV segmentation remains a challenge because of the large differences in cardiac structures in different research subjects. In this work, a network based on an encoder–decoder architecture for automatic LV segmentation of short-axis cardiac MR images is proposed. It combines UNet 3+ and Transformer to jointly predict the segmentation masks and signed distance maps (SDM). UNet 3+ can extract coarse-grained semantics and fine-grained details from full scales, while a Transformer is used to extract global features from cardiac MR images. It solves the problem of low segmentation accuracy caused by blurred LV edge information. Meanwhile, the SDM provides a shape-aware representation for segmentation. The performance of the proposed network is validated on the 2018 MICCAI Left Ventricle Segmentation Challenge dataset. The five-fold cross-validation evaluation was performed on 145 clinical subjects, and the average dice metric, Jaccard coefficient, accuracy, and positive predictive value reached 0.908, 0.834, 0.979, and 0.903, respectively, showing a better performance than that of other mainstream ones.

**Keywords:** left ventricle segmentation; UNet 3+; encoder–decoder; transformer; magnetic resonance imaging

**Citation:** Liu, Z.; He, X.; Lu, Y. Combining UNet 3+ and Transformer for Left Ventricle Segmentation via Signed Distance and Focal Loss. *Appl. Sci.* **2022**, *12*, 9208. https://doi.org/10.3390/app12189208

Academic Editors: Xinglong Zhang, Pengfei Jia and Yue Wu

Received: 30 August 2022
Accepted: 11 September 2022
Published: 14 September 2022

**Publisher's Note:** MDPI stays neutral with regard to jurisdictional claims in published maps and institutional affiliations.

## 1. Introduction

The World Health Organization (WHO) showed that in 2019, almost 17.9 million people died of cardiovascular diseases (CVDs), accounting for 32% of fatalities worldwide [1]. Early diagnosis of CVD can help improve cardiac function and reduce patient mortality [2]. Cardiovascular magnetic resonance (MR) imaging is harmless and has become the most commonly used technique for evaluating cardiovascular system structure and function. Left ventricle (LV) segmentation, as a key step in the treatment of CVD, can provide better visual aid during the diagnosis of CVD. Most CVDs affect the physiological shape of the cardiac LV, and LV dysfunction is the cause of many heart diseases, such as ventricular hypertrophy and myocardial infarction, making the examination of the LV an important prerequisite for determining whether the heart is diseased. LV segmentation can accurately delineate the boundaries of the LV on cardiac MR images so that physicians can better understand some clinical parameters, such as the patient's ventricular volume, ejection fraction, LV mass, and stroke volume [3,4].

In early clinical work, medical images are usually annotated by experts to mitigate the subjective bias caused by the level of a particular expert or possible negligence of subtle symptoms [5]. However, for most professional clinicians, manual segmentation is a cumbersome and time-consuming task. In general, it takes a clinician about 20 min to segment a patient's cardiac MR image. Additionally, in Figure 1, as you can see from the left ventricle's structure, the intensity and shape similarity between the LV and other organs, boundary inaccuracies, and the inherent noise of cardiac MR imaging have all posed obstacles to LV segmentation [6–8].

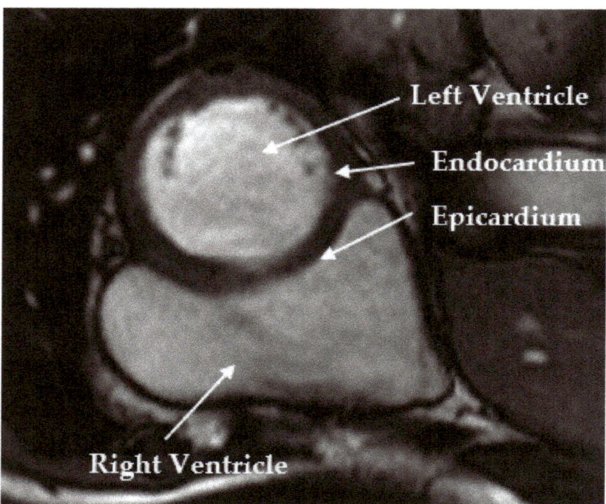


**Figure 1.** Short-axis cardiac MR anatomy image.

Some excellent segmentation algorithms spawned in the past decades have been caught in certain dilemmas that make them difficult to apply in clinical settings. Among them, many of the algorithms are traditional methods based on machine learning [9–11], such as thresholding [12], clustering [13], active contours [14,15], and split–merge. Figure 2 displays the thresholding, clustering, and active contours methods. Most of them use semi-automatic methods that heavily rely on the initialization step [16–19], resulting in their failure to achieve the desired results. Meanwhile, because of the rapid development of the accessibility of vast training data and computer hardware, medical image segmentation algorithms based on deep learning are becoming increasingly prevalent. In particular, Convolutional Neural Networks (CNNs) have achieved excellent results in various computer vision (CV) fields, such as image segmentation [20], object detection [21], and image classification [22]. In such a trend, CNN-based LV segmentation models such as Densely Connected Convolutional Networks (DenseNet) [23] and Fully Convolutional Neural Networks (FCN) [24] have been suggested, and have achieved good results in clinical trials.

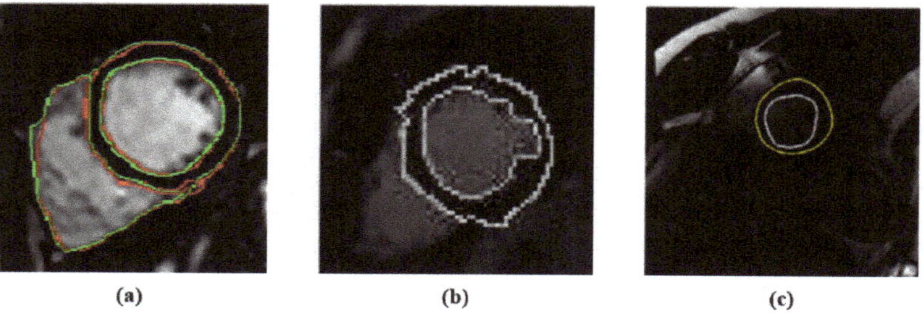

**Figure 2.** Traditional segmentation methods. (**a**) Thresholding. Automated contours (green) and manual contours (red). (**b**) Clustering. (**c**) Active contours.

Although these networks are representative, CNN-based approaches still have some limitations in capturing global information and recovering weak texture details. To overcome this limitation, some studies have suggested Transformer-based designs that use the self-attention mechanism to construct the contextual representations. Transformers, unlike traditional CNNs, can model global contextual information by modeling the relationships between spatially distant pixels [25]. However, both models have their drawbacks. CNN is insensitive to global features, and Transformer has the issues of high computation cost and a lack of ability to capture regional features, so the applicability of both models needs to be further improved.

To solve the above problems, this study proposes a fast and automated method of cardiac LV segmentation to help facilitate the diagnosis of CVD. The network has the following two main contributions: (1) The backbone uses a combined framework of UNet 3+ [26] and Transformer, which can efficiently acquire low-level spatial features as well as high-level semantic information and can also model the global context. (2) The shared backbone network is used to jointly predict the segmentation masks and signed distance maps (SDM) to study segmentation targets' different representations from different perspectives.

### 1.1. Traditional Segmentation Methods

Currently, the methods for LV segmentation can usually be divided into traditional segmentation methods and deep learning methods. However, the early traditional segmentation methods have some obvious drawbacks. Several of these algorithms can only obtain relatively accurate segmentation results when the pixel intensities of the LV and other tissues reach a high level of contrast. For example, the threshold-based segmentation method was used by Goshtasby et al. [27] in 1995 to extract the LV contour. It adaptively determines the threshold grayscale according to the global or local grayscale histogram of the image. Therefore, this method can only obtain better segmentation results on the condition that the grayscale of the region is significantly different from the background. Since its appearance, the K-Means clustering algorithm has been extensively applied in the fields of image analysis and data mining. However, because many regions in cardiac MR images are similar or even connected to the LV, the K-means clustering algorithm cannot achieve the expected results and needs further improvement. In 2006, Katouzian et al. [12] employed the idea of split–merge to segment the LV, which could successfully extract the epicardium and endocardium of the LV under the condition of manually annotating the epicardium and endocardium of the first slice. This manual process obviously increases the clinical application complexity. On the whole, these traditional methods have the characteristic of relying on manual design, which contradicts the idea of automatic segmentation.

### 1.2. Deep Learning

Unlike traditional segmentation methods, deep learning methods prefer to train with large-scale data to find the intrinsic patterns and representation levels in images and get more representative feature information [28]. They naturally describe image features without relying on the manual extraction of features, addressing the limitations of traditional segmentation methods. J. Long et al. [24] proposed a fully convolutional neural network (FCN) based on a CNN to recover the feature map to the original image size via transposed convolution, thus achieving image segmentation in 2015. P. V. Tran [7] segmented both the left and right ventricles in 2016 using FCN. However, the recovery of the LV contour was poor due to only a single upsampling. Ronneberger et al. [29] suggested the U-Net network based on FCN. U-Net effectively integrates low-resolution information and high-resolution information to learn better feature representation and improve generalization performance. However, the utilization of feature maps is poor and does not work well for object boundary segmentation. SegNet [30] proposed a decoder to perform nonlinear upsampling, the input feature maps of which were the maximum pooling index received from the corresponding encoder. In spite of avoiding the learning of upsampling and improving the precision of image boundary localization via this method, the segmentation accuracy is not high enough to satisfy the real-time requirement. On the other hand, despite the excellent representation capability of all these networks, CNN-based approaches have difficulty learning global semantic information because of the inherent locality of convolution operations. As a result, these networks usually yield weak segmentation performance, particularly for target structures that exhibit large inter-patient variations in size, shape, and texture.

### 1.3. Transformers

Transformer was first proposed by Vaswani et al. [31] as the main method for machine translation. It has now been introduced as a new model for image recognition [32], semantic segmentation [33], and many other computer vision tasks [34,35]. In contrast with previous CNN-based approaches, Transformer is powerful at modeling global contexts and demonstrates superior transferability for downstream tasks under large-scale pretraining. However, Transformer concentrates on modeling the global context at all stages, leading to generating low-resolution features. Further, due to the lack of detailed localization information for these low-resolution features, which cannot be efficiently recovered by direct upsampling to full resolution, the ultimate obtained segmentation results are coarse. In order to make the model focus on both regional and global features in segmentation tasks, recent studies have tended to combine CNN and Transformer, such as TransUNet [36] and TransFuse [37], which yield satisfactory performance in segmentation tasks. These works show that the combination model has great potential in the field of CV.

## 2. Method

The proposed network in this paper aims to automatically segment the LV in cardiac MR images, thus reducing the tedious manual segmentation and improving the disease diagnosis efficiency. The new medical image segmentation framework proposed in this work is shown in Figure 3, and the backbone part uses a combined architecture of UNet 3+ and Transformer. The network takes the same cardiac MR images as input and predicts both pixel probability maps and SDM. A loss function composed of two main components is designed to train the segmentation network. One is for computing the pixel probability map and the other one is for computing the SDM.

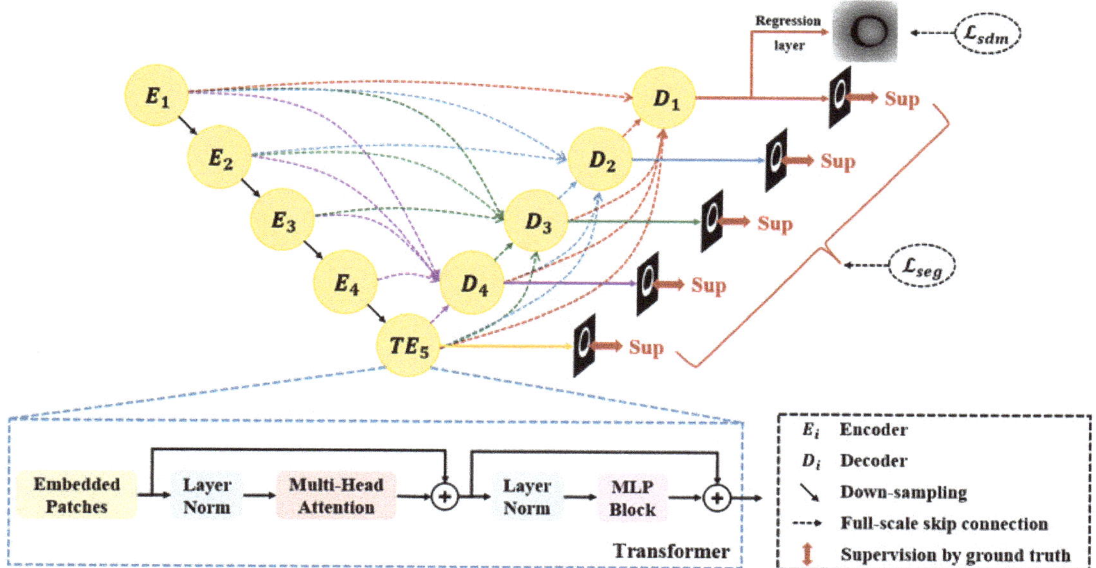


**Figure 3.** The proposed LV segmentation model is based on an encoder–decoder architecture. The network outputs pixel probability maps and SDM.

### 2.1. Segmentation Network

**Feature Extraction**—First, feature maps are generated for the input images using the encoder structure of UNet 3+ as a feature extractor. We are given an image $x \in \mathbb{R}^{H \times W \times C}$ with the spatial resolutions $H$, $W$ and $C$. The preprocessed image $x$ is fed into the network and the encoder applies a series of convolutional blocks to model the pixel-level contextual representations, with features progressively downsampled to $\frac{H}{16} \times \frac{W}{16}$.

**Transformer**—To extract the global features, the Transformer module is applied in the encoder design. Because Transformer is not as efficient as UNet 3+ when it comes to capturing regional features, patch embedding is directly used for patches generated from the UNet 3+ feature maps rather than from raw images. Then patches $x_p$ are mapped into a K-dimensional embedding space by a trainable linear projection. To preserve location information, specific position embeddings $E_{pos}$ are added to the patch embeddings when encoding the patch spatial information. Details are as follows:

$$z_0 = \left[ x_p^1 E; x_p^2 E; \cdots ; x_p^N E \right] + E_{pos}, \tag{1}$$

where $E \in \mathbb{R}^{(P^2 \cdot C) \times K}$ is the patch embedding projection, $E_{pos} \in \mathbb{R}^{N \times K}$ denotes the position embedding, and $x_p^1, \cdots, x_p^N$ are image patches.

Then, a stack of Transformer blocks consisting of alternating layers of multi-head self-attention (MSA) and multi-layer perceptron (MLP) blocks are used to learn the long-range context representation. Layer normalization (LN) is used before each block, and residual connections are used after each block. The following can be expressed as the output of the $i$-th layer:

$$z_i^{'} = MSA(LN(z_{i-1})) + z_{i-1}, \tag{2}$$

$$z_i = MLP\left( LN\left( z_i^{'} \right) \right) + z_i^{'}, \tag{3}$$

where $LN(\cdot)$ represents the layer normalization operator, $i \in \{1, 2, \cdots, L\}$ where $L$ is the number of Transformer layers and $z_i$ denotes the image representation of the encoded output of the $i$-layer Transformer.

**Decoder**—To generate segmentation masks and SDM in the raw image space, a decoder for UNet 3+ is introduced to perform feature upsampling. Since the output of the Transformer is sequential data, its output should first be recovered to spatial order. The encoded feature representation $z_i \in \mathbb{R}^{N \times K}$ is reshaped into $z_i \in \mathbb{R}^{\frac{H}{P} \times \frac{W}{P} \times K}$ where $p$ is the size of each patch, and then the channel dimension is reduced by a $3 \times 3$ convolution block. In addition, each decoder layer in the decoder combines the feature maps of all encoders. Additionally, the full-scale deep supervision proposed by UNet 3+ is used to learn hierarchical representations from the full-scale aggregated feature maps, and the output from each decoder stage is supervised by the ground truth (GT). To achieve deep supervision, a $3 \times 3$ convolution block, a bilinear upsampling, and a sigmoid function are added to the last layer of each decoder stage in the network. To generate the SDM, the network adds an SDM head at the last decoder stage, which consists of a convolution block and tanh activation.

### 2.2. Loss Function

Based on the above architecture design, the segmentation network generates pixel-level segmentation maps and SDM, and the following functions are introduced to convert GT to SDM. The SDM assigns each pixel a value, indicating its signed distance to the nearest boundary of the target object.

$$\mathcal{D}(a) = \begin{cases} 0, & a \in S \\ -\inf_{b \in C} \|a - b\|_2, & a \in C_{in} \\ +\inf_{b \in C} \|a - b\|_2, & a \in C_{out} \end{cases} \tag{4}$$

where $\|a - b\|$ is the Euclidean distance between pixels $a$ and $b$, $S$ represents the boundary of the target object, and $C_{in}$ and $C_{out}$ represent the region inside and outside of the target object, respectively. Typically, SDM takes negative values inside the target and positive values outside the target, with the absolute value indicating the distance from the point to the nearest point on the target object's surface.

In the network training, for the regression task branch, a $\mathcal{L}_2$ loss is used between the SDM of the network output $P_a$ and the transformed GT map $\mathcal{D}(Y)$.

$$\mathcal{L}_{sdm}(P_a, Y) = \|P_a - D(Y)\|^2, \tag{5}$$

where $Y$ denote the GT map.

For the segmentation task branch, the combination $\mathcal{L}_{seg}$ of dice loss and focal loss is applied as the loss function between the segmentation mask and the GT for each decoder output, and then their average is taken as the final segmentation loss.

$$\mathcal{L}_{seg}(P_b, Y) = \mathcal{L}_{Dice}(P_b, Y) + \mathcal{L}_{FL}(P_b, Y), \tag{6}$$

where $\mathcal{L}_{Dice}$ denotes the dice loss and $\mathcal{L}_{FL}$ denotes the focal loss. $P_b$ and $Y$ denote the prediction partition map and label, respectively.

The final loss is defined as:

$$\mathcal{L} = \mathcal{L}_{sdm}(P_a, Y) + \mathcal{L}_{seg}(P_b, Y). \tag{7}$$

## 3. Experiments

### 3.1. Datasets

The dataset from MICCAI 2018 is used to train and evaluate the proposed model [38]. The dataset contains 2900 short-axis cardiac MR images from 145 subjects at three hospitals attached to two healthcare centers (London Healthcare Center and St. Joseph's Healthcare). The age range of the study subjects is 16 to 97, with a mean of 58.9. In the 1.5625 mm/pixel mode, the pixel spacing of the MR images ranges from 0.6836 mm/pixel to 2.0833 mm/pixel. Pathologies such as myocardial hypertrophy, LV dysfunction, atrial septal defect, regional wall motion abnormalities, mildly enlarged LV, etc., are present. During the entire cardiac cycle, twenty frames are acquired for each subject. According to the standard AHA prescription, in each frame, the LV is divided into equal thirds (basal, medial, and parietal) perpendicular to the long axis of the heart. Before manual annotating GT, all cardiac MR images need to be performed with landmark labeling, rotation, ROI cropping, and resizing. After preprocessing, all images are cropped and resized to the dimension of $80 \times 80$ and normalized. After manual contouring, two experienced cardiac radiologists (A. Islam and M. Bhaduri) obtain the epicardial and endocardial boundary and perform a double examination. The labels of the data are approved by industry doctors.

### 3.2. Implementation Details

In the experiments, the backbone of the main framework is the combination of UNet 3+ and Transformer. The network is implemented in PyTorch (1.11.0), with the runtime platform processor of Inter(R) Core (TM) i9-10850K CPU, NVIDIA GeForce RTX 3080 Ti. All training and test images are uniformly adjusted to $80 \times 80$ dimensions. In the training stage, random rotation and flipping operations are applied as data augmentations. The model is evaluated and compared by the five-fold cross-validation. The dataset is split into five groups of 29 subjects each. One group (580 images) is selected for testing, and the remaining four groups (2320 images) are utilized as the training set. The final evaluation result is calculated using the average of five times this process. The proposed network is trained end-to-end with the Adam optimizer with a weight decay of $1 \times 10^{-5}$ and an initial learning rate of $2 \times 10^{-4}$. The model is trained in 100 epochs in a batch size of 16. The segmentation result utilized in testing is the output of the segmentation task branch.

### 3.3. Evaluation Metric

The goal of the network is making sure the model can accurately segment the LV from cardiac MR images. For objective evaluation of the proposed model, the region-based dice metric (DM) and Jaccard coefficient (JC) are employed as metrics. DM and JC are explained in detail as follows.

**Dice Metric**—DM calculates the overlap between the manual segmentation area and the automatic segmentation contour area obtained using the proposed method. DM lies in the [0, 1] range. The better the match between manual and predicted segmentation, the larger the DM value is. DM is defined as:

$$DM(A, B) = \frac{2|A \cap B|}{|A| + |B|},$$ (8)

where $A$ and $B$ represent the area of manual and automatic contour, respectively.

**Jaccard Coefficient**—The Jaccard coefficient, also named the Intersection over Union (IoU), is used to calculate the degree of dissimilarity between the manual segmentation area and the automatic segmentation contour area using the proposed method. Similar to DM, the JC is between [0, 1]. The larger the JC, the lower the similarity. The formula for JC is as follows:

$$J(A, B) = \frac{|A \cap B|}{|A \cup B|} = \frac{|A \cap B|}{|A| + |B| - |A \cap B|},$$ (9)

where $A$ and $B$ represent the area of manual and automatic contour, respectively.

In addition, we also use accuracy (ACC) and positive predictive value (PPV) to evaluate the results of pixel classification. They are defined by:

$$ACC = \frac{TP + TN}{TP + FN + FP + TN},$$ (10)

where $TP$, $TN$, $FP$, and $FN$ represent true positive, true negative, false positive, and false negative.

$$PPV = \frac{TP}{TP + FP},$$ (11)

where $TP$, $TN$, $FP$, and $FN$ represent true positive, true negative, false positive, and false negative.

## 4. Experimental Results

### 4.1. Performance of the Network

Figure 4 displays the predicted segmentation masks, GT, and contours for four subjects' cardiac MR images from the MICCAI 2018. It is clear from Figure 4a–d that the proposed method can accurately segment LV in cardiac MR images. In Figure 4d, it can be found that the automated segmentation contours marked by the red curve almost overlap on the GT (marked by the green curve), indicating that the model can accurately segment LV in diverse shapes. In summary, the model shows great potential for cardiac MR image segmentation with high accuracy, and thus may provide a visual aid to clinicians for qualitative diagnosis.

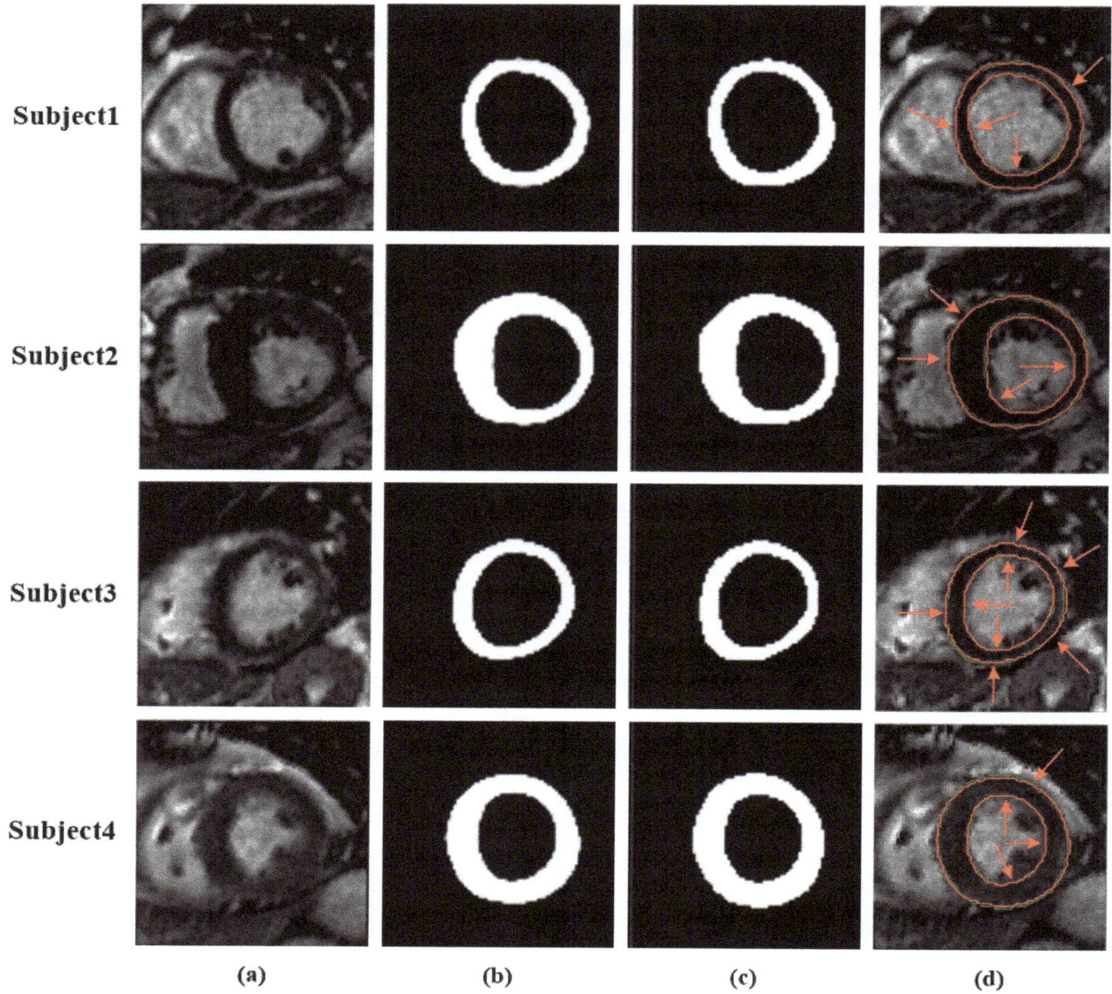

**Figure 4.** Segmentation results for four subjects. The arrows point to places where the proposed method can be seen to almost overlap with the GT manually delineated by the experts, indicating better segmentation. (**a**) Cardiac MR image. (**b**) Results of segmentation using the suggested method. (**c**) GT. (**d**) The segmentation contours obtained by the proposed method are marked by red curves, and the corresponding GT manually delineated by experts is marked by green curves.

### 4.2. Performance Comparison

On the MICCAI 2018 test set, comparison is made between the method and other prevalent segmentation methods such as FCN, Conv–Deconv [39], U-Net, Indices-JSQ [40], and SegNet to evaluate its effectiveness. As shown in Table 1, the DM and JC of the proposed method reach 0.908 and 0.834, respectively, with a certain extent of improvement compared to that of other segmentation methods. These results suggest that the model is capable of more accurately determining the class of each pixel point and then achieving higher segmentation performance.

**Table 1.** The comparison of the suggested method with several popular segmentation methods.

| Methods | DM | JC | ACC | PPV |
|---|---|---|---|---|
| FCN | 0.873 | 0.778 | 0.972 | 0.878 |
| SegNet | 0.843 | 0.728 | 0.964 | 0.845 |
| U-Net | 0.887 | 0.800 | 0.975 | 0.887 |
| Conv-Deconv | 0.846 | 0.733 | 0.965 | 0.840 |
| Indices-JSQ | 0.870 | $-$ [1] | $-$ | $-$ |
| Cardiac-DeepIED [41] | 0.890 | 0.801 | 0.976 | 0.891 |
| **Our Method** | **0.908** | **0.834** | **0.979** | **0.903** |

[1]—Indicates that the results are not provided.

*4.3. Ablation Studies*

In this section, ablation studies are performed on each component of the suggested method. As shown in Table 2, as the proposed modules are sequentially added on the UNet 3+ baseline, the model performance is gradually improved. It is clear from Table 2 that when Transformer is integrated into the UNet 3+ baseline, the DM reaches 0.907, with a 1.1% improvement compared to that of the baseline (0.896). This is because the incorporation of Transformer compensates for the inability of UNet 3+ to model the global context. Simultaneously, the combination also solves the problem of Transformer ignoring low-resolution detail features compared to using itself directly.

**Table 2.** Ablation experiments of the model on the MICCAI 2018 test set.

| UNet 3+ | Transformer | $\mathcal{L}_{seg}$ | $\mathcal{L}_{sdm}$ | DM |
|---|---|---|---|---|
| ✓ | × | ✓ | × | 0.896 |
| ✓ | ✓ | ✓ | × | 0.907 |
| ✓ | ✓ | ✓ | ✓ | 0.908 |

To further study the impact of the loss function in the model, an ablation study is conducted for the segmentation loss ($\mathcal{L}_{seg}$) and SDM loss ($\mathcal{L}_{sdm}$). By adding an additional regression head to the segmentation network's end and using the SDM loss, the dice metric value of the model is further enhanced to 0.908, with a slight increase of 0.1% compared to that of the model only with $\mathcal{L}_{seg}$. This result suggests that joint SDM training can implicitly force the model to learn shape information compared to traditional training using only segmented masks.

## 5. Conclusions

In this study, a network for automatic LV segmentation from cardiac MR images is proposed, providing an effective solution that allows physicians to diagnose CVDs. The proposed method extensively experimented on cardiac MR image data from 145 subjects, and the DM and JC reached 0.908 and 0.834, respectively, on the MICCAI2018 test set. The proposed module and loss function both improve the segmentation accuracy, as verified by ablation experiments. The method also outperforms the current mainstream methods in the comparison experiments suggesting that it can be considered an effective automatic LV segmentation task that will reduce the workload of radiologists during clinical diagnosis. In future research, more possibilities for Transformer application to medical image segmentation networks will be explored to provide a better technique for LV segmentation.

**Author Contributions:** Conceptualization, X.H.; methodology, Z.L. and X.H.; resources, Z.L. and X.H.; data curation, Z.L. and X.H.; writing—original draft preparation, X.H.; writing—review and editing, Z.L., X.H. and Y.L. All authors have read and agreed to the published version of the manuscript.

**Funding:** The APC was funded partly by the Natural Science Foundation of Chongqing, China (Grant Nos. cstc2019jcyj-msxmX0487, cstc2021jcyj-msxmX0605). The APC was funded partly by the National Natural Science Foundation of China (Grant Nos. 61971078, 61501070). The APC was funded partly by the Science and Technology Foundation of Chongqing Education Commission (Grant Nos. KJQN202001137, CQUT20181124). The APC was funded partly by the Scientific Research Foundation of Chongqing University of Technology (2020ZDZ026).

**Institutional Review Board Statement:** Not applicable.

**Informed Consent Statement:** Not applicable.

**Data Availability Statement:** Not applicable.

**Conflicts of Interest:** The authors declare no conflict of interest.

# References

1. WHO. *WHO Fact-Sheets Cardiovascular Diseases (CVDs)*; WHO: Geneva, Switzerland, 2021.
2. Shaaf, Z.F.; Jamil, M.M.A.; Ambar, R.; Alattab, A.A.; Yahya, A.A.; Asiri, Y. Automatic Left Ventricle Segmentation from Short-Axis Cardiac MRI Images Based on Fully Convolutional Neural Network. *Diagnostics* **2022**, *12*, 414. 12020414. [CrossRef] [PubMed]
3. Gessert, N.; Schlaefer, A. Left Ventricle Quantification Using Direct Regression with Segmentation Regularization and Ensembles of Pretrained 2D and 3D CNNs. *arXiv* **2019**, arXiv:1908.04181.
4. Tavakoli, V.; Amini, A.A. A survey of shaped-based registration and segmentation techniques for cardiac images. *Comput. Vis. Image Underst.* **2013**, *117*, 966–989. [CrossRef]
5. Petitjean, C.; Dacher, J.N. A review of segmentation methods in short axis cardiac MR images. *Med. Image Anal.* **2011**, *15*, 169–184. [CrossRef]
6. Dakua, S.P. Towards Left Ventricle Segmentation From Magnetic Resonance Images. *IEEE Sens. J.* **2017**, *17*, 5971–5981. [CrossRef]
7. Tran, P.V. A Fully Convolutional Neural Network for Cardiac Segmentation in Short-Axis MRI. *arXiv* **2016**, arXiv:1604.00494.
8. Xue, W.; Lum, A.; Mercado, A.; Landis, M.; Warrington, J.; Li, S. Full Quantification of Left Ventricle via Deep Multitask Learning Network Respecting Intra- and Inter-Task Relatedness. In *Lecture Notes in Computer Science, Proceedings of the International Conference on Medical Image Computing and Computer-Assisted Intervention, Quebec City, QC, Canada, 11–13 September 2017*; Springer: Cham, Switzerland, 2017.
9. Wernick, M.N.; Yang, Y.; Brankov, J.G.; Yourganov, G.; Strother, S.C. Machine Learning in Medical Imaging. *IEEE Signal Process. Mag.* **2010**, *27*, 25–38. [CrossRef]
10. Yabrin, A.; Amin, B.S.; Mir, A.H. A Comparative Study on Left and Right Endocardium Segmentation using Gradient Vector Field and Adaptive Diffusion Flow Algorithms. *Int. J. Bio-Sci. Bio-Technol.* **2016**, *8*, 105–120.
11. Li, W.; Ma, Y.; Zhan, K.; Ma, Y. Automatic Left Ventricle Segmentation in Cardiac MRI via Level Set and Fuzzy C-Means. In Proceedings of the International Conference on Recent Advances in Engineering & Computational Sciences, Chandigarh, India, 21–22 December 2015.
12. Katouzian, A.; Prakash, A.; Konofagou, E. A new automated technique for left-and right-ventricular segmentation in magnetic resonance imaging. In Proceedings of the International Conference of the IEEE Engineering in Medicine and Biology Society, New York, NY, USA, 30 August–3 September 2006; Volume 2006, pp. 3074–3077. [CrossRef]
13. Lynch, M.; Ghita, O.; Whelan, P.F. Automatic segmentation of the left ventricle cavity and myocardium in MRI data. *Comput. Biol. Med.* **2006**, *36*, 389–407. [CrossRef]
14. Zhang, Z.; Duan, C.; Lin, T.; Zhou, S.; Wang, Y.; Gao, X. GVFOM: A novel external force for active contour based image segmentation. *Inf. Sci.* **2020**, *506*, 1–18. [CrossRef]
15. Wu, Y.; Wang, Y.; Jia, Y. Segmentation of the left ventricle in cardiac cine MRI using a shape-constrained snake model. *Comput. Vis. Image Underst.* **2013**, *117*, 990–1003. [CrossRef]
16. Chakraborty, A.; Staib, L.H.; Duncan, J.S. Deformable boundary finding in medical images by integrating gradient and region information. *IEEE Trans. Med. Imaging* **1996**, *15*, 859–870. [CrossRef] [PubMed]
17. Lynch, M.; Ghita, O.; Whelan, P.F. Segmentation of the left ventricle of the heart in 3-D+t MRI data using an optimized nonrigid temporal model. *IEEE Trans. Med. Imaging* **2008**, *27*, 195–203. [CrossRef]
18. Wang, Y.; Zhang, Y.; Wen, Z.; Tian, B.; Kao, E.; Liu, X.; Xuan, W.; Ordovas, K.; Saloner, D.; Liu, J. Deep learning based fully automatic segmentation of the left ventricular endocardium and epicardium from cardiac cine MRI. *Quant. Imaging Med. Surg.* **2021**, *11*, 1600–1612. [CrossRef] [PubMed]
19. Xijing, Z.; Qian, W.; Ting, L. A novel approach for left ventricle segmentation in tagged MRI. *Comput. Electr. Eng.* **2021**, *95*, 107416.
20. Haque, I.R.I.; Neubert, J. Deep learning approaches to biomedical image segmentation. *Inform. Med. Unlocked* **2020**, *18*, 100297. [CrossRef]
21. Ren, S.; He, K.; Girshick, R.; Sun, J. Faster R-CNN: Towards Real-Time Object Detection with Region Proposal Networks. *IEEE Trans. Pattern Anal. Mach. Intell.* **2017**, *39*, 1137–1149. [CrossRef]
22. Baloglu, U.B.; Talo, M.; Yildirim, O.; Tan, R.S.; Acharya, U.R. Classification of myocardial infarction with multi-lead ECG signals and deep CNN. *Pattern Recognit. Lett.* **2019**, *122*, 23–30. [CrossRef]

23. Huang, G.; Liu, Z.; Laurens, V.; Weinberger, K.Q. Densely Connected Convolutional Networks. In Proceedings of the IEEE Computer Society, Honolulu, HI, USA, 21–26 July 2017.

24. Shelhamer, E.; Long, J.; Darrell, T. Fully Convolutional Networks for Semantic Segmentation. *IEEE Trans. Pattern Anal. Mach. Intell.* **2017**, *39*, 640–651. [CrossRef]

25. Shamshad, F.; Khan, S.; Zamir, S.W.; Khan, M.H.; Hayat, M.; Khan, F.S.; Fu, H. Transformers in Medical Imaging: A Survey. *arXiv* **2022**, arXiv:2201.09873.

26. Huang, H.; Lin, L.; Tong, R.; Hu, H.; Wu, J. UNet 3+: A Full-Scale Connected UNet for Medical Image Segmentation. In Proceedings of the ICASSP 2020—2020 IEEE International Conference on Acoustics, Speech and Signal Processing (ICASSP), Barcelona, Spain, 4–8 May 2020.

27. Goshtasby, A.; Turner, D.A. Segmentation of cardiac cine MR images for extraction of right and left ventricular chambers. *IEEE Trans. Med. Imaging* **1995**, *14*, 56–64. [CrossRef] [PubMed]

28. Wang, Z.; Peng, Y.; Li, D.; Guo, Y.; Zhang, B. MMNet: A multi-scale deep learning network for the left ventricular segmentation of cardiac MRI images. *Appl. Intell.* **2021**, *52*, 5225–5240. [CrossRef]

29. Ronneberger, O.; Fischer, P.; Brox, T. U-Net: Convolutional Networks for Biomedical Image Segmentation. *arXiv* **2015**, arXiv:1505.04597.

30. Badrinarayanan, V.; Kendall, A.; Cipolla, R. SegNet: A Deep Convolutional Encoder-Decoder Architecture for Image Segmentation. *IEEE Trans. Pattern Anal. Mach. Intell.* **2017**, *39*, 2481–2495. [CrossRef]

31. Vaswani, A.; Shazeer, N.; Parmar, N.; Uszkoreit, J.; Jones, L.; Gomez, A.N.; Kaiser, L.; Polosukhin, I. Attention Is All You Need. *arXiv* **2017**, arXiv:1706.03762.

32. Dosovitskiy, A.; Beyer, L.; Kolesnikov, A.; Weissenborn, D.; Zhai, X.; Unterthiner, T.; Dehghani, M.; Minderer, M.; Heigold, G.; Gelly, S.; et al. An Image is Worth 16x16 Words: Transformers for Image Recognition at Scale. *arXiv* **2020**, arXiv:2010.11929.

33. Zheng, S.; Lu, J.; Zhao, H.; Zhu, X.; Luo, Z.; Wang, Y.; Fu, Y.; Feng, J.; Xiang, T.; Torr, P.H.S. Rethinking Semantic Segmentation from a Sequence-to-Sequence Perspective with Transformers. In Proceedings of the Computer Vision and Pattern Recognition, Virtual, 19–25 June 2021.

34. Carion, N.; Massa, F.; Synnaeve, G.; Usunier, N.; Kirillov, A.; Zagoruyko, S. End-to-End Object Detection with Transformers. *arXiv* **2020**, arXiv:2005.12872.

35. Parmar, N.; Vaswani, A.; Uszkoreit, J.; Kaiser, L.; Shazeer, N.; Ku, A.; Tran, D. Image Transformer. In Proceedings of the Proceedings of the 35th International Conference on Machine Learning, Stockholm, Sweden, 10-15 July 2018; Volume 80, pp. 4055–4064.

36. Chen, J.; Lu, Y.; Yu, Q.; Luo, X.; Zhou, Y. TransUNet: Transformers Make Strong Encoders for Medical Image Segmentation. *arXiv* **2021**, arXiv:2102.04306. [CrossRef]

37. Zhang, Y.; Liu, H.; Hu, Q. TransFuse: Fusing Transformers and CNNs for Medical Image Segmentation. In *Lecture Notes in Computer Science, Proceedings of the International Conference on Medical Image Computing and Computer-Assisted Intervention, Strasbourg, France, 27 September–1 October 2021*; Springer: Cham, Switzerland, 2021.

38. Xue, W.; Islam, A.; Bhaduri, M.; Li, S. Direct Multitype Cardiac Indices Estimation via Joint Representation and Regression Learning. *IEEE Trans. Med. Imaging* **2017**, *36*, 2057–2067. [CrossRef] [PubMed]

39. Noh, H.; Hong, S.; Han, B. Learning Deconvolution Network for Semantic Segmentation. *arXiv* **2015**, arXiv:1505.04366. [CrossRef] [PubMed]

40. Du, X.; Tang, R.; Yin, S.; Zhang, Y.; Li, S. Direct Segmentation-Based Full Quantification for Left Ventricle via Deep Multi-Task Regression Learning Network. *IEEE J. Biomed. Health Inform.* **2019**, *23*, 942–948. [CrossRef] .

41. Du, X.; Yin, S.; Tang, R.; Zhang, Y.; Li, S. Cardiac-DeepIED: Automatic Pixel-Level Deep Segmentation for Cardiac Bi-Ventricle Using Improved End-to-End Encoder-Decoder Network. *IEEE J. Transl. Eng. Health Med.* **2019**, *7*, 1900110. [CrossRef] .

MDPI

St. Alban-Anlage 66

4052 Basel

Switzerland

Tel. +41 61 683 77 34

Fax +41 61 302 89 18

www.mdpi.com

*Applied Sciences* Editorial Office

E-mail: applsci@mdpi.com

www.mdpi.com/journal/applsci

www.ingramcontent.com/pod-product-compliance
Lightning Source LLC
LaVergne TN
LVHW071030170726
843515LV00006B/1253